ELECTROPHORESIS
AND
IMMUNOELECTROPHORESIS

SERIES IN LABORATORY MEDICINE
Leo P. Cawley, M.D., Series Editor

1. Electrophoresis and Immunoelectrophoresis
Leo P. Cawley, M.D.

2. The Diagnosis of Bleeding Disorders
Charles A. Owen, Jr., M.D., D.Sc., Ph.D.
E. J. Walter Bowie, B.M., B.Ch., M.A., M.S.
Paul Didisheim, M.D.
John H. Thompson, Jr., Ph.D.

3. Practical Manual of Blood Transfusion
Douglas W. Huestis, M.D.
Joseph R. Bove, M.D.
Shirley Busch, B.B. Spec. (ASCP)

4. Clinical Serum Enzymology
Paul L. Wolf, M.D.
Charlotte Bartczak, M.T. (ASCP)

Electrophoresis

AND

Immunoelectrophoresis

LEO P. CAWLEY, M.D.

Clinical Pathologist, Director of Laboratories, and Divisional Head of Special and Research Hematology, Wesley Medical Center; Scientific Director, Wesley Medical Research Foundation, Wichita

Little, Brown and Company
Boston

PUBLISHED IN GREAT BRITAIN
BY J. & A. CHURCHILL LTD., LONDON

BRITISH STANDARD BOOK NO. 7000 0141 7

PRINTED IN THE UNITED STATES OF AMERICA

To my father

whose stories of Finn MacCoul, epic hero of Erin,
stirred my imagination

PREFACE

A certain amount of information relating to hardware and software in the fields of electrophoresis and immunoelectrophoresis is contained in this book, but the emphasis is on the application of hardware and software for interpretation of the test results. The biology of medicine seems to be in need of more direct attention. In our own facility we have begun to think of this in the same vernacular as the computer sciences and have coined the word *bioware* to harmonize with the terms *hardware* and *software*. Although hardware and software must be involved in any laboratory, it is becoming increasingly evident that the application of bioware, or the biology of medicine, to the laboratory is without exception the most important function of the clinical pathologist and his laboratory staff. Thus we have become aware of the need to focus the attention of the laboratory on specific disease areas and to bring the enormous potential of the laboratory to aid in rapidly establishing a proper diagnosis or a "best answer."

The first duty of bioware must be to define tests in reference to specific diseases and to determine their diagnostic index singly and in sequence with other tests. Tests which exclude a disease are looked for and those which can give a simple yes or no answer are highly desirable. A series of tests which leads to a defined diagnostic goal constitutes a bioware package. Bioware packages drive software and hardware. It is too early to be sure, but it seems likely that bioware packages will prescribe both hardware and software, possibly in the same way that function dictates morphology.

It is our belief that the laboratory will play an expanding role in development of bioware packages. We have used the term *orchestration* to define a new dimension or function of the laboratory, directed at establishing a diagnosis. Too long has the laboratory been considered a passive entity, often fragmented and rarely unified into a single functional unit. By definition, orchestration is the process of arranging and harmonizing all the functions of the laboratory with clinical medicine in order to accelerate and improve the diagnosis of a specific disease. Proper orchestration requires that the laboratory have certain decision-making powers; that is, if test A is abnormal, test B, for confirmation, etc., is automatically ordered by the screening physician of the laboratory.

The laboratory in this day and age is a large, complex organization, and we have found it necessary to departmentalize the structure and appoint physicians to certain critical areas. These physicians are then in the position to monitor changes brought to their attention by the medical technology staff. Many of the problems of undetected diseases originate right at the bench with the technologist, where an unusual event, even though it may be noted and registered by the technologist on a laboratory slip, is not always followed up by the clinician because he may be unable to appreciate the subtleties of the report. The screening physician, on the other hand, is in a much more advantageous position in regard to laboratory tests and is attuned to the significance of unusual or positive laboratory tests. It is his function not to let this information go unnoticed. Moreover, he has the authority, with the cooperation of the attending physician, to set in order a sequence of tests which will arrive at a "best answer" or a diagnosis. Monitoring changes and keeping track of evidence of the success of the orchestration system can aid in development of advice rules for such procedures as diagnosis and follow-up therapy. Later such advice rules could become the nucleus of software information necessary for computerization of the process.

The objectives of orchestration related specifically to electrophoresis and immunoelectrophoresis are two: The first is to produce general diagnostic excellence in specific diseases by developing a logical system of laboratory tests with high diagnostic capabilities leading to a more rapid diagnosis (bioware). To this end the laboratory must often be reconstructed. Certain tests formerly done in chemistry must be done in the area in which a specific disease is under investigation. In other words, the laboratory must take on the aspects of a disease-oriented laboratory, which then alerts everyone in each department to all events which occur in his area. The second objective is automatically to provide a follow-up sequence of tests leading to either the most probable diagnosis or the "best answer" by on-line consultation of the screening physician based in the laboratory. It is hoped that the reader of this text will appreciate that much information which is not necessarily relevant to a particular problem may help in the interpretation of unusual events. In some areas, as in monoclonal peaks, the bioware package is well established, and the advice rules are beginning to become clear. In other areas the ability to offer on-line consultation is defective because information relating to hardware, software, or bioware is not advanced to a point of acceptance.

L. P. C.

Wichita

ACKNOWLEDGMENTS

It is with pleasure that I acknowledge the help, moral support, and scientific criticism received from the staff of the Wesley Medical Center laboratory and of the Wesley Medical Research Foundation. Many of the chapters represent personal attention from the technologists who are actually involved on a day-to-day basis in the performance of many of the tests used in this book.

Barbara Minard, M.T. (ASCP), head of the Department of Special Hematology, has contributed directly to the section dealing with methods. Lucile Eberhardt, M.T. (ASCP), in charge of Research Hematology, has been directly involved for several years in many areas dealing with enzyme procedures. William Goodwin, M.T. (ASCP), head of the Electron Microscopy Department, has contributed significantly to the procedures involved in continuous particle electrophoresis, disc electrophoresis, and ferritin-tagged antibodies. Paulette Dibbern, C.H. (ASCP), of the Department of Research Hematology, has contributed in the very interesting area of chemical coupling of compounds to proteins. Garry Millsap, of the Photography Department, has devoted himself to the job of developing better photographic read-out methods for electrophoresis and immunoelectrophoresis, and has been instrumental in preparing photographs for the text. The typing and final arrangement of many chapters have benefited from the helpful suggestions of Ronda Bippert, my secretary. Helen Stoskopf contributed many hours to typing and correcting the manuscript.

To several I owe a debt of gratitude for time and energy spent in reading certain chapters and offering their experience in making the subject matter more readable and more complete. These include Masashi Itano, M.D., Harold Bell, M.D., Carol Sleeper, M.D., Newton Ressler, Ph.D., and David Palmer, M.D.

L. P. C.

CONTENTS

Appendix

ELECTROPHORESIS
AND
IMMUNOELECTROPHORESIS

1

PRINCIPLES OF ELECTROPHORESIS AND IMMUNOELECTROPHORESIS

Migration of proteins in an electrical field was first perfected by Tiselius [17], who used a liquid medium. His technic of electrophoresis is known as "moving boundary electrophoresis" or "free electrophoresis." Moving boundary electrophoresis served for many years as the basis for classifying the electrophoretic mobility of proteins.

In zone electrophoresis a stabilizing medium serves as a matrix for the buffer in which the proteins travel and as a structure to which the proteins become attached after fixation. In other words, the *stabilizing medium* is used as a *trap* whereby proteins can be stained and examined after the electrophoretic phase. Paper, cellulose acetate, agar gel, and agarose are stabilizing media used to trap the migrating proteins. Zone electrophoresis became very popular because it did not require expensive equipment and because it permitted protein examination at leisure. Cellulose acetate has certain properties that make it better than paper; the same can be said of agarose and agar gel. For protein separation, paper requires about 18 hours, but cellulose acetate and agar gel require 30 to 45 minutes. Separation of proteins in agar gel or agarose is rapid and relatively complete, and the medium is optically clear. These advantages of agar gel and agarose as the media are most significant.

In the moving boundary technic and in zone electrophoresis, protein separation is based on electrophoretic mobility. Commonly, five protein bands are observable through separation based on electrophoretic mobility. The usual five bands are increased to about 25 bands in electrophoretic methods which use the "molecular sieve" principle. Starch [16] and acrylamide gels [4, 14] will trap relatively large molecules, but not the smaller ones. In the molecular sieve principle, separation of protein molecules is based not only upon electrophoretic mobility but also upon molecular size. Therefore, more protein molecules are separated by this system than is possible with zone systems. The 25 protein bands are complex to interpret, however. Also, the electrophoretic heterogeneity of albumin, for example, can be shown by the molecular sieve methods, but the additional bands of albumin travel in the same areas as other serum proteins, thus obscuring proper interpretation of the electrophoretogram. Molecular sieve methods are not suitable for the routine diagnostic laboratory.

Gel Electrophoresis

Agar-gel and agarose electrophoresis are well suited to the routine diagnostic laboratory [19], and they also serve as good illustrations of the principles of electrophoresis [9]. The gel structure is similar to that of a sponge, and this spongelike structure does *not* hinder particle migration. (Actually there is slight adsorption between the gel and certain serum proteins, but for this part of the discussion the gel is considered totally inert.) In other words, the pore size of the gel is relatively large compared with the size of the migrating protein. Examine the rough illustration, Figure 1-1. The meshwork of the agar gel serves as a carrier for the buffer solution. As shown in Figure 1-2, the buffer solution serves as the electrolyte. If a *negatively* charged particle were present in the agar gel, it would migrate toward the *anode*. Thus, the agar-gel or agarose system of electrophoresis has four basic components: a migrating protein, the buffer solution, an electrical field, and the stabilizing medium (agar gel or agarose). In most instances, agar gel itself has a charge—a negative charge. Thus, there would exist a tendency for the negative-(—)-charged agar gel to move toward the anode, but the agar gel cannot move toward the anode—it is fixed and is a gel. Since the agar gel really cannot move, the force vector is shown in Figure 1-2 as a broken-line arrow. Nevertheless, this force *does* exist, and its direction is toward the anode. Then according to Newton, if there is a force vector in the direction of the anode, one can expect an equal force vector in the direction of the cathode. This counterforce is made up of *buffer* and is illustrated by the solid arrow pointing toward the cathode (Fig. 1-3). The agar gel cannot move toward the anode, but a gradient of buffer

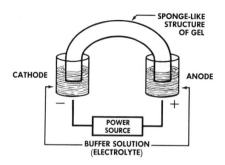

FIGURE 1-1

Simplified diagram of agar-gel electrophoresis apparatus

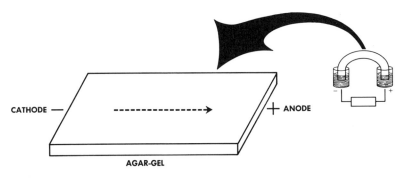

FIGURE 1-2

Diagrammatic view of agar gel serving as the electrolyte

moves toward the cathode. The "action" is fixed, but the "reaction" moves. The movement of buffer salts, called *electro-osmosis,* or endosmosis, is that force sweeping through the agar gel. With a negatively charged agar gel, electro-osmosis is toward the cathode. An increased negative charge on the agar gel results in an increased reaction of buffer movements, the magnitude of electro-osmosis being directly proportional to the charge on the agar gel—the greater the charge, the greater the electro-osmosis, whereas the smaller the charge, the less the electro-osmosis. It follows that electro-osmosis would be zero if and only if the agar-gel charge is zero, and the direction of electro-osmosis would be toward the cathode if the agar gel had a positive charge. Depending upon the charge of the agar gel, electro-osmosis may be either *weak* or *strong*. In most agar-gel preparations, there is considerable negative charge, whereas deionized agar gel has very little negative charge [3]. In the former case, electro-osmosis is strong, but in the latter, and with ionagar and agarose, electro-osmosis is weak.

If a negatively charged particle is placed in the agar gel under conditions of minimum electro-osmosis, the negative particle will advance toward the positive anode—that is, the negatively charged particle would advance to the *right* of the point of origin (Fig. 1-4). The negatively charged particle was attracted by the positive voltage at the anode, but one other force was at work—the buffer. Had the

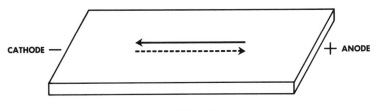

FIGURE 1-3

Direction of counterforce produced by buffer

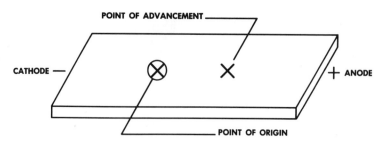

FIGURE 1-4
Movement of negatively charged particle in gel

flow of the buffer or *electro-osmosis* been somewhat greater, the point of advancement would not have been as far to the right of the point of origin as shown in Figure 1-4. The net mobility of the charged particle, therefore, is the difference between the movement based on electrostatic charge and the resistance offered by the flow of buffer or electro-osmosis. The magnitude of electro-osmosis depends on the charge of the agar.

Consider the electro-osmosis to be of a constant magnitude. A particle will migrate toward the anode if the electrostatic force is greater than the force of the electro-osmosis; that is, if the electrostatic force is sufficient to overcome the buffer force, the charged particle will migrate to the right of the point of origin. Conversely, if the electrostatic force is less than the buffer force, the charged particle will migrate to the left of the point of origin. Naturally, if the electro-osmotic force were exactly equal to the electrostatic force, net mobility of the charged particle would be zero, the charged particle remaining at the point of origin. Figure 1-5 shows the position of the gamma (γ) globulin with respect to the point of origin. The negatively charged γ-globulin appears to the left of the origin since the electrostatic force is insufficient to overcome electro-osmosis. With two negatively charged particles—for example, γ-globulin and beta (β) globulin, which has a greater negative charge than the γ-globulin—the two particles appear on opposite sides of the point of origin (Fig. 1-6). Although the nega-

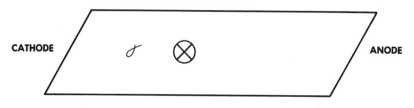

FIGURE 1-5
Position of γ-globulin in relation to point of origin

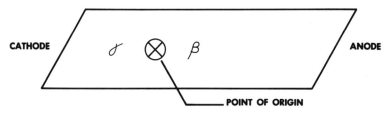

FIGURE 1-6

Electrophoretic mobility of γ-globulin and β-globulin

tively charged γ-globulin is attracted by the anode, it is overcome by the flow of buffer, whereas the β-globulin, with a greater negative charge, is attracted to the anode. It is thus possible for two negatively charged particles actually to migrate in opposite directions.

Electrical Charge of Proteins

Figure 1-7 shows the electrophoretic mobility of the five protein particles in a decreasing order of negative charge—that is, albumin (most negative), alpha₁ (α_1), alpha₂ (α_2), β, and γ (least negative). The electrophoretic mobility of the α-globulin is greater than the electrophoretic mobility of the β-globulin, the negative charge on the α-globulin being greater.

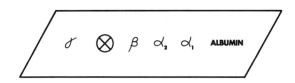

FIGURE 1-7

Electrophoretic mobility of five proteins

Proteins vary in size, shape, and *charge*. Being amphoteric, proteins, like peptides or amino acids, can have no charge, a negative charge, or a positive charge, depending upon the pH of the surrounding medium. The charge depends on the amine and carboxyl groups of the protein, peptide, or amino acids. In the protein group NH_3^+ —R—COO^-, for example, the positive (+) charge on NH_3^+ cancels the negative (—) charge on COO^-, leaving this protein with zero charge. This protein group is said to be at its *isoelectric point* where solubility, swelling, and viscosity are reduced, and it does not migrate. If the protein is put into an acidic medium, however, excess H ions combine with the protein molecule as follows:

$$\text{H}^+ + \boxed{\text{NH}_3^+ - \text{R} - \text{COO}^-} \longrightarrow \overbrace{\text{NH}_3^+ - \text{R} - \text{COOH}}$$

Now, the protein molecule has a positive charge—that is, putting the protein molecule into an acidic medium results in the protein molecule's taking on a positive charge. Similarly, if the protein molecule is put into an alkaline medium, it will take on a negative charge:

$$\boxed{\text{NH}_3^+ - \text{R} - \text{COO}^-} + \text{OH}^- \longrightarrow \overbrace{\text{NH}_2 - \text{R} - \text{COO}^-} + \text{H}_2\text{O}$$

Proteins are amphoteric in that the protein molecule takes on the characteristic of the buffer. A protein molecule in an excess OH⁻ (*negative*) medium results in the protein's acquiring a negative charge also, whereas a protein molecule in an excess H⁺ (*positive*) medium causes the protein to acquire a positve charge. Proteins can therefore have no charge, a positive charge, or a negative charge depending on the acidity or alkalinity of the buffer. If one considers a pH scale from 1 to 14, it is neither acidic nor alkaline at the isoelectric point (pH 7). Thus, the groups NH₃⁺—R—COO⁻ of this particular protein, when placed in buffer solution at pH 7, would continue to have no charge. But, as seen in Figure 1-8, buffer of pH greater than the isoelectric pH results in a negatively charged molecule. Conversely, the protein acquires a positive charge when in a buffer with pH less than the isoelectric point. Naturally, a protein molecule would be very insoluble at its isoelectric point—that is, in buffer with a pH of the isoelectric point. In our hypothetical protein equal numbers of NH₃⁺ and COO⁻ groups were present. In serum proteins this is not the case, and each protein has its own characteristic isoelectric point.

A commonly used buffer is Veronal (barbital), pH 8.6, which has been found by empirical testing to render proteins quite soluble, does not denature proteins, and imparts a satisfactory negative charge to the protein molecule; in the presence of an electrical field, the

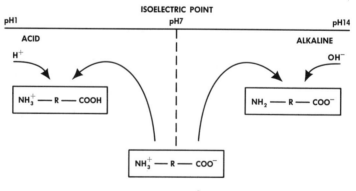

FIGURE 1-8

Effect of characteristic of buffer (i.e., acidity or alkalinity) on protein

negative protein molecule will migrate toward the anode. If a buffer with a pH less than 7 is used, the buffer will impart a positive charge to the protein molecule, and, in the presence of an electrical field, the positively charged protein will migrate toward the cathode.

Effect of buffer in electrophoresis

If, for example, the two tanks shown in Figure 1-1 contain barbital buffer (pH 8.6) and if agar gel, supported by glass or plastic, serves as a bridge between the two tanks, the current flows through the agar-gel bridge from one buffer tank to the other. Since the pH of the buffer is 8.6, protein molecules would take on negative charges, they would migrate toward the anode, and one would expect the protein molecules to separate.

Particle migration

The separation characteristics of the protein molecule depend upon the length of the agar gel from buffer surface to buffer surface. In Figure 1-9, the agar-gel length from buffer surface to buffer surface is 10 cm., power source is 130 volts, or

$$\frac{130\ v}{10\ cm.} = 13\ v\ per\ cm.$$

In agar-gel or agarose electrophoresis, use of less than 20 volts per centimeter can be thought of as low-voltage electrophoresis. Thus use of more than 20 volts per centimeter is considered high-voltage electrophoresis. Separation of the protein is more rapid in high-voltage electrophoresis. Therefore, the shorter the agar-gel length, the more rapid the separation. The distance of movement of charged particles through an agar strip is *inversely* proportional to the length of the agar gel, but directly proportional to the voltage. Naturally, time is also significant; if time is increased, the distance of movement of a charged particle will increase also. So, the distance of movement of a charged particle is directly proportional to both voltage and time,

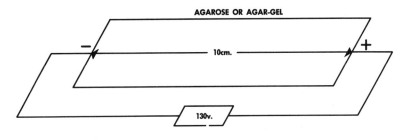

FIGURE 1-9
Length of gel from buffer surface to buffer surface

and inversely proportional to the length of the agar gel. The distance, d, of migration of a particular protein is expressed in centimeters. It can be computed by the following formula:

$$d = \frac{vt\,\mu}{l} \qquad (1)$$

in which d is the distance of migration in centimeters, v is voltage in volts, t is time in seconds, l is length of the agar gel in centimeters, and μ is the distance the protein will migrate within a given amount of time in an electrophoretic system; μ is an expression of *mobility*.

If equation 1 is solved for μ,

then $\qquad\qquad\qquad\qquad dl = vt\,\mu \qquad\qquad\qquad\qquad (2)$

and $\qquad\qquad\qquad\qquad \mu = \dfrac{dl}{vt} \qquad\qquad\qquad\qquad (3)$

Now, observe the units of measurement.

$$\left\{\frac{dl}{vt}\right\} \text{ means } \left\{\frac{cm \times cm}{volts \times seconds}\right\} = \frac{cm^2}{volts \text{ - } seconds}$$

Therefore, mobility is expressed in centimeters squared per volt-seconds. For example, the mobility of albumin in agar gel is about 6.5×10^{-5} cm.^{2}/v-sec. [1].

In practice, it is desirable to have rapid separation of proteins to preserve the sharpness of the separated protein bands. The longer the time of electrophoresis, the greater the radial diffusion of proteins, which results in less-well-defined protein bands. Therefore, as much as possible, it is desirable to shorten the time of electrophoresis. But how can the time of electrophoresis be reduced? Returning to equation 1 and solving for t,

$$t = \frac{dl}{v\mu} \qquad (4)$$

If one wishes the distance of protein migration to be 4 cm., but wants to shorten the time t, with μ being a constant for each protein group, one must either decrease length or increase voltage.

$$t = \frac{4l}{v\mu} \qquad (5)$$

If the length of the agar gel were decreased, there would be more volts per centimeter. Similarly, if the agar-gel length remained constant but the voltage source increased, volts per centimeter would increase. So, whether one changes length, l, or volts, v, the significant factor is volts per centimeter; that is, to decrease the required time and thereby increase the sharpness of separated protein bands, one must increase volts per centimeter. With a voltage increase, however, there is a corresponding current increase, and the resulting heat generated is a limiting factor. This may be controlled partially by lowering the ionic strength of the buffer in the agar gel and partly by surface cooling [18].

EFFECT OF PORE SIZE AND ADSORPTION ON MOBILITY

There are two factors affecting mobility that have not been mentioned: (1) pore size and (2) adsorption—that is, the adhesion of molecules to the surfaces of solid bodies or liquids with which they are in contact. The openings, or pores, in the meshwork of the agar gel are about 60 angstroms (Å). Since albumin has a diameter of approximately 30 Å, it will pass relatively uninhibited through the pores of the agar gel. Thus, molecules of albumin have a mobility in agar gel which is nearly identical to that in moving boundary electrophoresis. This is not necessarily true of the other protein molecules in serum. The outstanding example is the lipoproteins. Adsorption between the lipoproteins and the sulfonated polysaccharides of agar gel is often strong [18].

Adsorption phenomenon between protein and support medium remains an area of investigation. In general, as the pH is lowered for certain types of electrophoresis, particularly hemoglobin, adsorption becomes a significant feature. At pH's above 7, there is very little adsorption, but it does cause some pattern change. If an electrophoretic system were restricting protein mobility and one wanted the protein to move freely, one could try reducing the adsorption effect by increasing the pH, or one could choose a gel that had a pore size greater than the diameter of the protein. If, on the other hand, one wished to restrict the movement of the molecule, one would either select a gel with a pore size more nearly equal to the molecular diameter or, if such a gel were not available, increase the adsorption effect. In many ways the adsorption effect is similar to the technic of electrophoresis called "molecular sieve electrophoresis." The sieving principle does not operate strongly in agar gel, however, because of the large pore size, and the same is true for cellulose acetate [7]. Thus, in agar gel, with its pore size of approximately 60 Å, diffusion of proteins is reasonably uninhibited because the openings within the agar gel are sufficient to permit passage of most protein molecules [1, 18].

Protein Diffusion

Protein diffusion takes place in all directions. For example, a layer of 1 percent agar gel is placed on the bottom of a Petri dish, and a well is formed near the edge of the dish with a cork borer. If an antigen

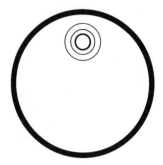

FIGURE 1-10

Diffusion pattern of albumin in circular well in Petri dish

such as human albumin is placed in the well, the proteins will diffuse in concentric rings (Fig. 1-10) much as waves spread when a pebble is thrown into water.

Suppose, instead of a circular well, a trench were cut into the agar-gel-covered Petri dish near the bottom. Then, as seen in Figure 1-11, the diffusion pattern depends on the geometry of the well, trench, etc. As will be seen later, the geometry of a γ-globulin in an electrophoretogram is different from that of albumin, and the diffusion patterns are different also, being similar to the respective geometries of the γ-globulin and albumin.

If both a well and a trench are cut into a Petri dish, the two diffusion patterns will eventually meet. However, if an *antigen* is placed in the well and an *antibody* in the trench, when the two patterns meet, a precipitate is formed (Fig. 1-12) owing to the interaction between the antigen and the antibody. (An antigen-antibody interaction can easily result in a structure larger than 60 Å.) This precipitate is trapped because (1) the structure is larger than 60 Å, further diffusion therefore being prevented, and (2) often the antigen-antibody combination

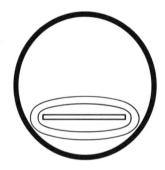

FIGURE 1-11

Diffusion pattern of albumin in a trench in a Petri dish

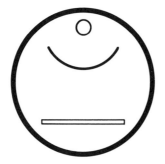

FIGURE 1-12

Diffusion pattern when antigen is placed in well (*top*) and antibody is placed in trench (*bottom*). Curved line indicates precipitin band.

is no longer soluble in the aqueous system. Precipitin bands form where the two patterns meet. The heavy curved line is indicative of a precipitin band.

The position of the precipitin band, which, as already stated, forms at the points where the two diffusion patterns meet, is dependent upon the size of the molecules and the concentration of the reactants. If the antibody is relatively large, it might diffuse more slowly since the openings in the agar gel are somewhat restrictive. In this case the precipitin band forms closer to the antibody. Abnormal *concentrations* of antigens affect the position of precipitin bands (discussed below), but in general the position of the precipitin bands depends upon two variables: (1) the molecular size and (2) the concentration of the antigen.

Immunodiffusion and Electrophoresis (Immunoelectrophoresis)

The formation of a precipitin band or immunoprecipitin band depends on interaction between the antigen and its antibody. Reactions between protein from tissue extract and albumin and hemoglobin can cause "spurious bands" which appear as authentic immunoprecipitins, and reactions between plant proteins and serum lipoprotein also appear as true gel-diffusion precipitin bands. If the antigen contains γ-globulin and the antibody contains anti-γ-globulin, the reaction will form an immunoprecipitin band; on the other hand, if the antigen contains γ-globulin and the antibody only anti-β-globulin, no precipitin band will form. When both antigen and antibody reactants are moving through agar gel, it is called *double gel diffusion* [12]. In 1953 Grabar and Williams [5] combined the technics of agar-gel electrophoresis and double gel diffusion to create *immunoelectrophoresis,* an additional dimension of protein analysis. Immunoelectrophoresis therefore consists of electrophoresis and immunodiffusion.

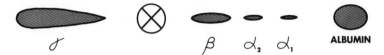

FIGURE 1-13

Electrophoretic pattern of normal serum proteins

IMMUNOELECTROPHORETIC TECHNIC

As discussed earlier, electrophoresis of serum proteins often results in five separate fractions: albumin, α_1-globulin, α_2-globulin, β-globulin, and γ-globulin. Figure 1-13 shows an antigen separated into these various fractions by agar-gel or agarose electrophoresis. The fractions are within agar gel or agarose, the anode being to the right and the cathode to the left; note, only the γ-globulin appears to the left of the point of origin. The strip is removed from the electrophoretic cell, and trenches are placed on each side and parallel to the line of electrophoresis, each trench being filled with antiserum (Fig. 1-14). In this technic of immunoelectrophoresis, if one assumes that antiserum containing antialbumin is placed in the top trench and allowed to diffuse, the pattern will be as shown in Figure 1-15. Note the respective identification and location of each globulin and the albumin. Figure 1-16 shows an electrophoretogram that resulted when an unknown antibody was applied against normal serum protein as an antigen. Observing from left to right, the unknown antibody contained the following specific antibodies: anti-γ-globulin, anti-α-globulin, and antialbumin.

OUCHTERLONY TECHNIC

Ouchterlony [12] found that he could utilize the system of double gel diffusion to compare a mixture of antibodies against a mixture of antigens in such a way that he might identify a specific antigen–anti-

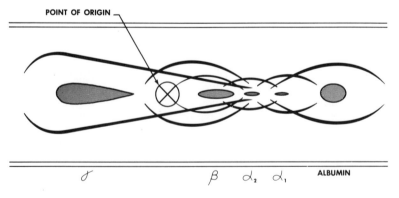

FIGURE 1-14

Pattern when gel electrophoresis is combined with immunodiffusion

body reaction out of a total mixture of antigens and antibodies. Using the Petri dish, he simply made three holes in the agar gel (Fig. 1-17). The top wells—the antigen wells—both contain purified human albumin, while the antialbumin and anti-γ-globulin, is placed in the lower well. The antiserum reacts with antigen from both wells, and these single precipitin bands unite to form a smooth junction (Fig. 1-18).

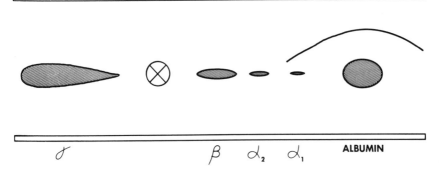

γ   β   α₂   α₁   **ALBUMIN**

FIGURE 1-15

Immunoelectrophoretic pattern with antialbumin in top trench

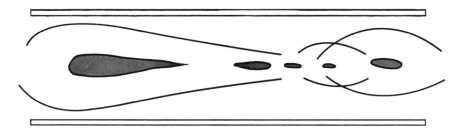

FIGURE 1-16

Immunoelectrophoretic pattern when unknown antibody (in trenches) is applied against normal serum proteins. See text.

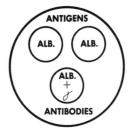

FIGURE 1-17

Ouchterlony technic

This reaction is called the "reaction of identity" and implies serologic identity; that is, the antigens in both upper wells share a common antigen site. If one places purified γ-globulin in the upper right well, however, the precipitin bands cross and do not unite (Fig. 1-19). The antigens do not share a common site, the reaction being called "nonidentity." Finally, if the upper right-hand well contains *nonhuman albumin*, the upper left-hand well contains human albumin, and the lower well has antiserum against human albumin and γ-globulin, precipitin bands form a partial union (Fig. 1-20)—that is, a "reaction of incomplete identity." This means there is a serologic relationship between the two sources of albumin, but *not* complete identity.

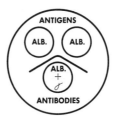

FIGURE 1-18

Ouchterlony technic showing reaction of identity

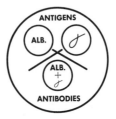

FIGURE 1-19

Ouchterlony technic showing reaction of nonidentity

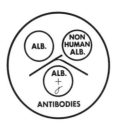

FIGURE 1-20

Ouchterlony technic showing reaction of incomplete identity

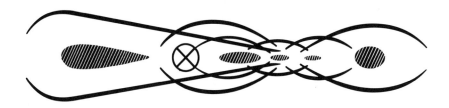

FIGURE 1-21

Immunoelectrophoretic pattern of normal serum protein to demonstrate characteristic γ precipitin band

COMPLEX PRECIPITIN BANDS

In more complex systems using polyvalent antiserum one obtains a great number of precipitin bands against one antigen and only a few against another. If both antigens are studied in the system outlined above, it is possible to relate the serologic properties of one antigen to the other.

Immunoelectrophoresis is very helpful where there are complex bands, since specific proteins produce specific characteristic precipitin bands. Observe, for example, a characteristic of the γ precipitin band: the right side of the band extends to the region of the α_2-globulin while the left side is on the cathode side of the γ-globulin (Fig. 1-21). This configuration is characteristic of the γ precipitin band and makes it one of the most easily identified bands in an immunoelectrophoretogram.

As many as 29 precipitin bands may be seen if polyvalent antisera and fresh serum for antigen are used (Fig. 1-22). Identification of each of these bands has been pursued with interest throughout the world. Their positions are relatively stable, and their shape is consistent. If one focuses attention on that area between the γ- and β-

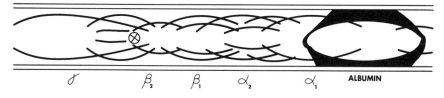

FIGURE 1-22

Immunoelectrophoretic pattern obtained with polyvalent antisera and fresh serum as antigen

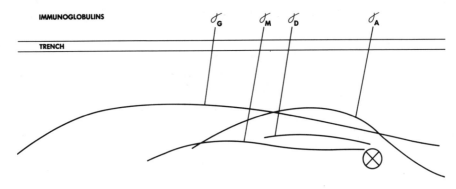

FIGURE 1-23

Precipitin bands of the immunoglobulins

globulins, one will note that there are four precipitin bands in this area (Fig. 1-23). These are the precipitin bands formed from immunoglobulins described by Heremans [6]. These immunoglobulins that travel electrophoretically in the zone between the β- and the γ-globulins have been designated as such because they carry antibodies—*immuno* —and also share important physical and chemical properties—*globulins*. In general, there are four immunoglobulins (Fig. 1-23): (1) γG, or IgG (immunoglobulin G); (2) γA, or IgA (immunoglobulin A); (3) γM, or IgM (immunoglobulin M); and (4) γD, or IgD (immunoglobulin D). As indicated, "γG" may be called "IgG" and so forth, both forms being acceptable [2]. The Svedberg units for each immunoglobulin are as follows:

Immunoglobulin	Svedberg Units
γG (IgG)	7
γA (IgA)	7
γM (IgM)	19
γD (IgD)	7

Obviously from the above the immunoglobulin γM could be differentiated from the other three by the use of an ultracentrifuge. The structure of these immunoglobulins is covered in Chap. 2.

ABNORMAL PRECIPITIN BANDS

As said earlier, the position of the bands is relatively stable, and their shape is consistent, and this rule applies to all the bands including the four immunoglobulin bands. Any deviation in the shape or position of the bands shown in Figure 1-23 is an *abnormal* condition and results from an *abnormal concentration of antigen*. Hence, one should expect any of the bands to be distorted when there is an excess of a particular immunoglobulin, for this would be an increased concentration of antigen. An excess protein concentration is often associated

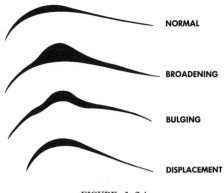

FIGURE 1-24
Distorted precipitin bands

with dysproteinemia and involves an excess of a particular immuno-globulin. Thus, if the concentration of γA immunoglobulin is higher than normal, a distorted γA precipitin band will result. If instead the γG precipitin band is distorted, the abnormal protein condition would be explained as an excess of γG immunoglobulin. If two immuno-globulins are in excess, both respective precipitin bands will appear distorted. Distortion of a precipitin band can be in the form of broad-ening, bulging, or displacement, as shown in Figure 1-24. The distorted band denotes an excess concentration of a specific protein.

One can identify the abnormal protein—paraprotein—by observing which immunoelectrophoretic band is distorted. It is especially im-portant, however, to note that immunoglobulin excess is not usually discernible by viewing just an electrophoretic pattern. Even if it were discernible by scanning the pattern for a narrow peak, the observation would be totally insufficient for complete identification because a para-protein peak can be either γG, γA, γM, or γD. Thus to identify the nature of a paraprotein peak—that is, to identify an abnormal protein—the diagnostic laboratory must rely upon immunoelectrophoresis.

OUCHTERLONY TECHNIC APPLIED TO IMMUNOELECTROPHORESIS

Electrophoretic heterogeneity of a serum protein or other protein system may or may not be due to separate proteins. In some instances the net charge difference is on the basis of carbohydrate difference, and in such cases removal of the carbohydrate moiety results in elec-trophoretic homogeneity. In other instances the heterogeneity is a reflection of different proteins. Resolution of multiple electrophoretic protein zones as to their respective serologic specificity can usually be done with immunoelectrophoresis by employing the same principles as Ouchterlony developed for double gel diffusion. In immunoelectro-phoresis the antigens are separated by electrophoresis and then

FIGURE 1-25
Reaction of identity

reacted by gel diffusion with a specific antiserum. In Figure 1-25 electrophoretic heterogeneity of protein with γ-globulin mobility is shown. In one antiserum trench is placed specific anti-γG. Note reaction of identity, showing that the proteins in the γ-globulin zone are serologically identical. This pattern is seen normally with spinal fluid transferrin.

The two other types of serologic reactions are shown in Figures 1-26 and 1-27: reactions of nonidentity and incomplete identity respectively.

In certain conditions several antigens which are not identical also have the same electrophoretic mobility; therefore the immunoprecipitin bands occur in the same area. If the amount of each antigen is different, the immunoprecipitin bands are found at different levels between the antigen well and the antiserum trench, as shown in Figure 1-28. This situation occurs principally in the α_2-globulin area with α_2-glycoprotein, haptoglobin, ceruloplasmin, and α_2-macroglobulin. Although the immunoprecipitin bands are in practice not as uniform and symmetrical as shown, the position of the immunoprecipitin bands is a guide to the identity of the serum antigen if their respective concentrations are within the normal range. Their respective positions are principally a function of their concentration. This pattern is com-

FIGURE 1-26
Reaction of nonidentity

ANTI ɣG

FIGURE 1-27
Reaction of incomplete identity

parable to an Ouchterlony double diffusion system with antiserum trench and antigen well as shown in Figure 1-12.

Osserman [11] employed a double gel diffusion superimposed on immunoelectrophoresis for identification of precipitin arcs. Electrophoresis is carried out as usual. Two trenches are made on either side of the line of electrophoresis, as in Figure 1-21. Antiserum is placed in one trench and a specific antigen to which the antiserum reacts in the other. The antiserum reacts with the separated protein and also with the antigen from the opposite trench. If the two proteins have the same antigen the immunoprecipitin band is clearly distorted.

Rarely one sees identical mobility and identical or nearly identical position of the immunoprecipitin band, as exemplified in an occasional observed situation involving γG immunoprecipitin band. Immunoglobulin γG is designated by its heavy-chain reaction, but the two

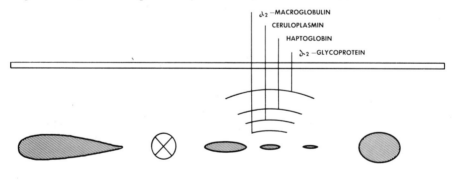

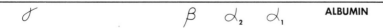

FIGURE 1-28
Antigens with essentially the same electrophoretic mobility but different concentration

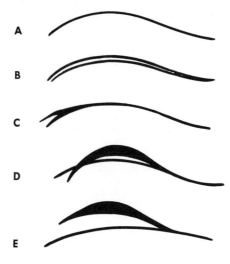

FIGURE 1-29

Types of "split" γG immunoprecipitin bands. A. Normal configuration. B. Double banding. C. Spur in a case of hypergammaglobulinemia. D and E. Multiple myeloma.

types of light chains, present on all immunoglobulins, are also available to react if the antisera, besides having antibodies directed at the heavy chain, also have antibodies directed at the two classes of light chain (ϰ and λ; see Chap. 2).

Figure 1-29 depicts four patterns of the split γG immunoprecipitin band that are most frequently encountered compared to the usual pattern at A. The patterns in B, C, D, and E suggest reaction of partial identity since each of the two immunoprecipitin bands appears to blend at the right. The blending is probably related to the fact that along the point of blending the antigens (light chains ϰ and λ) are in essentially the same concentration. The shifting of γG is usually associated with a condition of stress. Patterns shown are all from patients with specific diseases: B, hypogammaglobulinemia; C, hypergammaglobulinemia with unusual post-γ peak; and D and E, multiple myeloma.

Two-Dimensional Immunoelectrophoresis

Two-dimensional immunoelectrophoresis, as noted in Figure 1-30, consists of primary electrophoresis followed by electrophoresis at right angles against antiserum which, because of the differential electrophoretic mobility of antibody and antigen, cross each other during the second electrophoretic phase and react.

The principles of two-dimensional crossing electrophoresis were

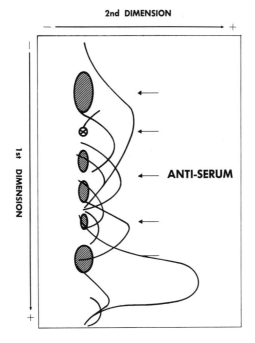

2nd DIMENSION

ANTI-SERUM

FIGURE 1-30

Diagram of technic of two-dimensional immunoelectrophoresis. The anti-serum is placed in agarose or buffer in the second dimension.

outlined by Nakamura et al. [10], who studied plant protein reaction with serum proteins. Ressler [15] incorporated antisera in the second stage. The method has been further refined by Laurell [8] for studies of serum protein heterogeneity. *Single gel diffusion* as depicted by Oudin [13] is similar in principle since antigen is moving through gel containing antiserum. Crossing-over electroprecipitin tests can be carried out in one dimension by placing antiserum just ahead of antigen so that during electrophoresis antibody and antigen will cross over each other and produce precipitin bands. This method has been employed primarily for quantitative studies.

References

1. Bodman, J. Agar Gel, Starch Block, Starch Gel and Sponge Rubber Electrophoresis. In Smith, I. (Ed.), *Chromatographic and Electrophoretic Techniques,* Vol. II, *Zone Electrophoresis.* New York: Interscience, 1963. Pp. 91–157.
2. Bulletin World Health Organization. Nomenclature for human immunoglobulins. *Bull. W. H. O.* 30:447, 1964.
3. Crowle, A. *Immunodiffusion.* New York: Academic, 1961.

4. Davis, B. J., and Ornstein, L. A new high resolution electrophoresis method. Delivered at meeting of the Society for the Study of Blood at the New York Academy of Medicine, March 24, 1959.

5. Grabar, P., and Williams, C. A., Jr. Méthode permettant l'étude conjuguée des propriétés électrophorétiques et immunochimiques d'un mélange des protéines: Application au sérum sanguin. *Biochim. Biophys. Acta* 10:193, 1953.

6. Heremans, J. F. Immunochemical studies on protein pathology: The immunoglobulin concept. *Clin. Chim. Acta* 4:639, 1959.

7. Kohn, J. Small-scale membrane filter electrophoresis and immuno-electrophoresis. *Clin. Chim. Acta* 3:450, 1958.

8. Laurell, E. -B. Antigen-antibody crossed electrophoresis. *Anal. Biochem.* 10:358, 1965.

9. McDonald, H. J. *Ionography, Electrophoresis in Stabilized Media.* Chicago: Year Book, 1955.

10. Nakamura, S., Tanaka, K., and Murakawa, S. Specific protein of legumes which react with animal proteins. *Nature* (London) 188:144, 1960.

11. Osserman, E. F. A modified technique of immunoelectrophoresis facilitating the identification of specific precipitin arcs. *J. Immun.* 84:93, 1959.

12. Ouchterlony, O. Antigen-antibody reactions in gels. *Acta Path. Microbiol. Scand.* 26:507, 1949.

13. Oudin, J. Immunologie. Méthode d'analyse immunochimique par précipitation spécifique en milieu gélifié. *C.R. Acad. Sci.* (Paris) 222:115, 1946.

14. Raymond, S., and Wang, Y. J. Preparation and properties of acrylamide gel for use in electrophoresis. *Anal. Biochem.* 1:391, 1960.

15. Ressler, N. Two-dimensional electrophoresis of protein antigens with an antibody containing buffer. *Clin. Chim. Acta* 5:795, 1960.

16. Smithies, O. Zone electrophoresis in starch gels: Group variations in the serum proteins of normal human adults. *Biochem. J.* 61:629, 1955.

17. Tiselius, A. A new apparatus for electrophoresis: Analysis of colloidal mixtures. *Trans. Faraday Soc.* 33:524, 1937.

18. Wieme, R. J. *Agar Gel Electrophoresis.* Amsterdam: Elsevier, 1965.

19. Wunderly, C. Immunoelectrophoresis: Methods, Interpretation, Results. In Sobotka, H., and Stewart, C. P. (Eds.), *Advances in Clinical Chemistry,* Vol. 4. New York: Academic, 1961. Pp. 207–270.

2

MONOCLONAL, POLYCLONAL, AND DYSCLONAL GAMMOPATHIES

Immunoglobulins: Immunochemical Characterization of Gammopathies

It is well established that malignant or excessive proliferation of plasma cells is associated with serum protein changes detected by electrophoresis [9, 14, 25]. These abnormalities of electrophoretic patterns have given rise to such descriptive terms as *hypergammaglobulinemia, M peaks, myeloma peaks,* and *gamma-globulin peaks.* The term *gamma globulin* is no longer precise enough for the protein molecules migrating electrophoretically in the γ- and β_2-globulin areas. Heremans [53] has introduced the term *immunoglobulins* to specify a group of biologically important proteins migrating in this area since they carry antibody activity. The three classes of proteins in this electrophoretic area were originally defined as $\gamma(7S)$, $\beta_2A(7S)$ and β_2M (19S) [53, 55, 79], but they are now designated γG, γA, and γM, or IgG, IgA, and IgM by international agreement [12]. A new immunoglobulin is γD [94, 119]. The electrophoretic and immunoelectrophoretic patterns related to densitometric scan are shown in Figure 2-1. γE, the most recently described immunoglobulin, is associated with reaginic activity [61, 62]. Immunoelectrophoretic patterns of the immunoglobulins are discussed in Chap. 1.

The purpose of this chapter is to demonstrate the role of serum protein zone electrophoresis (paper, cellulose acetate, and agarose) [14] and immunoelectrophoresis [48] in the clinical laboratory in categorizing various diseases involving plasma cells and lymphocytes [32, 33, 47, 54, 93, 98], and to summarize the relationship of these findings to recent information pertaining to the molecular structure of immunoglobulins [78, 79, 84, 110]. Molecular sieve electrophoresis (starch gel [104] or acrylamide gel [26, 27, 77, 86]) is not applicable to the diagnostic laboratory on a routine basis at this time [14]. However, for genetic studies such procedures are required [103]. In some instances heterogeneity of myeloma proteins can be detected only by molecular sieve electrophoresis [31]. Moving boundary electrophore-

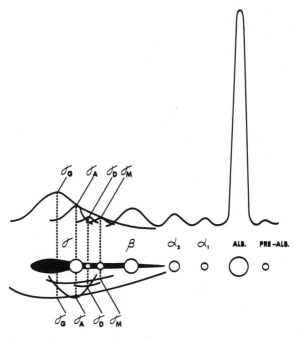

FIGURE 2-1

Drawing of electrophoretogram of serum proteins, densitometric scan, and immunoprecipitin bands of immunoglobulins γG, γA, γM, and γD. Drawing is designed to illustrate relative positions of immunoglobulins in an electrophoretogram and immunoelectrophoretogram. Anode is on the right.

sis, developed by Tiselius [109], also is not applicable for the diagnostic laboratory but does have a use in protein research laboratories.

Each electrophoretogram should be reviewed for abnormalities. When one is present in the γ-, β-, or α_2-globulin area, an immunoelectrophoresis is suggested to determine the serologic specificity of the abnormal protein. Within 24 to 36 hours after a serum sample has been obtained, it is possible to establish the serologic classification of the immunoglobulin responsible for a peculiar disturbance in the electrophoretic pattern [95].

Immunoglobulins and Dysproteinemia

The molecular structure of immunoglobulins consists of four polypeptide chains—two heavy chains (H) and two light chains (L)—and various carbohydrate side chains (Figs. 2-2 and 2-3) [84]. These molecules are manufactured by plasma cells and lymphocytes. Each immunoglobulin is thought to be synthesized by a specific clone. (A clone may be defined as a group of cells arising from a single an-

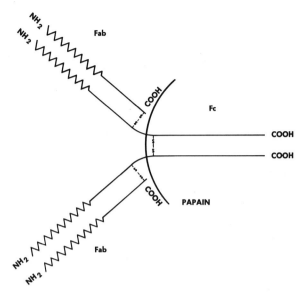

FIGURE 2-2

Diagram of immunoglobulin molecule showing position of 2 heavy (H) chains (*long curved lines*) and two light (L) chains (*short lines*). The variable portion of amino acid sequence is depicted in each chain as a wavy line. These specific areas are located on the N-terminal ends of both L and H chains. Two such antibody sites are shown for the molecule. The point where the heavy chain is broken by papain is shown. The position of breakage yields two Fab fragments. The H chains of the Fc fragment contain a disulfide linkage and remain together.

cestral cell.) If there is excessive production of molecules by a clone or an increase in number of cells of a clone, then there is an excess of a specific immunoglobulin, which electrophoretically is seen as a narrow zone of dense staining protein usually in the β or γ zone. A narrow protein zone in an electrophoretogram is a reflection of the homogeneity of the protein; that is, all of the molecules have essentially the same charge. A densitometric tracing of the electrophoretogram shows a peak or spike to match the dense protein zone and has led to the usage of such terms as *church steeple spike, M spike or peak,* and more recently *monoclonal peak* [115, 116].

The term *monoclonal peak* (Fig. 2-4) is more in keeping with the true nature of protein synthesis, and the disease state is referred to as monoclonal gammopathy [115, 116]. Lymphoproliferative disease, particularly multiple myeloma and macroglobulinemia, is an example of *monoclonal gammopathy.* The carbohydrate moiety of the immunoglobulin may differ without changing the net charge, at least as far as zone electrophoresis can detect electrophoretic heterogeneity

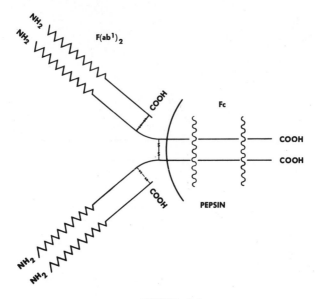

FIGURE 2-3

Diagram of immunoglobulin molecule identical to that in Figure 2-2, but demonstrating the point where pepsin attacks the H chains. Note that the Fc area is fragmented while the remaining H chains in combination with the two L chains yield one $F(ab^1)_2$ bivalent antibody unit.

[66]. Fahey has shown by starch-gel electrophoresis that there is considerable electrophoretic heterogeneity of some myeloma proteins; molecular heterogeneity has at least one more dimension, which is antigenic determinants such as described by Fessel for macroglobulins [34]. Carbohydrate heterogeneity of the immunoglobulin making up a monoclonal peak can be demonstrated by reaction between plant proteins and the immunoglobulin. The plant protein from the jack bean precipitates only those glycoprotein molecules with a specific terminal carbohydrate group—for example, N-acetyl glucosamine [66]. If more than one clone of cells participates, each making its respective immunoglobulin, multiple monoclonal peaks (biclonal, triclonal, etc.) may be found.

A polyclonal peak may be seen if there is a uniform blend of protein in the γ-globulin zone of the electrophoretogram. In the latter condition, referred to as *polyclonal gammopathy*, there is an increase of one or more immunoglobulins which are electrophoretically heterogeneous. This pattern is frequently observed in autoimmune disorders such as rheumatoid arthritis and lupus erythematosus [80]. If clones of cells normally present are absent, in reduced numbers, or dysfunctional, then a state of *aclonal* or *dysclonal gammopathy* develops. It is characterized by decrease of protein in the γ-globulin

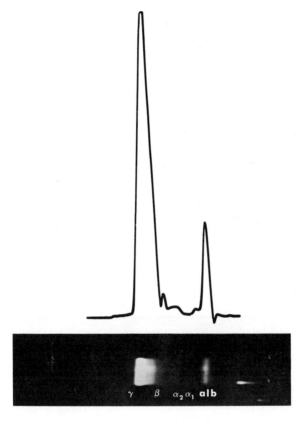

FIGURE 2-4

Photograph of an electrophoretic analysis of the "Check Sample" specimen of the Council on Clinical Chemistry, No. CC-34 [15]. The monoclonal peak in the γ-globulin region is enormous. Anode is on the right. (From L. P. Cawley [15].)

zone of the electrophoretogram. Examples of these changes are agammaglobulinemia (congenital or acquired) and acquired disturbances of the reticuloendothelial system such as lymphatic leukemia.

The structural formula of an immunoglobulin molecule as originally proposed by Porter delineated four peptide chains, two light chains and two heavy chains, bound together by disulfide bonds (Figs. 2-2 and 2-3) [84]. Selective cleavage of these bonds followed by isolation and purification led to several interesting conclusions. Light chains, the common structural subunits of all immunoglobulins, usually exist as dimers with a combined molecular weight of 50,000 (Table 2-1), have no carbohydrate content in the molecule, are related to or synonymous with Bence Jones protein, and exist in two immunologic forms, kappa ($\varkappa$) and lambda (λ). H chains are the spe-

TABLE 2-1

PROPERTIES AND MOLECULAR STRUCTURE OF THE IMMUNOGLOBULINS

Properties	γA	γG	γM	γD
Molecular weight	150,000	150,000	800,000	—
Carbohydrate content percent	10.5	2.5	12.0	—
Ultracentrifuge	7S (9,11,13S)	7S	19S (24S,32S)	7S
Molecular structure (2 heavy chains + 2 light chains)				
H chain (M.W. 50,000)	alpha (α)	gamma (γ)	mu (μ)	delta (δ)
L chain (M.W. 25,000)	lambda (λ)	λ	λ	λ
(Two serologic types)	kappa $(\varkappa)$	$\varkappa$	$\varkappa$	$\varkappa$
Complete formulas	$\alpha_2 \lambda_2$	$\gamma_2 \lambda_2$	$(\mu_2 \gamma_2)_5$	$\delta_2 \lambda_2$
	$\alpha_2 \varkappa_2$	$\gamma_2 \varkappa_2$	$(\mu_2 \varkappa_2)_5$	$\delta_2 \varkappa_2$

SOURCE: Modified from L. P. Cawley [16].

cific structural subunits of the immunoglobulin molecule, have (each) a molecular weight of 50,000, contain carbohydrate, bear no relationship to Bence Jones protein, and determine the class of immunoglobulins.

Clarification of the structure of immunoglobulins has resulted in a better understanding of the diseases which arise from abnormal synthesis of protein. Under usual circumstances all clones together manufacture 60 percent $\varkappa$ chains and 30 percent λ chains. The remainder are not typable [57]. Each clone manufactures either $\varkappa$ or λ chains but not both. In a malignant clone all of the light chains of the immunoglobulin produced are either $\varkappa$ or λ type. There is no mixture in the malignant state. This situation is consistent with the basic concept of the biology of tumors: tumors arise from single cells and thus represent a clone. As can be seen in Table 2-1, for the macroglobulins (γM) the unit formula $\mu_2\varkappa_2$ or $\mu_2\lambda_2$ is polymerized into five units, giving a total molecular weight of roughly 800,000, which is the approximate molecular weight of these macroglobulins [79, 84].

MONOCLONAL GAMMOPATHY

Several disease states of plasma cells can be precisely classified immunologically. In many instances the neoplastic cells of multiple myeloma manufacture a complete molecule of both heavy and light chains; however, in some cases there is excess production of light chains, which, being small, appear in the urine. These light chains may or may not exhibit the thermal properties characteristic of Bence Jones proteins. It is important to recognize that finding a monoclonal peak in an electrophoretogram is insufficient evidence to establish a

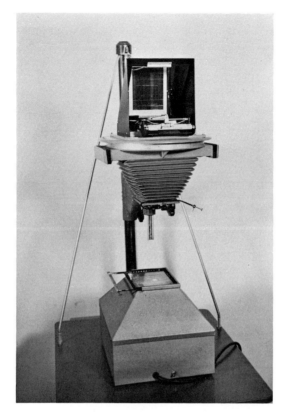

FIGURE 2-5

Device developed in our laboratory known as the Immuno-Glo [16], which is used in conjunction with a Polaroid MP-3 or, as shown here, a special camera, the Solar Immuno-Glo camera, to photograph immunoelectrophoretograms [18]. The Solar Immuno-Glo camera uses Polaroid film. As described in Figure 2-4 serum is subjected to immunoelectrophoresis, intensified in tannic acid for seven minutes, and photographed using the Immuno-Glo camera set up as shown to produce the photographic readout in Figure 2-6.

specific diagnosis of myeloma or macroglobulinemia. A definitive diagnosis can be established only with the use of immunoelectrophoretic analysis employing specific antisera (Figs. 2-4, 2-5, 2-6, and 2-7), bone marrow examination, and clinical evidence of disease. Rapid reports from the laboratory are speeded by using photographic readout of immunoelectrophoretograms (Figs. 2-5 and 2-7) [16]. By immunoelectrophoresis of serum and urine with specific antisera all known gammopathies of myeloma type and also macroglobulinemia of Waldenström can be defined. Table 2-2 lists the findings necessary to establish a definitive diagnosis of malignant dysproteinemia: elec-

TABLE 2-2

ABNORMAL IMMUNOLOGICAL SYNTHESIS AND ASSOCIATED CLINICAL AND LABORATORY FINDINGS

| Monoclonal Peak | | Bence | Skeletal | Bone | Molecular | Diagnosis |
Serum	Urine	Jones	Lesions	Marrow	Structure	
X	X	X	X	Positive	$\alpha_2\lambda_2$; $\alpha_2\varkappa_2$ (γA) $\gamma_2\lambda_2$; $\gamma_2\varkappa_2$ (γG) $\delta_2\lambda_2$; $\delta_2\varkappa_2$ (γD) L chains in urine	Multiple myeloma
X	O	O	X	Positive	As above without L chains	γA, γG, or γD type
X	X	O	O	Lymph nodes Lymphosarcoma	γ-H chains	H chain disease
X	O	O	O	Positive	$\mu_2\lambda_2$ or $\mu_2\varkappa_2$ (γM)	Macroglobulinemia
O	X	X	X	Positive	λ or $\varkappa$ L chains	B.J. myeloma
O	X	X	O	Negative	λ or $\varkappa$ L chains	Amyloid
X	O	O	O	Negative	$\alpha_2\lambda_2$; $\alpha_2\varkappa_2$ (γA) $\gamma_2\lambda_2$; $\gamma_2\varkappa_2$ (γG)	Monoclonal gammopathy of unknown etiology
O	X	X	O	Negative	λ or $\varkappa$ L chains	Lymphosarcoma

SOURCE: Modified from L. P. Cawley [16].

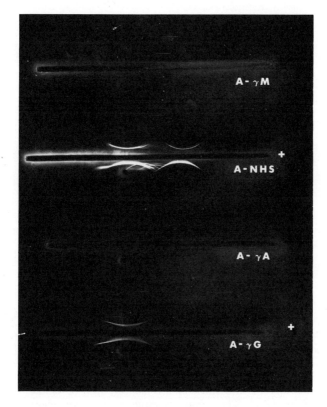

FIGURE 2-6

Immunoelectrophoretogram from the Council on Clinical Chemistry "Check Sample" No. CC-34 [15], showing the monoclonal peak to have serologic specificity of γG. This is determined by noting distortion of the γG immunoprecipitin band, particularly with specific anti-γG serum. This sample was analyzed some three months before test material was available. In the interim γM became elevated, and the patient thus had a biclonal gammopathy in which both γG and γM were increased. (From L. P. Cawley [15].)

trophoretic analysis of serum protein and urine, Bence Jones determination in urine, skeletal lesions, bone marrow findings, and the possible formula of the immunoglobulin involved. There are three serologic types of multiple myeloma, γA, γG, and γD, which may exist with or without Bence Jones proteinemia. H chain disease, recently described [79], was predicted on a theoretical basis. In this disorder the abnormal clone manufactures excess H chain or exclusive H chain molecules. The H chain dimer is found in the urine and serum in H chain disease. The H chain molecule is small enough to appear in the urine, but the plasma is not cleared of all H chains. Since the H chain molecule contains the specific serologic marker of the immunoglobulins, a

FIGURE 2-7

View looking down on the Immuno-Glo illumination box showing the position of the 5″ by 7″ film holder with agarose-covered plastic sheet with immunoelectrophoretogram [16, 22]. The lighting system directs light at an oblique angle to the surface of the gel, permitting intense illumination of the bands. Photographic reproduction with Polaroid film is swift, and immediate use of the information for interpretation is possible [16]. Immunoelectrophoretograms are subsequently soaked free of tannic acid, dried, and stained for filing. The same principle is used for photographing all immunoelectrophoretic and immunodiffusion studies [19].

specific anti-H chain was used to define this peculiar protein in the urine. Thus far, only γ, the H chain of γG, has been found [79]. The γ chain, unlike Bence Jones proteins, does not have thermal properties. Patients with H chain disease have lymphosarcoma without skeletal involvement. L chain production in pathologic conditions is relatively common compared to H chain disease. L chains, being of small size, readily pass through the renal glomeruli and appear in the urine in multiple myeloma, amyloid disease, and unusual lymphoproliferative disease with only Bence Jones protein production.

It should be recognized in reviewing Table 2-2 that a combination of electrophoretic analysis of urine and serum, Bence Jones protein determination of urine, skeletal survey, and bone marrow findings is necessary for a definitive diagnosis. It must be stressed that electrophoretic and immunoelectrophoretic analysis alone are insufficient to establish the diagnosis of a malignant dysproteinemia. In approximately 10 to 20 percent of cases with monoclonal peaks no clinical, malignant syndrome can be identified [79].

Types of monoclonal gammopathies (monoclonal peaks in serum and/or urine) found in patients studied in our laboratories during 35 months (Table 2-3) are similar to those reported by Osserman et al. [79] and Ritzmann et al. [87] in most respects (Table 2-4) [101]. γG is, in each instance, the most prevalent immunoglobulin produced in malignant monoclonal gammopathies. The implication is that there are more γG clones in the body to become malignant than there are clones producing γA, γD, or γM immunoglobulins. None of these

TABLE 2-3

MONOCLONAL GAMMOPATHIES (MG) IN SERUM AND URINE
Classification of 57 Cases

Type of MG	Cases	%
MG with plasma cell proliferative disease		
Multiple myeloma		
γG	5	8.9
γG + B. J.	5	8.9
γA	1	1.7
γA + B. J.	0	0.0
B. J. only	2	3.5
	(13)	(23.0)
H chain (Franklin's) disease	0	0
H chains of G		
Macroglobulinemia of Waldenström	(2)	(3.5)
γM		
MG with lymphoproliferative disease		
γG	1	1.7
B. J. only	1	1.7
	(2)	(3.5)
MG with epithelial malignancy		
γG—bladder	1	1.7
γA—uterus	1	1.7
Untyped—kidney	1	1.7
	(3)	(5.0)
MG of undetermined etiology[a]		
Associated with specific diseases		
Adequate diagnostic work-up	6	10.5
Insufficient diagnostic work-up	12	21.0
	(18)	(31.0)
Idiopathic (no apparent disease)		
Adequate diagnostic work-up	6	11.0
Insufficient diagnostic work-up	13	23.0
	(19)	(34.0)

[a]See Table 2-5 for further breakdown of this category.

TABLE 2-4

RELATIVE FREQUENCY OF VARIOUS TYPES OF MONOCLONAL GAMMOPATHY

Type of MG	Ritzmann's Composite Survey	Ritzmann	Osserman	Wesley
γG	54%	58%	58%	75%
γA	20%	17%	16%	9%
γM	19%	18%	11%	5%
L chain	6%	5%	15%	7%
H chain, untypable, other	1%	2%	1%	4%
Total cases	2727	339	397	57

studies can be expected to give the absolute frequency of the specific types of monoclonal gammopathies present and developing in the population since all were performed on hospitalized patients and not on the population at large. As Osserman et al. [79] surmised, the increased use of electrophoresis in hospitals and physicians' offices has resulted in the detection of numerous occult dysproteinemias, many of which prove to be caused by malignant plasma cells or lymphoproliferative disease. A good number of these occult dysproteinemias do not, of course, fit into our concept of clinical disease and are classified as monoclonal gammopathy of unknown etiology, first suggested by Waldenström [114, 115, 116], or essential monoclonal gammopathy [51]. According to the findings of our study, monoclonal gammopathies of unknown etiology make up a significant number of the total (Table 2-3) [101].

The monoclonal gammopathies in the group with no evidence of malignant plasma cell or lymphoproliferative disease represent a sizable proportion of the total cases. Monoclonal gammopathies not related to these diseases are listed in Table 2-5 as being associated with specific disease or as idiopathic (no specific disease). The size of the monoclonal peak may be of future value in identifying patients as candidates to be followed for development of malignancy. The small peaks predominate in those patients with autoimmune disorders and in apparently healthy persons.

Longitudinal studies under way in many laboratories will undoubtedly clarify the natural history of the development of dysproteinemia in malignant plasma cell and lymphoproliferative diseases. Current studies are inadequate to justify a firm prediction on the outcome in patients with monoclonal peaks found in their serum. Osserman et al. [79] suggest that many such patients will eventually develop clinical evidence consistent with plasma cell malignancy. Hallen [51], in a study of 150 subjects without myelomatosis, found no definite conclusion as to whether M components are related to diseases other

TABLE 2-5

MONOCLONAL GAMMOPATHY OF UNDETERMINED ETIOLOGY
37 Cases or 65%

Associated with specific diseases—16 cases or 32%

Adequate diagnostic work-up	Cases	%		
γG + B. J.				
Asthma	1	1.7		
γG				
Subacute sclerosing leukoen-cephalitis	1	1.7		
Polycythemia vera	1	1.7		
Hypertension	1	1.7		
B. J. only				
Acute nephritis + CVA	1	1.7		
γG early + B. J. later				
Rheumatoid arthritis + agranulocytosis	1	1.7		
	(6)	(10.5)		
Insufficient work-up			Cases	%
γG				
Asthma			1	1.7
Cholecystitis			1	1.7
Rheumatic heart disease			1	1.7
Arteriosclerotic heart disease			5	8.9
Thrombocytopenia			1	1.7
Thrombophlebitis + pulmonary embolism			1	1.7
Arteriosclerosis obliterans			1	1.7
γM				
Pemphigus			1	1.7
			(12)	(21.0)

Idiopathic (no specific disease apparent)—16 cases or 32%

Adequate diagnostic work-up	Cases	%		
γG	5	8.9		
γA	1	1.7		
	(6)	(10.5)		
Insufficient work-up				
γG			11	19.0
γA			1	1.7
Untyped			1	1.7
			(13)	(22.4)
Totals	12	21%	25	43%

than myelomatosis, macroglobulinemia of Waldenström, or chronic lymphatic leukemia. Monoclonal peaks were found in 3 percent of patients over 70 years of age. Although Hallen believes there is strong evidence of the existence of benign essential monoclonal gammopathy, the borderline between benign conditions and malignancy is diffuse. A single observation is unquestionably not definitive. Norgaard [74] reported cases which progressed to clinical myeloma in from 6 to 17 years after initial detection of "idiopathic" monoclonal gammopathy. Because of the uncertain nature of monoclonal peaks a follow-up program is deemed mandatory. Our own study, although short, suggests that large monoclonal peaks carry a more dire probability of eventual malignancy than small peaks, particularly those that are 1 cm. or less in height. Kyle et al. [64] suggest that the ratio of height to width of a peak at half peak height is a useful criterion in determining whether a peak is monoclonal. Peaks characteristic of products of monoclonal proliferation do have narrow width compared to height, and in some instances it may be necessary to define a monoclonal peak by measurement. We have not had to resort to measurement in order to differentiate a monoclonal peak from another type of peak. Immunoelectrophoresis with specific antisera is much more to the point since it is a direct immunochemical test of the homogeneity of the proteins in a given area. We are primarily interested in establishing that homogeneity is the main feature of a given band or peak in an electrophoretogram. In fact, biclonal or triclonal peaks (Fig. 2-8) cannot be properly defined except by specific antisera in immunoelectrophoretograms. Electrophoretic heterogeneity versus immunochemical heterogeneity can be resolved only by immunoelectrophoresis. Examples of monoclonal peaks can be seen in Figure 2-8.

In view of the seriousness of the implied malignant potential in a finding of monoclonal peak in a patient (regardless of size), guidelines for assuring the patient of adequate follow-up are important. The patients who are undergoing periodic monitoring in our study are examined at six-month intervals for two years and then yearly. The examination includes systems review, physical examination, and laboratory tests. A serum protein electrophoresis is compared with the original for change in size and shape of the monoclonal peak. The rate of change in the concentration of the monoclonal peak is ascertained by measuring peak height. Decrease in albumin and of globulins other than the globulin comprising the abnormal protein is looked for. Urine protein electrophoresis and test for Bence Jones protein are made. A hemogram is also part of the laboratory study. The patient is reassured if all is well. If the height of the monoclonal peak has increased, the patient is requested to undergo a more extensive study including x-ray examination. A bone marrow examination is not performed until the patient either has x-ray evidence of disease, shows increase of monoclonal peak, complains of bone pain, or develops anemia.

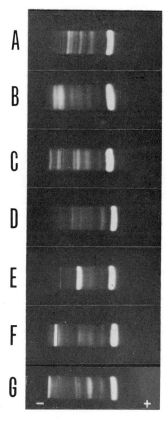

FIGURE 2-8

Seven electrophoretic patterns from patients with various types of monoclonal gammopathies. A. From a patient with no disease at this time, but
showing three very small monoclonal peaks, which by immunoelectrophore-
sis are designated as γG, γA, and γM. The bright band just anodic to the
point of application is γM. B. Triclonal gammopathy kindly furnished to
us by Harold Bell, M.D., Edmonton, Canada. Two monoclonal peaks show
clearly. By immunoelectrophoresis γG, γA, and γM were demonstrated.
C. Biclonal peaks (or two monoclonal peaks from same clone?). By immunoelectrophoresis both peaks are γG. The patient at this time has converted to frank myeloma. D. A small γG peak is seen just behind the β_2
band. Although a very delicate band, by immunoelectrophoresis it was
shown to be γG. The patient also shows hypogammaglobulinemia. He had
a plasmacytoma in the right femur detected following a pathologic fracture.
E. Monoclonal gammopathy of γA specificity. The patient has myeloma.
F. γM monoclonal peak in a patient with macroglobulinemia of Waldenström. G. Monoclonal peak of γG type. The patient died before diagnosis.
In E, F, and G the three major immunoglobulins are shown. γA characteristically travels on the anodic side of the point of application, whereas γM
may travel at the point of application or on either side of the point of application. γG is deceptive and may travel in many areas.

A very enlightened discussion of this subject is contained in a paper by Hobbs [56]. Central to the effectiveness of detecting malignant monoclones by demonstration of monoclonal peak in electrophoretograms are the amount of protein produced each day, the sensitivity of detection, and the quantity of protein produced per cell. Hobbs [56] correctly points out that clinically evident multiple myeloma usually is detected when there is about 4.3 gm. per 100 ml. of abnormal protein. In mice with plasma cell tumors the serum level of abnormal protein increases linearly with the volume of tumor. Hobbs states that with cellulose acetate he can detect 0.2 gm. per 100 ml. of pure paraprotein added to an otherwise normal serum. If 12 months is used as a median doubling time of tumor tissue (assuming equal performance of new cells as to production of protein), it would take five years for serum protein levels to reach 4.3 gm. per 100 ml. from a starting level of 0.2 gm. per 100 ml. Hobbs also notes that, if one assumes origin of tumor from a single cell with constant doubling time, then extrapolation to zero suggests that the tumor started many years before.

It is evident from the studies of Hobbs [56] and Hallen [51] that rate of increase of abnormal protein (monoclonal peak) is indicative of benignancy or malignancy of the clone manufacturing the abnormal protein. A monoclonal peak that remains unchanged in height for a year or more or disappears altogether is unquestionably the product of a benign clone. If the doubling rate of a tumor is taken at the longest figure found by Hobbs (25 months), a significant change in the concentration of abnormal protein (height of monoclonal peak) should be evident within 12 months and certainly by two years.

A direct approach to the problem of defining the nature of small monoclonal peaks merits investigation. Since immunoglobulins produced by malignant plasma cells usually carry no antibody specificity, particularly as related to amount of protein produced, malignancy might be excluded if specific antibody activity could be demonstrated in the small monoclonal peak. Antibody activity per milligram of antibody would test the affinity of the antibody; however, as mentioned, monoclonal proteins, by and large, do not have detectable antibody activity. A few instances of cold agglutinin activity particularly in macroglobulinemia of Waldenström have been described, and it seems worth considering that there is an antigen in nature to which the monoclonal protein would react. The question of whether these monoclonal proteins are nonsense or inert molecules—i.e., cell production of protein is fully expressed but no antibody specificity is evident—cannot now be answered.

DIAGNOSTIC INDEX OF ELECTROPHORESIS

Many laboratory tests when abnormal indicate a high probability of a specific disease. For example, elevated blood glucose bears a high correlation with diabetes. It has been of some interest to us to determine first the incidence of monoclonal peaks per electrophoretogram in our institution and to develop some understanding of the significance of a monoclonal peak and a specific disease. We have chosen to use the term *diagnostic index* or *probability index* to describe the relationship of monoclonal peak and malignant plasma cell disease.

The incidence of monoclonal peaks found per electrophoretogram during almost three years was 0.45, 0.75, and 1.50 percent respectively for 1965, 1966, and 1967 (Table 2-6) [101]. If the figure of 1.5 mono-

TABLE 2-6

INCIDENCE OF MONOCLONAL GAMMOPATHY

Year	Electro-phoretograms	Monoclonal Peaks	% MG/ Electro.
1965	2350	11	0.46
1966	3200	25	0.78
1967 (Jan. through Nov.)	2000	30	1.50

clonal peaks per 100 electrophoretic analyses can be accepted as a relatively stable finding, then the fraction of these that were from malignant plasma cell disease is a figure which would be helpful in establishing an understanding of the natural history of the disease. Over the three-year period of our experience (1965, 1966, and 1967), 37 percent of patients with monoclonal peaks proved to have malignant plasma cell disease, lymphoproliferative disease, or epithelial malignancy. It should be apparent from the foregoing discussion that additional cases will convert to frank myeloma from this group during the next 48 months or longer. Recognizing these limitations one can approximate the number of monoclonal peaks found in any period that are from overt or occult malignancy. In 1967, 30 monoclonal peaks were found. About 11, or 37 percent of 30 cases, should have or develop clinical multiple myeloma. Thus far, 9 of the 30 cases with monoclonal peaks have been diagnosed as having multiple myeloma. Superficially one might say that the finding of a monoclonal peak carries a 37 percent risk of being a result of a malignant condition. Tempting as such generalities are, it has been our experience that the picture is not as neat as all that. Furthermore, the greater the amount of abnormal proteins, the greater the chance of malignancy. If the figure of 37 percent risk is coupled with other findings, more reliability is possible. Greater exactness is possible if instead we determine the rate of increase, say at three and six months. Assuming maximum doubling time of 25 months for a tumor, a 12.5 percent increase in six months would be serious, particularly if the increase of abnormal protein continued.

CHROMOSOMAL ABNORMALITIES

Chromosomal changes in lymphoproliferative disorders, particularly myeloma [13, 67] and macroglobulinemia of Waldenström [6, 8], have been reported. There are two types of chromosomal aberrations frequently detected in all types of monoclonal gammopathies [58]. The first is a large supernumerary chromosome in the AB size range with variable centromeric position, designated the MG chromosome. The second is an abnormality in pair 12 of group C which may involve a missing chromosome, an extra chromosome, or a structural defect. Chromosomal karyotyping will undoubtedly become commonplace in many diagnostic studies of lymphoproliferative disease associated with dysproteinemia in the near future.

POLYCLONAL GAMMOPATHY

The development of the concept of clones of lymphocytes and plasma cells and their uniqueness from the standpoint of protein synthesis make possible a more basic approach to interpretation of serum protein electrophoretograms. Diseases known to give rise to polyclonal gammopathy were previously referred to as hypergamma-globulinemias. Agarose or agar-gel electrophoresis, as pointed out by Wieme [118], and Papadopoulos and Kintzios [83], provides better resolution than that available through paper and has added some refinement in the evaluation of electrophoretograms. Gross inspection of the stained electrophoretogram or this in combination with densitometric scan is sufficient to establish whether or not an individual pattern falls into the category of monoclonal, polyclonal, or dysclonal pattern (Figs. 2-8, 2-9, 2-10).

Polyclonal peaks from a gross inspection standpoint may be divided into two broad groups: (1) symmetrical or diffuse, in which the γ-globulin fraction appears to approach a theoretical distribution and a scan shows a theoretical curve; and (2) nonsymmetrical or skewed, a pattern which is most frequently associated with skewing toward the cathodic side. Two divisions of the latter type are frequently encountered: In the first the β-γ linking [28] is characterized by tailing toward the anode into the region of the α_2 area and is associated with increase of γG and γA immunoglobulins; in the second there is no linking of the β-γ area, and the pattern is usually associated with γG increase. The symmetrical type is most frequently associated with autoimmune conditions and often with increase of α_2-globulin. The nonsymmetrical type, particularly the β-γ linking, is likely to be found in serum of patients with Laennec's cirrhosis, whereas the pattern with no β-γ linking is frequently associated with infectious hepatitis [76, 83, 87, 88].

Polyclonal peaks are seen in many conditions, and Table 2-7 outlines the breakdown of the types that are usually associated with polyclonal gammopathy, with classification of the pattern usually asso-

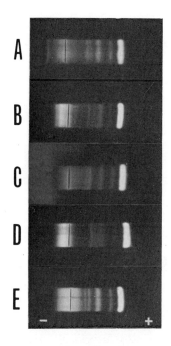

FIGURE 2-9

Series of electrophoretograms showing polyclonal peaks. A. From the patient described in Chap. 4 who had long-standing rheumatoid arthritis, showing elevation of γG globulin. In addition, a small post-γ monoclonal peak of γG specificity is shown. This later disappeared. B. From a small girl with chronic cirrhosis, showing two discrete monoclonal peaks within the substance of the γG. The two peaks are not clearly shown in the photograph. C. Cirrhosis with slight elevation of the γ-globulin. Notice the bridging between the γ-globulin fraction and the β-globulin. D. Hepatitis. Note the skewing of γ-globulin to the cathodic side. E. Long-standing cirrhosis. Notice the bridging of γ-globulin zone with the β-globulin.

ciated with diseases—that is, symmetrical or nonsymmetrical—and the immunoglobulins most frequently found to be elevated. By all odds, liver disease is most often associated with polyclonal gammopathy, and decreased albumin is an additional hallmark of the disease [76, 80, 87, 88, 106]. In autoimmune disease, particularly rheumatoid arthritis, there is a symmetrical increase in the γ-globulin fraction and an increase of α_2-globulin [76, 106]. Chronic infections of bacterial or viral origin are frequently associated with an increase of α_2- as well as γ-globulin [83, 87, 88, 106].

Although increase of γ-globulin is nonspecific, the increased resolution available through agarose electrophoresis has improved the usefulness of gross impressions, which are now more meaningful than

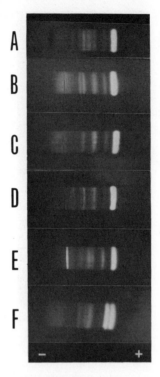

FIGURE 2-10

Electrophoretic patterns of normal, unusual, and hypogammaglobulinemic types. A. Hypogammaglobulinemia with extra band traveling between the α_1 and the α_2. In some instances this band represents hemoglobin from hemolyzed erythrocytes. B. A normal pattern. Notice the six zones. C. Complete absence of the β_2-globulin which is normally occupied by complement. D. Hypogammaglobulinemia with decreased α_1. An extra component cathodic to the point of application is present. Its nature is unknown. E. Serum from a patient with 80 percent third-degree burns showing hypogammaglobulinemia. Notice the extra band lying between α_1 and α_2, which in this case was hemoglobin. Also, notice the very dense band traveling in front of the point of application, which is the position usually taken by fibrinogen and should not be confused with a monoclonal peak. F. Bisalbuminemia of the A/B type. (Furnished by L. W. Kleppe, M.D., Beloit, Wisconsin).

formerly. In time, information will accumulate that will permit more precise pattern recognition. Tissue injury characterized by α_1 increase is reasonably well documented [83, 118]. Changes in reference to tissue injury are frequently observed in malignancy of epithelial or of the lymphoid system [83, 87, 106]. In either instance there is also an increase in γ-globulin, usually of the symmetrical type, associated

TABLE 2-7

POLYCLONAL GAMMOPATHIES AND TYPES OF DISEASE

Disease	Type of Peak	Predominant Immunoglobulin	Associated Findings
A. Autoimmune			
a. Lupus erythematosus	Symmetrical	γG	↓Albumin
b. Rheumatoid arthritis	Symmetrical	γG γA γM	↓Albumin
c. Hemolytic anemia	Symmetrical	γG	↓Haptoglobin
d. Cirrhosis	Skewed β/γ fusion	γG γA	↓Albumin
B. Infectious			
a. Bacterial tuberculosis	Symmetrical	γG	Normal α_2-globulin
b. Viral			
1. Hepatitis	Skewed	γG	↑β-lipoprotein, ↓ albumin
2. Infectious mononucleosis	Skewed	γG γM	↓Albumin
c. Fungal			
1. Actinomycosis	Symmetrical	γG γA γM	Normal α_2-globulin
C. Malignancy			
a. Epithelial with metastases	Symmetrical	γG	↑α_2-globulin, ↑α_1-globulin
b. Hodgkin's	Symmetrical	γG	↑α_2-globulin
c. Monocytic leukemia	Symmetrical	γG γA γM	↑α_2-globulin
D. Other			
a. Sarcoid	Symmetrical	γG γA γM	Normal α_2-globulin

with an increase of α_2- and sometimes of α_1-globulins. It is not yet possible to utilize the electrophoretic polyclonal shape to establish the presence of a specific disease (Fig. 2-9). Cirrhosis, infectious hepatitis (frequently associated with an increase of β-lipoprotein), rheumatoid arthritis, and autoimmune diseases are diseases or disease groups in which the shape of the polyclonal peak or association with other changes is reasonably reproducible. More detailed breakdown of gross patterns is impossible at this time. Polyclonal gammopathy has been described in the mink and is referred to as Aleutian mink disease [75]. In some animals arthritis develops, and its severity is correlated with high levels of γ-globulin [113]. The disease develops spontaneously and is of viral etiology [52]. Occasionally monoclonal gammopathy with Bence Jones proteinuria appears.

By immunoelectrophoresis the immunoglobulins in various polyclonal gammopathies can be determined and their concentration estimated. This approach offers additional information and permits somewhat more discrimination of the nature of polyclonal gammopathy. The polyclonal peak represents extreme heterogeneity, in which many clones contribute their products. It becomes apparent that there is a considerable number of plasma cell clones stimulated by a host of antigens to produce their respective protein, which appears somewhere in the electrophoretic area of the γ-globulin. Not infrequently a polyclonal peak with a small, very narrow zone typical of a monoclonal peak is observed. This probably represents an isolated clone product. In some instances we have observed the same thing in the post-γ position (Fig. 2-9A). Figure 2-9 shows a number of patterns of polyclonal gammopathy. One (B), from a young girl with liver disease, discloses two small monoclonal peaks within the framework of the diffuse hypergammaglobulinemic pattern. Familial hypergammaglobulinemia has been studied by Waldenström, and he has found symptomatic and asymptomatic members of families [114, 115]. Seligmann et al. [100] described a familial β_2-macroglobulinemia (γM).

DYSCLONAL OR ACLONAL GAMMOPATHIES

These terms are used in our laboratory to describe conditions in which the γ-globulin is low or absent as far as the gross observation of the electrophoretogram is concerned. This can be discerned either by visual observation of the stained pattern or such observation in combination with the densitometric scan. Quantitative immunoassay in addition to immunoelectrophoresis is necessary to classify the condition into a scheme related to the antibody-deficiency syndrome (Table 2-8). Rosen and Janeway [90] recently reviewed the antibody-deficiency syndromes and suggested a system of classification based on information obtained by many investigators. The various types of antibody-deficient syndromes are shown in Table 2-8. The hypogammaglobulinemic syndrome may be transient or permanent and is

TABLE 2-8

CLASSIFICATION OF DYSCLONAL OR ACLONAL GAMMOPATHIES

Antibody-Deficient Syndrome	Findings
Transient hypogammaglobulinemia	Infants, very low γG, γA, γM, for first 9–15 months
Congenital agammaglobulinemia	
Sex-linked	Infants, extremely low γG, γA, γM, male only
Autosomal	Infants, very low γG, γA, γM, males and females
Acquired agammaglobulinemia	
Primary	Adults, sudden depression of γG, γA, and γM
Secondary	Adults, children, depression of γG, γA, and γM by systemic disease
Swiss-type agammaglobulinemia[a]	Infants, lymphopenia, thymic aplasia, autosomal
Dysgammaglobulinemia (congenital and acquired)	
Type I	Low γG and γA, increased γM, sex-linked
Type II	Low γA and γM, increased γG
Type III	Absence of γA

[a]Hereditary thymic aplasia.

45

broken down into transient hypogammaglobulinemia, agammaglobulinemia congenital or acquired, Swiss-type agammaglobulinemia (hereditary thymic aplasia), and dysgammaglobulinemia congenital and acquired. As will be seen, it is important to have adequate knowledge of the normal concentration of immunoglobulins in infants to properly interpret concentration of immunoglobulins. Values of serum γG, γA, and γM immunoglobulins in children are covered in Chap. 15 under Quantitation of Immunoglobulins.

In the transient type of hypogammaglobulinemia the infants usually recover at 9 to 15 months of age. In the stage of defective protein synthesis they are susceptible to infections involving skin, lung, meninges, and respiratory tract.

Congenital agammaglobulinemia is sex linked, occurring only in males [43]. Male infants afflicted are particularly susceptible to bacterial infections, which may be persistent until proper therapy is administered. Viral diseases are not unusually detrimental to these infants. Good et al. [44] called attention to the fact that approximately one-third to one-half of these children have a condition similar to rheumatoid arthritis which disappears when substitute therapy is instituted. Atopic, allergic, and hypersensitivity manifestations occur with increased frequency in agammaglobulinemic children. Protein synthesis is disrupted, and the serum γG is less than 100 mg. per 100 ml. Serum γA and γM are usually present in less than 1 percent of normal values. Both $\varkappa$ and λ chains are formed. Defective or poor protein synthesis is easily demonstrable since antigenic stimulus to any number of antigens fails to cause an antibody response. The isoagglutinins are low, and the Schick test is usually not satisfactory. The basic defect appears to be the absence of the plasma cells from lymph nodes, spleen, intestine, and bone marrow [42]. In addition to the above, there is disorganization of the lymphoid tissue. Normal follicular organization is lacking, and the tissue lymphocytes appear diminished in number.

Sporadic or autosomal agammaglobulinemia has been reported. We have seen a family in which the only children, two male infants, died from abscesses 18 months apart at the age of 2. Serum on the last infant taken at autopsy was subjected to electrophoresis, and an aclonal pattern with complete absence of γ-globulin was found. Immunoelectrophoresis and immunodiffusion established no appreciable quantities of γG, γA, or γM (10, 0, and 0 mg. per 100 ml. respectively). The father did not have adequate quantities of γG, γA, or γM (10, 16, and 31 mg. per 100 ml. respectively), although he was not known to suffer from unusual infection. The mother had normal amounts of γG, γA, and γM. In the child, lymphoid tissue was scanty and was deficient of lymphocytes. The thymus was small but did contain Hassall's corpuscles.

Acquired agammaglobulinemia is distinguished from the above in that the hypogammaglobulinemia or agammaglobulinemia develops later in life. Although a well-established familial trend is not present, some suggestion along this line has been made. The condition is autosomal. Clinically the patients are susceptible to pyogenic infections, particularly involving the lung, and bronchiectasis may develop. A complication of acquired agammaglobulinemia is a spruelike syndrome [97], characterized by diarrhea, steatorrhea, and at times protein loss. The immunoglobulins in the serum are decreased. γG is usually less than 500 mg. per 100 ml. γA and γM are present but usually decreased. Lymphoid tissue shows a lack of plasma cells. Patients with primary acquired agammaglobulinemia have an unusual proneness to lymphoreticuloneoplasia. Gafni and his co-workers noted the frequent association of thymomas with primary acquired agammaglobulinemia [41, 63, 68] and became interested in the thymus and immunologic competence.

In secondary acquired agammaglobulinemia defective synthesis of γ-globulin is associated with neoplasia of the lymphoid system, particularly reticulum cell sarcoma, lymphosarcoma, and Hodgkin's disease. These diseases may show polyclonal peaks on electrophoresis. Patients with secondary agammaglobulinemia or dysclonal peaks probably are susceptible to a number of infections because of their basic disorder. Persons with monoclonal peaks are placed in this category because the synthesis of the other immunoglobulins is depressed.

Swiss-type agammaglobulinemia has been labeled hereditary thymic aplasia by Rosen and Janeway [90] because of the genetic transmission of the defect and the failure of the embryonic differentiation of the thymus gland. The defect is transmitted as an X-linked characteristic in some families and is autosomal in others. Frequent infections by gram-negative enteric bacilli is also a feature of the disease. Death usually results from *Pseudomonas* infection of the lung. The infants are unduly susceptible to vaccination, developing severe and often fatal vaccinia [89, 91]. γG, γA, and γM are usually absent in the serum, although some cases have been recorded in which some γA or γG or γM has been found. The lymphocyte count of peripheral blood usually is less than 1500 per cu. mm. Delayed hypersensitivity is absent. In all the other conditions described before, blood lymphocytes, when challenged by extract of *Phaseolus vulgaris*, will undergo transformation with development of mitosis. In the Swiss type of agammaglobulinemia the lymphocytes fail to respond to the stimulus of the phytohemagglutinin. The lymphoid tissue in afflicted infants is almost absent. Lymph nodes contain stroma or reticular cells but no lymphocytes and no plasma cells. The thymus is usually small, and no Hassall's corpuscles are found.

The pathogenesis of agammaglobulinemia of the Swiss type is of considerable interest. Miller [70] called attention to the importance of the thymus in the development of immune response. Thymectomy of mice in the neonatal period gives rise to a syndrome which resembles the condition just described. These animals lack small lymphocytes in the follicles of their lymph nodes and spleen, cannot manifest delayed hypersensitivity, and are incapable of rejecting a homograft. Patients with Swiss-type agammaglobulinemia eventually develop a situation resembling the runting syndrome of the graft-versus-host reaction. The immunologic defect can be repaired by transplantation of thymic tissue in experimental animals and even by thymic tissue inside of Millipore chambers. From these experiments it has been concluded that the thymic gland plays a central role in the maturation and distribution of lymphocytes whose function is crucial to the delayed hypersensitivity and homograft rejection phenomena. It is not yet entirely clear whether lymphocytes originate within the thymus and migrate from there to the lymphoid centers or whether the lymphocytes originate in the bone marrow and subsequently migrate to the thymus where they receive some instruction that renders them immunologically competent. Although the thymus appears to elaborate a hormone which imparts function to the lymphocytes, no lymphocytotropic hormone has as yet been isolated.

The above described conditions involve defective synthesis of all three of the major immunoglobulins. There are, however, a number of other conditions in which one or two of the immunoglobulins are defective. These have been labeled dysgammaglobulinemias. Three types are generally recognized. Type I, the most frequent, is characterized by decrease of γA and γG and increased concentration of γM. Although the γM immunoglobulin level may reach 150 to 1000 mg. per 100 ml., a monoclonal peak is rarely observed in the electrophoretogram. It, therefore, appears to be made up of many molecules giving rise to electrophoretic heterogeneity. The condition may be acquired or congenital [91]. In addition to being susceptible to infections, some patients develop thrombocytopenia, neutropenia, renal lesions, aplastic anemia, or hemolytic anemia, possibly related to an autoimmune disorder. The disorder, apparently can be transmitted as an X-linked phenomenon; however, it also occurs in girls. In the hereditary form of the disease, the histopathology of lymphoid tissue is noteworthy since no follicular formation is evident. In the spleen and other lymphoid tissue the L-P cells characteristic of Waldenström's disease are found.

In type II dysgammaglobulinemia γG is increased or normal, whereas γA and γM are decreased. In agammaglobulinemia, the γG molecule present has antibody activity. In type I dysgammaglobulinemia the γM immunoglobulin also has antibody activity, but the γG in type II appears to be inert [6].

In type III dysgammaglobulinemia γA is decreased or absent. A small proportion of the normal population [3] have a decrease of γA.

A high proportion of patients with a hereditary ataxia telangiectasia have a decrease of γA [120]. Patients with type III dysgammaglobulinemia have frequent respiratory infections.

Biological Significance of Immunoglobulins

STRUCTURE AND FUNCTION OF THE IMMUNOGLOBULINS

The general structure of the immunoglobulin molecule has been characterized in a relatively short period of time by several investigators. At this time the three-dimensional structure is not known in full detail, since pure crystalline antibody has not been made in quantities adequate for crystallographic studies. Summation of the concepts of the structure of immunoglobulin molecules as currently conceived from the work of Porter [84, 85], Edelman and Gally [29], Nisonoff et al. [73], and Franklin [38] is outlined in Figures 2-2 and 2-3. The immunoglobulin molecule is bivalent—the active sites being composed of the end amino acids of H and L peptide chains. L and H chains are held together by disulfide bonds and covalent bonds [57, 73, 84, 85]. In the diagram the antibody active sites are at the N-terminal of each pair of L and H chains, which is a part of the variable portion of the molecule as far as the sequence of amino acids is concerned. The remaining portion of both the L and H chains represents the coded sequence of amino acids depicting the specific serologic type of the basic molecule, the L chain, that is, being either $\varkappa$ or λ and the H chain being either γ, α, μ, or δ. A sequence of only about 12 amino acids at the active site of L and H is specifically designed for antigen recognition. About 107 amino acids constitute the variable section of both H and L chains. The overall measurements of the immunoglobulins are 35 Å in diameter and 240 Å in length [29], the molecular weight is approximately 150,000, and there are about 1400 amino acid residues [57, 73, 84, 85].

Since the molecule is bivalent, it is capable of combining with two separate antigens or attaching to the antigen at two separate sites. The variability of only a small portion of an antibody molecule is unique among proteins in that the overall molecule is very similar to others in the same class, for example, two γG immunoglobulins. Both may be serologically identifiable as γG immunoglobulins and yet each has its own specific reactive site directed toward a different antigen.

Treatment of the molecule with papain produced fragments (Fig. 2-2). Three fragments are derived from papain digestion—one fragment that can be crystallized (Fc), and two identical antibody fragments (Fab), which do not crystallize [84, 85]. The ab stands for antigen binding. The Fc fragment does not have antibody ability, whereas the other two behave as incomplete antibodies [57, 84, 85]. They are incapable of forming precipitates but are able to combine with antigen. The Fc fragment apparently is tied up biologically with the properties of passive cutaneous anaphylaxis, the ability of an antibody to cross the placenta, rate of catabolism (biologic half-life), and fixation of complement in an antigen-antibody reaction. Pepsin digestion of antibodies produces a bivalent antibody fragment of about 100,000 molecular weight and a number of small fragments which have no activity [73]. The bivalent fragment is designated $F(ab^1)_2$ (Fig. 2-3). It can be converted into two univalent fragments similar to Fab by reduction with propionic acid [73]. Table 2-9 summarizes some important features of immunoglobulins.

TABLE 2-9

IMMUNOGLOBULINS

Immunoglobulin	
Molecular weight	150,000
Amino acid residue	1,400
Size	19 Å × 35 Å × 280 Å
Peptide chains	2 heavy and 2 light
Heavy chains	
Molecular weight	55,000
Amino acid residue	430
Serologic types	γG, γA, γM, γD, γE
Variable amino acids	25% (N-terminal)
Light chains	
Molecular weight	22,000
Amino acid residue	214
Serologic types	ϰ, λ
Variable amino acids	50% (N-terminal)
Fc fragment	
Molecular weight	50,000
Amino acid residue	430
Peptide chain	2 portions of heavy chain
Percent variable amino acids	None
Fab fragment	
Molecular weight	50,000
Amino acid residue	430
Peptide chain	1 light chain and fragment of heavy chain
Antibody activity	Univalent
$F(ab^1)_2$ fragment	
Molecular weight	100,000
Amino acid residue	840
Peptide chains	2 light chains and 2 fragments of heavy chain equivalent to 2 Fab fragments
Antibody activity	Bivalent

HETEROGENEITY OF HEAVY CHAIN

Grey et al. [49] have described antigenic heterogeneity of the heavy chains of γG. Ne, We, Vi, and Ge, the four types delineated by Grey et al., now are referred to respectively as γ_{2a}, γ_{2b}, γ_{2c}, and γ_{2d} [108], or more conveniently as γG_1, γG_2, γG_3, and γG_4. These are found in all normal sera and are analogous to the $\varkappa$ and λ subtypes of the light chain.

ALLOTYPES

Antigenic determinants of the immunoglobulins that differ among normal individuals, referred to as allotypes, were originally detected by Grubb [50]. He found that a protein factor in some human rheumatoid serum would clump erythrocytes coated with incomplete anti-Rh$_0$ (D). He could distinguish differences in individual γG immunoglobulins by incorporating the test immunoglobulin in a system utilizing rheumatoid factor and erythrocytes coated with incomplete anti-Rh$_0$ (D). If the incorporated serum inhibited the reaction between the rheumatoid factor and the coated erythrocytes, the subject was said to be Gm positive. If the test serum did not inhibit the reaction, the subject was said to be Gm negative. Some individuals injected with γG from another individual in the same species may produce antibodies reacting with the donor's γG [1, 112]. The ability of one individual to respond by producing antibodies against another's γG depends upon whether they have had the same or different antigenic variants of immunoglobulin. Besides the Gm system, there is the Inv allotypic system. The determinants of the Gm system are numerous, and the inheritance is complex [72]. The Gm antigenic character is located on the heavy chain of γG and thus is not present in γA, γM, or γD [65, 82]. The Gm system involves only γG_1 [85]. The Inv system is confined to the light chains of the $\varkappa$ type and hence appears in all immunoglobulins [57, 65, 82].

BIOLOGIC ACTIVITIES OF IMMUNOGLOBULINS

As might be suspected, the immunoglobulin classes differ in their biologic properties. The concentrations of the immunoglobulins in the peripheral blood listed in Table 2-10 point out the great abundance of γG and the very small amount of γD. This difference apparently is not related to difference in synthesis but is more related to differences in clearance [7, 57], in particular of γG [59]. About 5 percent of the intravascular γG is catabolized daily. γM and γA are cleared two or three times or more of that rate [10], which, however, is independent of serum concentration [99] (Table 2-10). The catabolism of immunoglobulins resembles the clearance mechanism for removing such serum enzymes as lactate dehydrogenase (LDH) [81], glutamic pyruvic transaminase (GPT) [36], and glutamic oxylacetic transaminase

TABLE 2-10

SERUM IMMUNOGLOBULINS

Immuno-globulin	Concentration (mg./100 ml.)	Biologic Half-life (days)	Rate of Synthesis (gm./day)
γG	800–1600	25	2.3
γA	140–420	6	2.7
γM	50–200	5	0.4
γD	0.1–0.7	2.8	—
γE	0–0.1	—	—

(GOT) [36]. The clearance or catabolism of proteins such as enzymes is covered in Chaps. 3 and 7.

Only γG of the immunoglobulins passes the placenta to reach the fetus. The passage is due to an active transport process of the immuno-globulin across the trophoblastic cells of the placenta. The Fc part of the heavy chain is responsible for this property of γG [57]. γG also arises as a secondary response rather than a primary response follow-ing immunization.

γA is unique among the immunoglobulins since it is the only one secreted in colostrum, saliva, intestinal juices, and the respiratory secretions in great amount [23]. The concentration of γA in the secre-tions is greater than that of γG whereas in serum γG is 10 times as prevalent as γA. The process of secretion is not passive but active, and the γA in these body fluids differs from that in the serum by possessing a "secretor" piece, which is a small protein with 11S ultra-centrifuge characteristics [57]. The transport of the γA through epithelium such as the salivary gland has been demonstrated by immunofluorescent methods [57]. Some serum γA is made in plasma cells in the intestinal wall. γA monoclonal gammopathy is frequently associated with lipids and skin disease (see Chap. 4).

γM antibodies are the first to form after contact with antigen, read-ily combine with specific sites on cell membranes, and are effective in activating complement to produce immune lysis. It has been shown by electron microscopy that immune hemolysis causes small perforations in the cell membrane similar to those seen in erythrocyte membranes following treatment with saponin [5, 20]. γM antibodies are more than 100,000 times more effective in lysis of cells than molecules of γG anti-bodies [57]. It might appear that γM antibodies are adapted for deal-ing with particulate antigens such as bacteria in the blood stream. Also γM antibodies are large and are thus confined to the intravascular spaces.

The shift in antibody production from γM type to γG (7S) in the secondary phase has suggested that γM antibodies perform some important preliminary function but long-term immunologic protection follows the appearance of γG (7S) antibodies. A possible biologic function of γM antibodies has been described by Weir [117] in a hypothesis relating to immunologic consequences of cell death. This hypothesis suggests that following cell death the particulate cellular components are taken up by phagocytic cells, resulting in the formulation of γM autoantibodies. The soluble components of the cell which are not phagocytized induce immunologic tolerance by a yet unknown influence on lymphoid cells. A combination of the γM autoantibodies with the tissue antigen and complement releases a chemotactic agent which attracts neutrophils. Weir notes the experimentally supported observation that serum albumin rendered free of aggregates by ultracentrifugation will not call forth an antibody response when it is injected into a rabbit, but rather will induce tolerance. The tolerance, however, may be prevented by pretreatment of the animal with bacterial products, which causes an exaggerated antibody response. Although an established tolerance against proteins cannot be broken by this mechanism, it may play an important role in causing an animal to form autoantibodies to soluble products, whereas under noninfectious conditions it would produce tolerance.

Fudenberg [40], in an article reviewing the various hypotheses, reasons that autoimmune disease, rather than being an effect of increased synthesis of γ-globulins, is more likely to be a consequence of a specific immunologic defect of antibody synthesis and type of lymphocytes. Individuals with the disorder are, therefore, incapable of defending against an infectious agent which damages only certain organs of the body (tropism of viruses), releasing sequestered antigens or modified cellular antigens that stimulate proliferation of "forbidden" lymphocytes to react directly with site of cell injury or to produce autoantibodies. The "forbidden" lymphocytes, as proposed in the clonal selection theory of antibody formation, are mutants which are removed by other lymphocytes.

The finding in many areas that autoantibodies are present in normal individuals and in high concentration among relatives of persons with autoimmune disease has caused a number of investigators to consider that these autoantibodies are secondary to the basic disorder or possibly a physiologic response [21]. Grabar called attention to this possible explanation of autoantibodies in 1963 [46]. In this regard, as proposed by Weir [117], γM autoantibodies appear to perform a useful function by combining with tissue antigen and complement to release a factor causing attraction of neutrophils to the site of injury. In lupus erythematosus low-weight γM antinuclear autoantibodies have been described [92]. Stobo and Tomasi [105] found low-weight 7S autoantibodies of the γM type in the serum from hereditary ataxia telangiectasia, disseminated lupus, and Waldenström's macroglobulinemia. Only 10–15 percent of the total γM molecules were of the 7S type, and they were found in 17 percent of 52 patients with disseminated lupus erythema-tosus. The significance of these low-weight γM antibodies is unknown.

The skin-sensitizing antibodies have been referred to as reagins and have the unusual property of becoming fixed to tissues and re-maining attached for relatively long intervals of time [57]. The Prausnitz–Küstner reaction is a clinical demonstration of skin-sensi-tizing or skin-fixing antibodies. The challenging dose in such instances may be held for several days, but the antibody remains at the site of injection and is available for reaction when the challenging dose of antigen is injected. Reagin-type antibodies are of the γE type [61, 62].

ULTRASTRUCTURE OF IMMUNOGLOBULINS

Electron microscopic analysis of γG molecules to synthetic antigen shows a triangular arrangement suggesting that three immunoglobulins have shared three antigens (in this case, antigen has identical anti-genic groups at each end of the molecule). Since pepsin did not abolish the ability of the antibody to function, the Fc fragment is probably not required for immunologic reaction to take place. The final pattern resembles three Y's facing inward ⋏⋏ and joining to form a triangle △ , antibodies at the corners and Fc portion directed outward [85]. Electron microscopic study of purified Waldenström's macro-globulins and normal rabbit γM immunoglobulin revealed a spider-like structure with five legs. These preparations were not combined with antigen [107]. The structure resembles a circle with five spokes ✩ .

IMMUNOGLOBULINS AND ANTIBODIES

From the standpoint of antibody production, the best evidence suggests that the gene controlling the variable portion of each immuno-globulin—that is, the gene that codes for specificity of the antibody—is available at the time of contact of the antigen with the antibody-forming cells. What is not clear is why a clone becomes committed to production of a single specific antibody. The actual presence of immunoglobulins is demonstrable in many plasma cells by the fluores-cent antibody method of Coons [24]. The technic can be made dis-criminating by using a fluorescent-labeled antiserum specific for γG, γA, or γM, or specific for light chain or other antigenic variants. Used on frozen tissue sections, it has shown that individual plasma cells contain both a heavy and a light chain, but any one cell is restricted to the class, subclass, type, or allotype of immunoglobulin that is produced. If a molecule is γG in a particular plasma cell, it has ϰ or λ chains, but not both. Individuals heterozygous for the genes determin-ing allotype specificities show plasma cells which produce immunoglobulins carrying only one of the allotypic Gm or Inv specificities. It appears that there is a restricted capability in the individual plasma cell in production of a single immunoglobulin. Thus, lymphoid tissue maintains versatility in the range of molecular variants it can produce. As already mentioned, the clone that becomes malignant in lympho-proliferative diseases of the myeloma and macroglobulinemic types is committed to the production of a single class of molecules.

The antibody-forming cells appear to respond to only a single or at most two antigens, by undergoing rapid proliferation and eventual formation of plasma cells which produce and release a specific im-munoglobulin with specific antibody characteristic. Macrophages may be involved in the production of antibody [35]. A preliminary step in antibody production may involve digestion of the antigen by macro-phages. The digested antigen possibly complexes with ribonucleic acid (RNA) after release from the macrophage and upon contact with receptive clonal cells stimulates specific antibody formation by binding to membrane markers. A peptide-oligonucleotide complex, presumably an RNA antigen with immunogenic properties, has been isolated from antigen-exposed macrophages [45]. It may be that production of nonsense or inert immunoglobulins—that is, immunoglobulins with no antibody specificity—can be brought about by a variety of unknown stimuli, as is manifested in malignant transformation of plasma cells. At this time we have no knowledge as to what sets off the overproduc-tion of plasma cells in multiple myeloma. Although in some instances the myeloma proteins have been shown to have cold agglutinin activ-ity, this is not the usual situation. In general, myeloma proteins do not have antibody specificity. The stimulus to form immunoglobulins may well center on a surface membrane reaction to an unknown chemical mediator, possibly along the lines of the mitogens of plants, in partic-ular the extract of the kidney bean, *Phaseolus vulgaris*, which causes lymphocytes to undergo a series of changes, including blast formation [60, 69], and production of immunoglobulins [29, 30].

Thus, specific antibody production requires more than just stimu-lation of immunoglobulin production. The part that the macrophage plays in supplying additional information is thus very appealing. The antigen or antigen coupled to RNA from a macrophage may operate not only to stimulate the proliferating of lymphocytes and also im-munoglobulin synthesis but to supply the code for the variable portion of the immunoglobulin to render it specific for a certain antigen.

Metabolism of Immunoglobulins

SYNTHESIS

The rate of immunoglobulin synthesis is regulated by type and amount of antigen [99]. Very low levels of immunoglobulins in germ-free animals and newborn infants support the importance of antigen.

In contrast, high rates are found in hyperimmunized animals and in patients with chronic infection. The rates of synthesis of immunoglobulins differ in normal, hyperimmune, and germ-free mice. The highest rate occurs in hyperimmune animals, demonstrating the importance of antigen in determining the rate of synthesis of immunoglobulins. In a summary of data from several sources on immunoglobulins and the metabolic behavior of immunoglobulins in the human, Schwartz shows the rate of synthesis of γG and γA to be almost equal although their biologic half-lives are quite different [99]. As he points out, the serum concentration and the rate of synthesis of immunoglobulins (Table 2-10) are unequal, and the antigen must, therefore, play some role in controlling the synthesis of a specific immunoglobulin. For a number of sources it appears that the physical condition of the antigen has a distinct part in antibody response. The particulate type of antigen calls forth γM whereas the soluble type is more likely to call forth γG antibodies [99]. The weight of the antigen is similarly related. Both of these mechanisms bear on tolerance. Antigen cleared by high-speed centrifugation or some suitable method of removing all particulate matter surprisingly will not induce antibody formation in the experimental animal when injected but rather, as noted elsewhere, will induce a state of tolerance. Phagocytosis in induction of antibody synthesis is apparently an important step [35].

Although antigen is apparently intimately involved in the regulation of the synthesis of antibody, the antibody itself is likewise an important regulator of antibody synthesis. In 1909 Smith demonstrated that excess of antibody in a mixture of diphtheria toxin and antitoxin prevented an immune response to the toxin [102]. This is the principle currently employed in prevention of Rh immunization [39]. Uhr showed clearly that synthesis of antibody is under the influence of feedback; that is, the higher the serum antibody concentration, the greater the suppression of antibody production [111]. Each specific antibody influences its own synthesis and thus prevents overimmunization and excessive multiplication of a particular cell line. In animals not previously immunized, infusion of an antibody prevents immunization altogether. Small amounts of antibody injected into animals already producing antibody is inhibitory, and this effect is demonstrated within 24 hours after infusion of the antibody [96]. Of specific interest is the observation that specific γG antibodies have been shown to inhibit the production of corresponding γM antibodies, and the inhibition is 100 or more times the inhibiting efficiency of the γM antibodies [71]. Thus, the appearance of γG antibodies in the secondary phase or following the primary production of γM reduces the influence of the γM immunoglobulins since synthesis is suppressed by the γG antibody. Feedback restriction on antibody production may be defective in certain disorders characterized by excessive hyperglobulinemia (macroglobulinemia or benign hypergammaglobulinemia) [99].

CATABOLISM

Control of serum immunoglobulin levels by catabolic breakdown is important only for γG. The biologic half-life of γG is reduced in hypogammaglobulinemia and is increased if the serum level is high. The catabolic rate of γG is specifically related to the serum level and not to its rate of synthesis or to the concentration of the other immunoglobulins [59]. In contrast, breakdown of γM and γA is independent of their serum concentrations. The dominant factor controlling the synthesis of γG is antigen, not serum concentration [99]. The biologic half-life of γG is unrelated to its function as an antibody but is related to the Fc portion of the molecule [99].

References

1. Allen, J. C., and Kunkel, H. G. Antibodies to genetic types of gamma globulin after multiple transfusions. *Science* 139:418, 1963.
2. Bach, F., and Hirschhorn, K. Gamma globulin production by human lymphocytes *in vitro*. *Exp. Cell Res.* 32:592, 1963.
3. Bachmann, R. Studies on serum γ-A-globulin level: III. Frequency of A-γ A-globulinemia. *Scand. J. Clin. Lab. Invest.* 17:316, 1965.
4. Bachmann, R., and Laurell, C. B. Electrophoretic and immunologic classification of M-components in serum. *Scand. J. Clin. Lab. Invest.* 15:11, 1963.
5. Bangham, A. D., and Horne, R. W. Action of saponin on biological cell membranes. *Nature* (London) 196:952, 1962.
6. Barandun, S., Stampfli, K., Spengler, G. A., and Riva, G. Die Klinik des Antikorpermangel Syndroms. *Helv. Med. Acta* 26:163, 1961.
7. Barth, W. F., Wochner, R. D., Woldmann, T. A., and Fahey, J. L. Metabolism of human gamma macroglobulin. *J. Clin. Invest.* 43:1036, 1964.
8. Benirschke, K., Brownhill, L., and Ebaugh, F. G. Chromosomal abnormalities in Waldenström's macroglobulinaemia. *Lancet* 1:594, 1962.
9. Berlin, N. I., Merwin, R., Potter, M., Fahey, J. L., Carbone, P. P., and Cline, M. J. Neoplastic plasma cell. Combined Clinical Staff Conference at the National Institutes of Health. *Ann. Intern. Med.* 58:1017, 1963.
10. Birke, G., Norberg, R., Olhagen, B., and Plantin, L. -O. Metabolism of human gamma macroglobulins. *Scand. J. Clin. Lab. Invest.* 19:171, 1967.
11. Bottura, C., Ferrari, I., and Veiga, A. A. Chromosome abnormalities in Waldenström's macroglobulinaemia. *Lancet* 1:1170, 1961.
12. Bulletin World Health Organization. Nomenclature for human immunoglobulins. *Bull. W.H.O.* 30:447, 1964.
13. Castoldi, G. L., Ricci, N., Puntururi, E., and Bosi, L. Chromosomal imbalance in plasmacytoma. *Lancet* 1:829, 1963.
14. Cawley, L. P. Changing concepts in protein electrophoresis. *J. Kansas Med. Soc.* 64:470, 1963.
15. Cawley, L. P. Molecular pathology of dysproteinemias. Clinical Chemistry Check Sample No. CC-34, Commission on Continuing Education. Chicago: American Society of Clinical Pathologists, 1965.

16. Cawley, L. P. *Workshop Manual on Electrophoresis and Immuno-electrophoresis* (Rev. ed.). Commission on Continuing Education, Council on Clinical Chemistry. Chicago: American Society of Clinical Pathologists, 1966.

17. Cawley, L. P., and Eberhardt, L. Simplified gel electrophoresis: I. Rapid technique applicable to the clinical laboratory. *Amer. J. Clin. Path.* 38:539, 1962.

18. Cawley, L. P., Eberhardt, L., and Schneider, D. Simplified gel electrophoresis: II. Application of immunoelectrophoresis. *J. Lab. Clin. Med.* 65:342, 1965.

19. Cawley, L. P., Eberhardt, L., and Wiley, J. L. Double immunodiffusion with agar-coated plastic film base. *Vox Sang.* 10:116, 1965.

20. Cawley, L. P., and Goodwin, W. L. Electron microscopic investigation of the binding of bromelin by erythrocytes. *Vox Sang.* 13:393, 1967.

21. Cawley, L. P., Riner, A., Houser, C., and Huaman, A. M. Thyroid and autoimmune disorders; thyroid auto-antibodies in hospital patients. *J. Kansas Med. Soc.* 67:263, 1966.

22. Cawley, L. P., Schneider, D., Eberhardt, L., Harrouch, J., and Millsap, G. A simple semi-automated method of immunoelectrophoresis. *Clin. Chim. Acta* 12:105, 1965.

23. Chodirker, W. B., and Tomasi, T. B., Jr. Gamma-globulins: Quantitative relationships in human serum and nonvascular fluids. *Science* 142:1080, 1963.

24. Coons, A. H., Leduc, E. H., and Connolly, J. M. Studies on antibody production: I. Method for histochemical demonstration of specific antibody and its application to study of hyperimmune rabbit. *J. Exp. Med.* 102:49, 1955.

25. Crowle, A. L. *Immunodiffusion.* New York: Academic, 1961.

26. Davis, B. J. *Disc Electrophoresis,* Part II. Rochester, N.Y.: Distillation Products Industries, 1961.

27. Davis, B. J., and Ornstein, L. A New High Resolution Electrophoresis Method. Delivered at meeting of the Society for the Study of Blood, at the New York Academy of Medicine, March 24, 1959.

28. Demeulenaere, L., and Wieme, R. J. Special electrophoretic anomalies in the serum of liver patients: A report of 1145 cases. *Amer. J. Dig. Dis.* 6:661, 1961.

29. Edelman, G. M., and Gally, J. A. A model for the 7S antibody molecule. *Proc. Nat. Acad. Sci. U.S.A.* 51:846, 1964.

30. Elves, M. W., Roath, S., Taylor, G., and Israels, M. C. G. The in vitro production of antibody lymphocytes. *Lancet* 1:1292, 1963.

31. Fahey, J. L. Heterogeneity of myeloma proteins. *J. Clin. Invest.* 42:111, 1963.

32. Fahey, J. L. Antibodies and immunoglobulins: I. Structure and function. *J.A.M.A.* 194:183, 1965.

33. Fahey, J. L., and McLaughlin, C. Preparation of antisera specific for 6.6 s gamma-globulins, beta 2A-globulins, gamma-1-macroglobulins, and for type I and II common gamma-globulin determinants. *J. Immun.* 91:484, 1963.

34. Fessel, W. J.   Multiple antigenic determinants on macroglobulins. *Proc. Soc. Exp. Biol. Med.* 113:446, 1963.
35. Fishman, M., and Adler, F. L.   Antibody synthesis in x-irradiated recipients of diffusion chambers containing nucleic acid derived from macrophages incubated with antigen. *J. Exp. Med.* 117:595, 1963.
36. Fleisher G. A., and Wakim, K. G.   The fate of enzymes in body fluids—an experimental study: I. Disappearance rates of glutamic pyruvic transaminase under various conditions. *J. Lab. Clin. Med.* 61:76, 1963.
37. Fleisher, G. A., and Wakim, K. G.   The fate of enzymes in body fluids—an experimental study: III. Disappearance rates of glutamic–oxalacetic transaminase II under various conditions. *J. Lab. Clin. Med.* 61:98, 1963.
38. Franklin, E. C.   Structural units of human 7S gamma globulin. *J. Clin. Invest.* 39:1933, 1960.
39. Freda, V. J., Gorman, J. G., and Pollack, W.   Rh factor: Prevention of isoimmunization and clinical trial on mothers. *Science* 151:826, 1966.
40. Fudenberg, H. H.   Are autoimmune diseases immunologic deficiency states? *Hosp. Pract.* 3:43, 1968.
41. Gafni, J., Michaeli, D., and Heller, H.   Idiopathic acquired agamma-globulinemia associated with thymoma: Report of two cases and review of literature. *New Eng. J. Med.* 263:536, 1960.
42. Good, R. A.   Studies on agammaglobulinemia: II. Failure of plasma cell formation in bone marrow and lymph nodes of patients with agammaglobulinemia. *J. Lab. Clin. Med.* 46:167, 1955.
43. Good, R. A., Kelly, W. D., Rotstein, J., and Varco, R. L.   Immunological deficiency disease. *Progr. Allerg.* 6:187, 1962.
44. Good, R. A., Rotstein, J., and Mazzitello, W. F.   Simultaneous occurrence of rheumatoid arthritis and agammaglobulinemia. *J. Lab. Clin. Med.* 49:343, 1957.
45. Gottlieb, A. A.   Studies on a peptide-oligonucleotide complex with immunogenic properties from antigen-exposed macrophages. *J. Clin. Invest.* 46:1062, 1967.
46. Grabar, P.   The problem of autoantibodies: An approach to a theory. *Texas Rep. Biol. Med.* 15:1, 1963.
47. Grabar, P., Fauvert, R., Burtin, P., and Hartmann, L.   Études sur les protéines du myéloma. II. L'analyse immuno-électrophorétique des sérums de 30 malades. *Rev. Franc. Étud. Clin. Biol.* 1:175, 1956.
48. Grabar, P., and Williams, C. A.   Méthode permettant l'étude conjuguée des propriétés électrophorétiques et immunochimiques d'un mélange de protéines. Application au sérum sanguin. *Biochim. Biophys. Acta* 10:193, 1953.
49. Grey, H. M., and Kunkel, H. G.   H chain subgroups of myeloma proteins and normal 7S gamma globulin. *J. Exp. Med.* 120:253, 1964.
50. Grubb, R.   Agglutination of erythrocytes with incomplete anti-Rh by certain rheumatoid arthritic sera and some other sera: The existence of human serum groups. *Acta Path. Microbiol. Scand.* 39:195, 1956.
51. Hällén, J.   Discrete gammaglobulin (M-) components in serum.

Clinical study of 150 subjects without myelomatosis. *Acta Med. Scand.* Suppl. 462, p. 8, 1966.

52. Henson, J. B., Gorham, J. R., and Leader, R. W. Hypergamma-globulinemia in mink initiated by a cell-free filtrate. *Nature* (London) 197:206, 1963.

53. Heremans, J. F. Immunochemical studies on protein pathology: The immunoglobulin concept. *Clin. Chim. Acta* 4:639, 1959.

54. Heremans, J. F., and Heremans, M. T. Studies on "abnormal" globulins (M-components) in myeloma, macroglobulinemia and related disease: Immunoelectrophoresis. *Acta Med. Scand.* 170 (Suppl. 367): 27, 1961.

55. Hermans, P. E., McGuckin, W. F., McKenzie, B. F., and Bayrd, E. D. Electrophoretic studies of serum proteins in cyanogum gel. *Proc. Staff Meet. Mayo Clin.* 35:792, 1960.

56. Hobbs, J. R. Paraproteins, benign or malignant? *Brit. Med. J.* 3:699, 1967.

57. Holborow, E. J. An ABC of modern immunology. I. Links between the old and the new. *Lancet* 1:833, 1967; II. Antibodies anatomised. *Lancet* 1:890, 1967; III. Immunoglobulin physiology. *Lancet* 1:942, 1967; IV. Cellular immune faculties, competence and memory. *Lancet* 1:995, 1967; V. The immunological capability of small lymphocytes. *Lancet* 1:1049, 1967; VI. Cell-mediated immunity. *Lancet* 1:1098, 1967; VII. The genesis of immunological competence. *Lancet* 1:1148, 1967; VIII. Defects and evasions. *Lancet* 1:1208, 1967.

58. Honston, E. W., Ritzmann, S. E., and Levin, W. C. Chromosomal abnormalities common to three types of monoclonal gammopathies. *Blood* 29:214, 1967.

59. Humphrey, J. H., and McFarlane, A. S. Rate of elimination of homologous globulins (including antibody) from the circulation. *Biochem. J.* 57:186, 1954.

60. Hungerford, D. A., Donnelly, A. J., Nowell, P. C., and Beck, S. The chromosome constitution of a human phenotypic intersex. *Amer. J. Hum. Genet.* 2:215, 1959.

61. Ishizaka, K., Ishizaka, T., and Hornbrook, M. M. Physiochemical properties of human reaginic antibody: IV. Presence of a unique immunoglobulin as a carrier of reaginic activity. *J. Immun.* 97:75, 1966.

62. Ishizaka, K., Ishizaka, T., and Hornbrook, M. M. Physiochemical properties of reaginic antibody: V. Correlation of reaginic activity with gamma E-globulin antibody. *J. Immun.* 97:840, 1966.

63. Jacox, R. F., Mongan, E. S., Hanshaw, J. B., and Leddy, J. P. Hypogammaglobulinemia with thymoma and probable pulmonary infection with cytomegalovirus. *New Eng. J. Med.* 271:1091, 1964.

64. Kyle, R. A., Bayrd, E. D., McKenzie, B. F., Heck, F. J. Diagnostic criteria for electrophoretic patterns of serum and urinary proteins in multiple myeloma. *J.A.M.A.* 174:245, 1960.

65. Lawler, S. O., and Cohen, S. Distribution of allotypic specificities on the peptide chain of human gamma globulin. *Immunology* 8:206, 1965.

66. Leon, M. A. Concanavalin A reaction with human normal immuno-globulin G. *Science* 158:1325, 1967.

67. Lewis, F. J. W., MacTaggert, M., Crow, R. S., and Wills, M. R. Chromosomal abnormalities in multiple myeloma. *Lancet* 1:1183, 1963.

68. MacLean, L. D., Zak, S. J., Varco, R. L., and Good, R. A. Thymic tumor and acquired agammaglobulinemia: Clinical and experimental study of immune response. *Surgery* 40:1010, 1956.

69. Marshall, W. H., and Roberts, K. B. The growth and mitosis of human small lymphocytes after incubation with phytohaemagglutinin. *Quart. J. Exp. Physiol.* 48:146, 1963.

70. Miller, J. F. A. P. Thymus and development of immunologic respon-siveness. *Science* 144:1544, 1964.

71. Moller, G., and Wigzel, H. Antibody synthesis at the cellular level: Antibody-induced suppression of 19S and 7S antibody response. *J. Exp. Med.* 121:969, 1965.

72. Muir, W. A., and Steinberg, A. G. On the genetics of the human allotypes, Gm and Inv. *Seminars Hemat.* 4:156, 1967.

73. Nisonoff, A., Wissler, F. C., and Lipman, L. N. Properties of the major component of a peptic digest of rabbit antibody. *Science* 132:1770, 1960.

74. Norgaard, O. Recherches sur l'évolution préclinique du myélome multiple. *Acta Med. Scand.* 176:137, 1964.

75. Obel, A. L. Studies on a disease in mink with systemic proliferation of the plasma cells. *Amer. J. Vet. Res.* 20:384, 1959.

76. Ogryzlo, M. A., Maclachlan, M. J., Dauphinea, J. A., and Fletcher, A. A. The serum proteins in health and disease, filter paper electro-phoresis. *Amer. J. Med.* 26:596, 1959.

77. Ornstein, L. *Disc Electrophoresis*, Part I. Rochester, N.Y.: Distilla-tion Products Industries, 1961.

78. Osserman, E. F., and Lawlor, D. Immunoelectrophoretic character-ization of the serum and urinary proteins in plasma cell myeloma and Waldenström's macroglobulinemia. *Ann. N.Y. Acad. Sci.* 94:93, 1961.

79. Osserman, E. F., and Tahatsuki, K. Plasma cell myeloma: Gamma globulin synthesis and structure; a review of biochemical and clinical data with the description of a newly recognized and related syn-drome, "Hr-2-chain (Franklin's) disease." *Medicine* 42:357, 1963.

80. Osserman, E. F., and Tahatsuki, K. The plasma proteins in liver disease. *Med. Clin. N. Amer.* 47:679, 1963.

81. Palmer, D. L., and Cawley, L. P. Factors governing the hemostatic regulation of LDH isoenzymes in human serum. *Amer. J. Clin. Path.* 47:367, 1967.

82. Papadopoulos, N. M., and Kintzios, J. A. Differentiation of path-ological conditions by visual evaluation of serum protein electro-phoretic patterns. *Proc. Soc. Exp. Biol. Med.* 125:927, 1967.

83. Polmar, S. H., and Steinberg, A. G. Dependence of a Gm (b) anti-gen on the quaternary structure of human gamma globulin. *Science* 145:928, 1964.

84. Porter, R. R. Structure of Gamma Globulin and Antibodies. In Gill-horn, A., and Hirschberg, E. (Eds.), *Basic Problems in Neoplastic*

Disease. New York: Columbia University Press, 1962. Pp. 177–194.

85. Porter, R. R.   The structure of antibodies. *Sci. Amer.* 217:81, 1967.
86. Raymond, S., and Nakowichi, M.   Electrophoresis in synthetic gels: 1. Relation of gel structure to resolution. *Anal. Biochem.* 1:23, 1962.
87. Ritzmann, S. E., and Levin, W. C.   II. Polyclonal and monoclonal gammopathies. *Lab. Synopsis* 2:9, 1967.
88. Riva, G.   *Das Serumeiweissbild.* Bern and Stuttgart: Humber, 1957.
89. Rosen, F. S., and Janeway, C. A.   Danger of vaccination in lympho-penic infants. *Pediatrics* 33:310, 1964.
90. Rosen, F. S., and Janeway, C. A.   The gamma globulins: III. The antibody deficiency syndrome. *New Eng. J. Med.* 275:709, 769, 1966.
91. Rosen, F. S., Kevy, S., Merier, E., Janeway, C. A., and Gitlin, D.   Recurrent bacterial infections and dysgammaglobulinemia, deficiency of 7S gamma-globulins in presence of elevated 19S gamma globulin: Report of two cases. *Pediatrics* 28:182, 1961.
92. Rothfield, N. F., Frangione, B., and Franklin, E. C.   Slowly sedi-menting mercaptoethanol-resistant anti-nuclear factors related anti-genically to M immunoglobulins (gamma-1 M-globulin) in patients with systemic lupus erythematosus. *J. Clin. Invest.* 44:62, 1965.
93. Roulet, D. L. A., Spengler, G. A., and Hassig, A.   Immunoelectro-phoretical studies in paraproteins. *Vox Sang.* 7:281, 1962.
94. Rowe, D. S., and Fahey, J. L.   A new class of human immunoglob-ulins. I. A unique myeloma protein. *J. Exp. Med.* 121:171, 1965.
95. Ruggles, J. G., and Cawley, L. P.   Molecular pathology: Newer concepts of molecular pathology in dysproteinemias. *J. Kansas Med. Soc.* 67:72, 1966.
96. Sahiar, K., and Schwartz, R.   Inhibition of 19S antibody synthesis of 7S antibodies. *Science* 145:395, 1964.
97. Sanford, J. P., Favour, C. B., and Tribeman, M. S.   Absence of serum gamma globulin in adult. *New Eng. J. Med.* 250:1027, 1954.
98. Schultze, H. E., and Heremans, J. F.   *Molecular Biology of Human Proteins.* Vol. 1, *Nature and Metabolism of Extracellular Proteins.* New York: American Elsevier, 1966. Chap. 1, Sec. 1.
99. Schwartz, R. S.   Immunoglobulin metabolism. *Med. Clin. N. Amer.* 50:1487, 1966.
100. Seligman, M., Danon, F., and Fine, J. M.   Immunological studies in familial β 2-macroglobulinaemias. *Proc. Soc. Exp. Biol. Med.* 114: 482, 1963.
101. Sleeper, C. A., and Cawley, L. P.   Detection and diagnosis of mono-clonal gammopathy. *Amer. J. Clin. Path.* Vol. 51, 1969.
102. Smith, T.   Active immunity produced by so-called balanced or neu-tral mixtures of diphtheria toxin and antitoxin. *J. Exp. Med.* 11:241, 1909.
103. Smithies, O.   Zone electrophoresis in starch gels: Group variations in the serum proteins of normal human adults. *Biochem. J.* 61:629, 1955.
104. Smithies, O.   An improved procedure for starch-gel electrophoresis. Further variations in the serum proteins of normal individuals. *Bio-chem. J.* 71:585, 1959.
105. Stobo, J. D., and Tomasi, T. B.   A low molecular weight immuno-

globulin antigenically related to 19S IgM. *J. Clin. Invest.* 46:1329, 1967.

106. Sunderman, F. W., Jr. Recent Advances in Clinical Interpretation of Electrophoretic Fractionations of the Serum Proteins. In Sunderman, F. W., and Sunderman, F. W., Jr. (Eds.), *Serum Proteins and the Dysproteinemias.* Philadelphia: Lippincott, 1964. Pp. 323–345.

107. Svehag, S. -E., Chesebro, B., and Bloth, B. Ultrastructure of gamma M immunoglobulin and alpha macroglobulin: Electron microscopic study. *Science* 158:933, 1967.

108. Terry, W. D., and Fahey, J. L. Subclasses of human $gamma_2$ globulin based on differences in the heavy polypeptide chains. *Science* 146:400, 1964.

109. Tiselius, A. A new apparatus for electrophoresis: Analysis of colloidal mixtures. *Trans. Faraday Soc.* 33:524, 1937.

110. Tomasi, T. B., Jr. Human gamma globulin. *Blood* 25:382, 1965.

111. Uhr, J. W., and Baumann, J. B. Antibody formation: I. The suppression of antibody formation by passively administered antibody. *J. Exp. Med.* 113:935, 1961.

112. Vierucci, A. Gm groups and anti-Gm antibodies in children with Cooley's anemia. *Vox Sang.* 10:82, 1965.

113. Wagner, B. M. Aleutian disease of mink. *Arthritis Rheum.* 6:396, 1963.

114. Waldenström, J. Studies on conditions associated with disturbed gamma globulin formation (gammopathies). *Harvey Lect.* 56:211, 1961.

115. Waldenström, J. Hypergammaglobulinemia as a clinical hematological problem: A study in the gammopathies. *Progr. Hemat.* 3:266, 1962.

116. Waldenström, J. Monoclonal and polyclonal gammopathies and the biological system of gamma globulins. *Progr. Allerg.* 6:320, 1962.

117. Weir, D. M. The immunological consequences of cell death. *Lancet* 2:1072, 1967.

118. Wieme, R. J. *Agar Gel Electrophoresis.* Amsterdam: Elsevier, 1965.

119. Young, R. R., Austen, K. F., and Moser, H. W. Abnormalities of serum gamma-1-A globulin and ataxia telangiectasia. *Medicine* 43:423, 1964.

120. Zawadzki, Z. A., and Rubini, J. R. D-myeloma: Report of two cases. *Arch. Intern. Med.* (Chicago) 119:387, 1967.

3

LACTATE DEHYDROGENASE
ISOENZYMES

lthough determination of serum lactate dehydrogenase (LDH)
has increased the diagnostic acumen of the physician, more precise information has been obtained by the electrophoretic separation of LDH into fractions (isoenzymes) [50, 69, 72, 85]. By definition
[40], isoenzymes are multiple molecular forms of an enzyme with closely
related substrate specificities [45]. They share additional physical and
chemical characteristics but differ not only in electrophoretic mobility
but also in Michaelis constant, variable substrate specificity, immunologic characteristics, thermal stability, and inhibition by pyruvate
excess [14, 37, 55, 61, 72, 85]. Clinical diagnostic tests which take
advantage of these dissimilarities are the LDH heat stability test [68,
85] and hydroxybutyric dehydrogenase enzyme assay [29, 61, 82].
Such tests reflect the properties of the isoenzyme fractions elevated in
myocardial infarction and hemolytic anemias to withstand inactivation by heat and to have selective affinity for a substrate other than
lactate. More precise information for differentiation of disease is possible by agarose electrophoretic separation of the LDH isoenzymes,
histochemical localization with tetrazolium, and quantification by
densitometry [15, 17, 48, 69, 70, 81, 84].

Nomenclature and Genetic Variants

There are five major mammalian LDH isoenzymes separable by
electrophoresis [45, 71, 85]. They are designated 1 through 5 in order of
descending mobility, such that the fraction closest to the anode is 1 and
the fraction closest to the cathode is 5 (Fig. 3-1). This is the opposite of
the designation used by some investigators but is in keeping with
historical precedent [77] and the classic method of lettering and
numbering the globulins as α_1, α_2, β_1, β_2, and γ. In Figure 3-1 the relative mobilities of the various serum proteins and LDH fractions are
shown. LDH_1 travels in the α_1-globulin zone and partially overlaps
the trailing edge of albumin. The major portion of LDH_2 travels in
the same zone as the α_2-globulins, and LDH_3 travels in the β-globulin
zone. LDH_4 and LDH_5 are found within the γ-globulin zone. It should

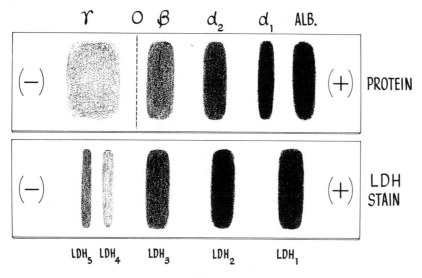

FIGURE 3-1

Relative mobilities of the various LDH fractions in relation to the serum proteins. (From E. J. Wright, L. P. Cawley, and L. Eberhardt [84].)

be noted that the LDH isoenzymes are not attached to the serum proteins, with the possible exception of β-lipoprotein.

Additional zones of LDH isoenzyme activity [30] shown by starch-gel electrophoresis of mouse tissue have been reported after treatment of sample with 2-mercaptoethanol: LDH_5—five zones; LDH_4—four zones; LDH_3—three zones; LDH_2—two zones; and LDH_1—just one zone. Fritz and Jacobson [30] suggested that the 2-mercaptoethanol removed bound nicotinamide-adenine dinucleotide (NAD) from a portion of A subunits (M), but not from B subunits (H). A different explanation of this effect is that offered by Houssais [38], who suggested from his studies that the extra zoning of LDH bands represents polymerized units from the five major LDH bands. Gelderman and Peacock [32] found increased activity of rabbit muscle LDH_5 by molecular sieving on Sephadex G-200 in the presence of low concentrations of β-mercaptoethanol. This activated enzyme contains no nicotinamide-adenine dinucleotide reduced-X (NADH-X) which is present in an inhibitor separated from LDH_5 by the Sephadex G-200. LDH_1 was not affected by the β-mercaptoethanol. Ressler [59], using starch-gel electrophoresis, described subbands of LDH similar to those described by Fritz and Jacobson [30]. The difference was in method of treatment of material and mobility of LDH subbands. Ressler used extracts of tissue fixed in formaldehyde. The additional subbands of LDH had greater mobility than did the usual LDH band. As concentration of formaldehyde increased, the LDH subbands with faster mobility became predominant and the slower subbands disappeared. The interpretation of the phenomenon is believed to rest with the reaction between M monomers and formaldehyde resulting in some tetramers with a greater net negative charge. LDH_5 would give rise to five subbands, one with the same mobility as the usual LDH band and four with more mobility than the original band. Saturation of all M units with formaldehyde would give only one subband. In the study by Fritz and Jacobson [30] the subbands had less mobility than original, suggesting that 6-mercaptoethanol combined with M monomers to produce lesser net negative-charged tetramers.

The two methods of producing subbands of LDH isoenzymes are shown in Figure 3-2. The soundness of the concept of the tetrameric structure of LDH isoenzyme, covered in detail later in this chapter, has been questioned in recent reports from Fritz and Jacobson [31], Stambough and Buckley [64], and Steglink and Vestling [67]. Specific immunochemical testing, when available, will aid in clarifying the nature of subbands of LDH isoenzymes.

A consistent finding is the X band in testis and sperm which is between LDH_3 and LDH_4 [7, 33]. A variant of human erythrocyte LDH consisting of multiple bands for LDH_1, LDH_2, LDH_3, and LDH_4 was reported by Boyer et al. [12]. The authors concluded that the variant was probably a mutant allele at the genetic loci producing a different LDH_1 with less negative charge. A heterozygote would have the pattern found in the Nigerian where five zones of LDH_1, four zones of LDH_2, three zones of LDH_3, and two zones of LDH_4 were shown by starch-gel electrophoresis. These findings were not altered

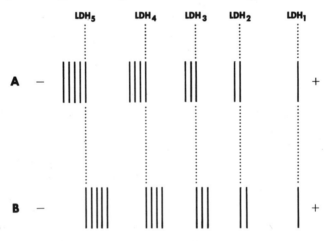

FIGURE 3-2

Diagram of subbands of LDH isoenzymes in starch-gel electrophoretograms after sample was reacted with β-mercaptoethanol (A) and with formaldehyde (B)

by 2-mercaptoethanol. Nance et al. described an LDH variant in erythrocytes of a Brazilian family. There were two zones in LDH_2 and LDH_3 [49]. Beckman goes into many other reports of erythrocyte LDH variations and discusses the genetics involved [5]. Marked predominance of LDH_3 has been seen in serum of two patients in our laboratory, but the significance of this is not known [15, 53].

Structure of LDH Isoenzymes

In 1950 Meister [47], using moving boundary electrophoresis, split beef heart LDH into two fractions. Since then a variety of electrophoretic methods have been used to define the LDH isoenzymes. These methods include electrophoresis on agar-gel or agarose [15, 69, 70, 84], paper [56, 62], cellulose acetate [3], starch gel [22, 43, 72], and polyacrylamide gel [34]. Localization includes stepwise elution of electrophoretic strips and spectrophotometric determination of concentration [22, 85]; scanning for fluorescence [28] or lack of fluorescence [6], properties of NADH and NAD respectively; and staining by tetrazolium salts [3, 22, 48, 56, 69, 70, 81, 84].

It is possible to separate chemically the molecules of purified LDH_1 into homogenous monomers with identical amino acid residues, sedimentation constant, diffusion coefficient, and electrical charge [2, 14, 21, 30, 43]. LDH_5 also yields homogenous monomers with identical features of diffusion, sedimentation, electrical charge, and amino acid composition. The features are distinct from those of the monomers of LDH_1. The monomers from LDH_1 have been designated as B and those from LDH_5 as A by Markert [43] (H and M respectively by Cahn et al. [14]). The letters H and M will be used in this text for the monomers of LDH_1 and LDH_5, respectively, because of their selective increase in heart (H) and muscle (M). Disruption of the isoenzymes of intermediate electrophoretic mobility yields a combination of the two monomers, thus accounting for the variable electrical charge, cross-immunity, relative heat stability, substrate affinity, and substrate inhibition which characterize the physical chemical heterogeneity of the LDH isoenzymes. Freezing equal amounts of purified LDH_1 and LDH_5 in 1M saline solution produces complete disruption of the isoenzymes into monomers. After recombination, electrophoretic separation yields all five fractions in the ratio of 1:4:6:4:1, which is what should be expected if random reassociation of subunits into tetramers occurs [43]. This type of information suggests that LDH exists as a tetramer composed of four monomers, and the five isoenzymes of LDH represent all the possible tetrameric combinations of two unlike monomers (Table 3-1). By varying the amount of LDH_1 and LDH_5, recombination may be skewed to yield any number of ratios, some of which resemble the characteristic profiles obtained from various tissue extracts [43].

TABLE 3-1

CHARACTERISTICS OF LDH

Class	Subunits	Band	M.W. (10^3)	Metab.
Monomer	H		34	Aerobic
	M		34	Anaerobic
Tetramer	HHHH	LDH_1	135	Aerobic
	HHHM	LDH_2	135	
	HHMM	LDH_3	135	Intermediate
	HMMM	LDH_4	135	
	MMMM	LDH_5	135	Anaerobic

As mentioned earlier, several recent experiments with extracts of animal muscle suggest that the molecular structure of the LDH may be more complex than that described above because of the isoenzyme of spermatozoa and testis and the resolution of the five major LDH isoenzymes into subbands. Steglink and Vestling [67] showed that rat liver and bovine heart LDH contained two polypeptide chains per subunit. Stambough and Buckley [64] propose that if the H and M subunits in rabbits also have two polypeptide chains, then the X_4 isoenzyme of sperm, the subbands, and their physical chemical properties might be explained on bases of hybridization of the two chains. These authors propose the following system:

1. H subunits consist of 2 chains, one enzymatically active h and the other enzymatically inactive A.
2. M subunits consist of 2 chains, one enzymatically active m and the other enzymatically inactive C.
3. Crossover hybridization between 4 chains. A (anode) chain determines fast electrophoretic mobility and C (cathode) chain determines slow electrophoretic mobility.
4. LDH_1 and LDH_5 each, therefore, would have 5 bands which are listed below in decreasing order of enzyme activity and slowest mobility.

LDH_1	LDH_5
$H_4-1 = (hA)_4$	$M_4-1 = (mC)_4$
$H_4-2 = (hA)_3 (mA)$	$M_4-2 = (mC)_3 (hC)$
$H_4-3 = (hA)_2 (mA)_2$	$M_4-3 = (mC)_2 (hC)_2$
$H_4-4 = (hA) (mA)_3$	$M_4-4 = (mC) (hC)_3$
$H_4-5 = (mA)_4$	$M_4-5 = (hC)_4 = X_4$

5. X_4 by several physical chemical tests including electrophoretic mobility appears to be identical to $M_4-5 = (hC)_4$, the weakest-staining LDH_5 subband. The presence of X_4 is usually associated with loss of those bands having M chains such as H_4-2, H_4-3, H_4-4, H_4-5, M_4-1, M_4-2, M_4-3, and M_4-4, and also H_4-1 is intensified.

The authors suggest that molecular structure of LDH is of great interest to all biologists, particularly to those in the clinical laboratory since hints of subbands are occasionally seen in serum LDH zymograms. Palmer and Cawley [53] noted "peaking" of LDH_1 in early myocardial infarction. The LDH_1 band in such cases appeared to show subbanding. It is too early to say whether the proposed structures of LDH can be established unequivocally. In the meantime the more widely used nomenclature will be used in this text.

Clinical Application

Serum LDH isoenzyme values from healthy nonhospitalized individuals are relatively constant (Table 3-2). Figure 3-3 shows three zymograms with tracings of myocardial infarction, hepatitis, and a normal pattern. Figure 3-4 is a diagrammatic representation of typical densitometric scan of LDH zymogram in normal serum and in serum of patients with myocardial infarction. The normal patterns are fairly constant, and the relative proportionality of the LDH isoenzymes remains nearly the same with varying amounts of sample [84].

TABLE 3-2

PERCENTAGE OF LDH ISOENZYMES IN NORMAL SERUM

Band	Mean	S.D.	Range
LDH_1	22.0	4.6	15.0–30.0
LDH_2	39.0	4.0	33.0–51.5
LDH_3	27.0	4.3	20.0–35.5
LDH_4	7.5	1.0	2.5–13.5
LDH_5	4.5	0.8	0.0– 8.5

Figure 3-5 diagrammatically summarizes the types of pathologic serum LDH zymograms found in a clinical laboratory service [15, 84]. A normal serum pattern is present in the upper left-hand corner. Malignancies, including acute myelogenous leukemia and carcinoma of the lung, breast, thyroid, gallbladder, pancreas, bronchus, stomach, and colon, present a rather characteristic zymogram pattern with absolute increases in all fractions and significant relative increases in LDH_3. Figure 3-6 presents a pattern and Analytrol scan of serum LDH zymogram on a patient with γA-type myeloma [16]. Total LDH was 160 international units (I.U.) (lactate $\rightarrow$ pyruvate; normal less than 60 I.U.) [75]. Percentages from the Analytrol scan are as follows: LDH_1, 28; LDH_2, 33.5; LDH_3, 32; LDH_4, 5; and LDH_5, 1.5. Figures 3-7 and 3-8 are a serum electrophoretogram with monoclonal peak and an immunoelectrophoretogram showing bulging and increase of γA immunoprecipitin band.

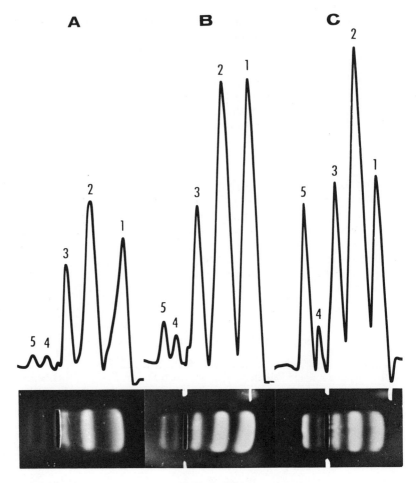

FIGURE 3-3

Photographs of three serum LDH zymograms with Analytrol scans representative of (A) normal, (B) myocardial infarction, and (C) infectious hepatitis. In B, LDH_1 and LDH_2 are both elevated. In C, LDH_5 is clearly increased. (From E. J. Wright, L. P. Cawley, and L. Eberhardt [84].)

In hepatic metastasis, LDH_5 is also elevated; it is not increased in patients with carcinoma without metastasis to the liver. The composite LDH profile produced by hepatocellular damage includes hepatitis, jaundice secondary to chlorpromazine, postcholecystectomy, and hepatic contusion. There is usually an absolute increase in all fractions; however, the distinctive feature of this profile is a marked increase in LDH_5. Skeletal muscle damage, as in crushing trauma, hip fractures, and extensive surgical procedures involving muscle, also causes a decided but transient rise in LDH_5. The profile of biliary obstruction

NORMAL LDH ZYMOGRAM

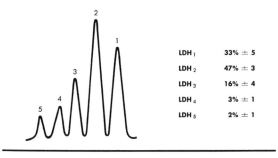

LDH₁	33% ± 5
LDH₂	47% ± 3
LDH₃	16% ± 4
LDH₄	3% ± 1
LDH₅	2% ± 1

TYPICAL LDH ZYMOGRAM IN INFARCTION

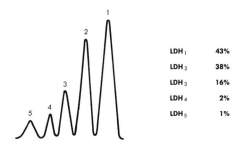

LDH₁	43%
LDH₂	38%
LDH₃	16%
LDH₄	2%
LDH₅	1%

FIGURE 3-4

Diagram of serum LDH isoenzymes in healthy patient and following myo-cardial infarction. Note shift of LDH isoenzymes to LDH_1 (From M. E. Blough, E. W. Crow, and L. P. Cawley, *J. Kansas Med. Soc.* 67:545, 1966.)

discloses an absolute increase in LDH_4 and LDH_5 with LDH_4 in excess of LDH_5. This is in contrast to the marked relative increase in LDH_5 with hepatocellular damage. Untreated pernicious anemia, paroxysmal nocturnal hemoglobinuria, sickle cell anemia, autoimmune hemolytic anemia, and hemolytic anemia of undetermined etiology present almost equal absolute increases in LDH_1 and LDH_2 and a concomitant re-duction (absolute or relative) of the other fractions. Acute myocardial infarction consistently shows marked increase in LDH_1 and LDH_2 with LDH_1 at least equal to, and usually in excess of, LDH_2. There is a relative and in some instances an absolute reduction in the remaining bands. In congestive failure a rise in LDH_5 may occur, presumably because of anoxic changes in the liver, although, as discussed later, clearance of LDH_5 by the reticuloendothelial (RE) system may be

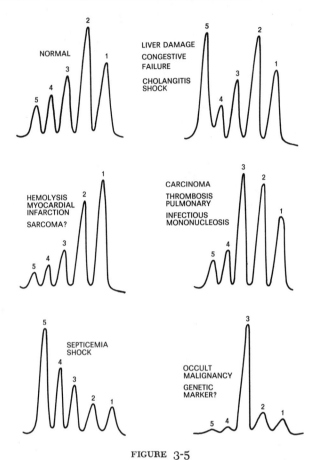

FIGURE 3-5

Representative Analytrol scans of serum LDH zymograms from various
diseases (see text). (From L. P. Cawley [15].)

temporarily impaired by material from the infarcted myocardium. The
elevation of LDH₅ is frequently transient; there may be a return to
normal within 12 hours or less. Narrowness of LDH₁ resulting in a
peak higher than LDH₂, but not quantitatively containing more enzyme, has been seen early in myocardial damage from myocarditis and
infarction [54]. Vascular collapse (shock) shows an absolute increase
in all fractions with nearly equal distribution. The LDH pattern in
the lower right of Figure 3-5 is most unusual and has been seen in
only two patients in over 10,000 determinations [15]. One patient had
a gastric ulcer, and the other complained of chronic weakness. Test
of a son of the patient with gastric ulcer was normal, and although this
finding does not exclude genetic influence, it does mediate against it.
Lundh [42] reported a macromolecular serum lactate dehydrogenase

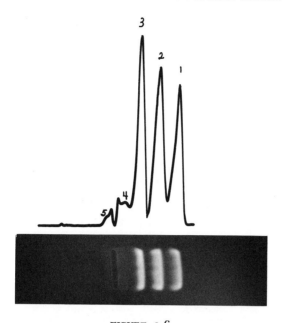

FIGURE 3-6

Serum LDH zymogram and Analytrol scan showing increase of LDH₃

activity which trailed just behind LDH₃. The band by Sephadex G-200 fractionation was shown to be a macromolecule. Special tests excluded its attachment to γA, γG, γM, α₁-antitrypsin, β₁C globulin, β-lipoprotein, α₂-macroglobulin, and C-reactive protein. The band was also not related to the X band of sperm. Two other unpublished cases were cited by the author. Each patient had a different disease, and even though Lundh's patient had leukemia, he felt that the association was coincidental.

Production, Release, and Catabolism

The differing amino acid composition of the two monomers implies two separate genes responsible for their production [43]. Random combination at the cellular level would then be dependent upon the ratio of monomers present at the time of assemblage and would result in the distinctive patterns obtained from any one tissue. In tissues, the genetic control of synthesis of LDH monomers is influenced by the type of metabolism prevailing at the cellular level. In anaerobic states the hybrids are skewed toward LDH₅ (MMMM) while in aerobic states LDH₁ (HHHH) predominates [14]. If metabolism is the underlying primary factor determining the ratio of H and M monomers produced and the subsequent isoenzyme pattern, it is perhaps through this parameter that genetics [49], hormones [35], and disease [56, 70,

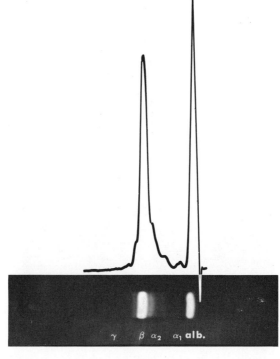

FIGURE 3-7

Electrophoretogram and Analytrol scan showing monoclonal peak in the γ-globulin area. Note decrease of γ-globulin. (From L. P. Cawley [15].)

80, 85] effect the change in the isoenzyme profiles of the tissue. Once in the serum, the enzyme is removed over a period of time but is not excreted in the urine or bile.

Elevations due to tissue necrosis with subsequent cellular leaching probably have no more than transient influence on the total LDH isoenzyme pattern because of rapid plasma clearance of the enzyme [15]. Other mechanisms must be sought as an explanation for the prolonged elevations seen in disease, such as myocardial infarction with persistent elevation of serum LDH for as long as two weeks. It is estimated [36] that the marked serum LDH elevation seen in pernicious anemia could be accounted for only by assuming a daily intravascular hemolysis of 16–18 percent of the red cells, which would correspond to the unlikely red cell survival of only several days. Since H and M monomers are not formed intravascularly or extracellularly, the serum level is a balance between their release from tissues and their clearance or removal from the serum by the RE system (Figs. 3-9 and 3-10). It is unknown whether or not the RE system releases

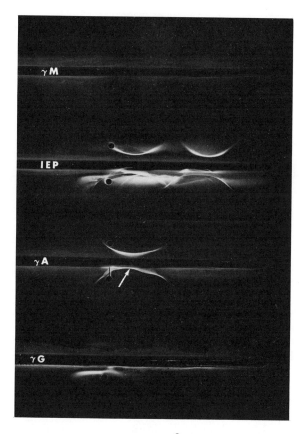

FIGURE 3-8

Immunoelectrophoretogram with anti-γG, anti-γA, anti–human serum (immunoelectrophoresis precipitin [IEP]), and anti-γM against patient's plasma. *Upper*, antigen well in each set diluted 1:10 with saline. *Bottom*, antigen well, undiluted serum. The γA immunoprecipitin bands correspond to the monoclonal peak of Figure 3-7. Note bulge of γA band (*arrow*). The low concentration of γG and the absence of γM are evident. Anode is on the right. (From L. P. Cawley [15].)

back into the serum some of the monomeric or tetrameric structures. The concentration of inhibitors and accelerators in the serum also influences the serum level. In isoenzyme studies these substances are probably separated from the enzyme by the electrical field and are not as influential as they are in the total LDH. The influence of binding of LDH isoenzymes by other proteins or particles on the serum levels of LDH may play an important role in both isoenzyme studies and total LDH determination, as shown by Cohen et al. [20]. If the monomers were individually released, one might expect a hybrid

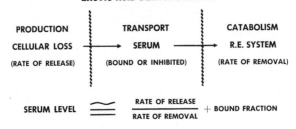

FIGURE 3-9

Relationship of serum LDH to production and catabolism

formation in the serum to result in a ratio of LDH isoenzymes approaching the theoretical values of 1:4:6:4:1, if equal units of H and M were available. In human serum the ratio is approximately 1:2:6:16:12, which is skewed toward LDH_1 (HHHH) and LDH_2 (HHHM). Either LDH_1 and LDH_2 are indicative of selective tissue destruction, such as erythrocytes, or the rate of removal of the LDH_5, LDH_4, and LDH_3 is more rapid [54]. The erythrocyte, because of rapid destruction, may be responsible for the prevalence of LDH_1 and LDH_2 in the serum if one takes the view that tissue destruction is the primary factor in determining the LDH isoenzyme profile in serum.

In the rat the erythrocyte contains predominantly LDH_5 [54, 71]. The serum profile also shows almost exclusively LDH_5. This seems to support the direct relationship between the tissue undergoing most rapid destruction and the serum profile of LDH isoenzymes. In the horse LDH_3 accounts for 46 percent of the total LDH, followed by LDH_2, LDH_1, LDH_4, and LDH_5 in that order of magnitude [19]. Yet the analysis of tissues from various organs of the horse does not reveal

PRODUCTION, TRANSPORT, AND CATABOLISM OF LDH

SERUM LEVELS INFLOW—OUTFLOW	PRODUCTION INTRACELLULAR	TRANSPORT BOUND TO PROTEIN	CATABOLISM MACROPHAGES
NORMAL	AGING & TRAUMA ⟶	ALL COMPARTMENTS FILLED ⟶	RATE OF REMOVAL (NORMAL)
INCREASE A	RELEASE — FACTORS CAUSING INJURY, INFARCTION, VITAMIN DEF., ANOXIA, HEMOLYSIS ⟶	FREE UNBOUND LDH PREDOMINATES ⟶	NORMAL
INCREASE B	NORMAL OR ⟶ AS IN A	NORMAL ⟶	RATE OF REMOVAL BLOCKED BY VIRUS, BACTERIA, OR RBC STROMA
INCREASE C	NORMAL OR ⟶ AS IN A	INCREASE OF ⟶ BOUND FRACTION	NORMAL

FIGURE 3-10

Summary of three causes of increase of serum LDH

any organ with predominant content of LDH$_3$. The tissue profiles are very similar to those of corresponding human tissues [19]. It is assumed that the erythrocytes of the horse have a similar life-span relative to those of other animals. In other words, the release of LDH from tissues in the horse, if identical to that in humans, should result in a similar LDH profile providing that the clearance of the RE system is not a factor. In the dog LDH$_5$ is predominant in the serum while the red cell has a pattern similar to that of human red cells, a relationship which does not conform to the basic thesis mentioned above [19]. In the rabbit the serum LDH isoenzyme profile is analogous to that in the human. The red cell LDH isoenzyme pattern is also similar to that in the human. In rabbits given intravenous injections of rat hemolysates (LDH$_5$), the rat LDH$_5$ is cleared within one hour and cannot be detected by LDH zymogram studies. If the rabbit receives an injection of his own hemolysate followed 15 minutes later by an injection of rat hemolysate, rat LDH$_5$ can be detected in rabbit serum for 24 hours.

Taken together, these factors suggest that not only is the source of the LDH important in determining the serum profile but so is the mechanism of excretion or clearance. In human serum it is difficult to account for the preferential concentration of LDH$_2$ if one assumes that the rate of clearance of the five LDH isoenzymes is identical. Although a few human organs have LDH$_2$ predominantly, the increase of LDH$_2$ relative to that of other bands is not great.

The survival of individual LDH isoenzymes based on their independent rate of clearance may be an important factor in explaining the serum LDH isoenzyme profile. The immunoglobulins γG, γA, γM, and γD have similar rates of synthesis but different rates of catabolism [4, 60]. It is this difference in rates of catabolism that is responsible for their serum levels. γG has a long survival time whereas γM has a short survival time and thus a low serum level. If one assumes that the serum level of the LDH isoenzymes is a reflection of their individual catabolism as well as production, then the serum LDH pattern suggests that the RE system catabolizes LDH$_5$ rapidly whereas LDH$_1$ is catabolized less rapidly and possibly by a different set of macrophages [53]. In that case tissue release of LDH$_5$ is considerably higher than is reflected in the amount of LDH$_5$ found in the serum by current technics. The hybrids then will be removed by either of the two sets of macrophages. The efficiency of this removal, however, will be considerably less, such that LDH$_2$ is not cleared as rapidly as LDH$_1$ presumably because it has an M monomer whose influence, although small, may affect the clearance rate. The basis of this hypothesis is that catabolism in the RE system is based on recognition of M and H monomers.

Clinically, some cases support this type of reasoning. For example, in heart failure LDH$_5$ elevation is very transient, in some instances disappearing in the first 24 hours after development. In septicemia LDH$_5$ increases rapidly [15, 84] and thus may reflect liver damage. On the other hand, the quick rise may indicate blockage of the RE system. It is known that the Riley virus produces permanent increase of LDH by RE blockage [51]. Glutamic oxalacetic transaminase (GOT) and malic dehydrogenase were also elevated by the viremia, but alkaline phosphatase, which is removed by separate macrophages, was not affected. In viral pneumonia we have on occasion noted marked increase of LDH$_1$, LDH$_2$, and LDH$_3$. Elevations of these isoenzymes in viral pneumonia may reflect interference of normal clearance by blocking the RE system (Fig. 3-10); however, subclinical myocarditis cannot be excluded. Boyd [10] has demonstrated that LDH$_5$ and LDH$_1$ from sheep muscle and heart, respectively, have different rates of disappearance from plasma. He found that LDH$_5$ had a more rapid rate of clearance than LDH$_1$ and calculated that LDH$_5$, in order to match the plasma level of LDH$_1$, would have to be released 7.5 to 15 times faster than LDH$_1$. This conclusion tends to support the concept of a specialized system of clearance, as discussed before. Coffman et al. [18] found no difference in clearance rates in dogs after beef LDH$_1$ and rabbit LDH$_5$ were injected. However, in the dog, the rabbit and beef enzyme acted as a foreign protein, and therefore both H and M structures were removed indiscriminately. When dog LDH$_1$ and LDH$_5$ were injected into dogs, LDH$_5$ was removed more rapidly than LDH$_1$.

Other factors obviously play a role in regulation of serum LDH levels. These include influence of steroids on both the production and the RE clearance mechanism. Inhibitors and accelerators normally found in serum also are involved in regulation of LDH in the serum. Not to be overlooked is the amount of hemolysis induced by venipuncture, which would give rise to elevation of LDH. Loeb et al. [41] showed that in the dog serum LDH and phosphohexose isomerase are directly related to amount of hemolysate present as measured by plasma hemoglobin.

In a more direct approach to the problem of inhibition and clearance Boyd [11] labeled sheep muscle LDH$_5$ with 2-C^{14}-iodoacetate. The radioactive carboxymethylated isoenzyme was shown to be active, to have essentially the same electrophoretic mobility as untagged LDH, and to react with rabbit anti–sheep LDH$_5$. The radioactive LDH$_5$ was given intravenously to sheep, and the rate of disappearance of enzyme activity compared to disappearance of radioactivity was studied. About 90 percent of the injected C^{14}-labeled LDH$_5$ had disappeared from the circulation within 24 hours. The rapid loss of LDH$_5$ activity represents actual disappearance of the isoenzyme from plasma. No significant amount appeared in the urine, and the enzyme did not become inactivated and continued to circulate, since no evidence of it was found with rabbit anti–sheep LDH$_5$.

Pathologic alteration of the serum LDH profile is generally, but not necessarily, accompanied by an increase in total LDH. Because of the diverse mechanisms which apparently influence the profile (Fig. 3-5), it should be emphasized that only by quantifying the fractions can the maximum information be obtained from the determination of lactic dehydrogenase. Agarose electrophoresis with histochemical localization of LDH isoenzymes is a satisfactory and useful laboratory procedure for quantification of LDH isoenzymes [84].

Isoenzymes of Lymphocytes, Granulocytes, Erythrocytes, and Platelets

Various reports [9, 25, 26, 58, 65, 73] on the LDH isoenzyme profiles of extracts of blood cells make note of the central criterion for valid results, namely, proper isolation of the cell type in relatively pure form. Attempts to evaluate the LDH isoenzymes of cell types have been complicated by lack of pure cell populations to work with. The technic of glass-bead chromatography for separation of cell types as outlined by Rabinowitz et al. [58] is appealing from the standpoint that, theoretically, complete separation of lymphocytes from granulocytes is possible. Lymphocytes and erythrocytes come off the column together and must be separated by other methods. Differential-gradient centrifugation employed by Starkweather et al. [65] was found to be the best technic for their purposes. A more exacting approach for preparation of pure cell types is cell electrophoresis [1], which is now suitable for the preparation of adequate amounts of cells for study. The procedure is very gentle and does not change cell detail. Besides the use of glass beads for separation of various cells, Rabinowitz et al. [58] utilized disc electrophoresis as a method of determining LDH isoenzyme profiles. Our experience with disc electrophoresis [15] causes us to be cautious in our interpretation of determinations made by this method. However, Dietz and Lubrano [23], in a very complete study of LDH isoenzyme separation by disc electrophoresis, found that by using sucrose as a nonionic diluent mixed with serum in a ratio of one to three parts of 40 percent sucrose they obtained reproducible LDH zymograms. LDH_5 was regularly demonstrated. This is the method used by Rabinowitz and Dietz [57] in their study of LDH in lymphocytes. Starkweather et al. [65, 66] used agarose-gel electrophoresis, and Dioguardi et al. [24, 25, 26] used DEAE column chromatography and starch-gel and cellulose acetate electrophoresis for demonstration of LDH isoenzymes in blood cells. There is some difference of reported results in the LDH isoenzymes found for lymphocytes.

In immature tissues LDH_3 is the predominant band [65] (Fig. 3-11). As maturation occurs, the pattern of LDH isoenzyme becomes

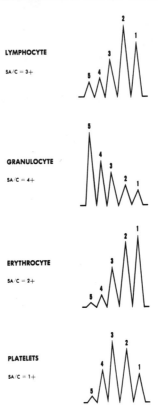

FIGURE 3-11

Diagrammatic representation of LDH isoenzyme scans of blood cells. Each cell type is unique.

skewed toward either LDH_1 or LDH_5. In general, in malignant conditions LDH_3 is increased.

LYMPHOCYTES

Starkweather et al. [65] found that normal mature peripheral lymphocytes and lymph node lymphocytes have an LDH isoenzyme pattern similar to that of granulocytes. Rabinowitz et al. [58] and Dioguardi et al. [24] found that LDH_1, LDH_2, and LDH_3 predominated, with LDH_2 being the highest (Fig. 3-7). The amount of enzyme per cell is high. Histochemically, the enzyme appears to be located in the cytoplasm at discrete sites. Nuclear localization is questionable. In chronic lymphatic leukemia there is a shift to LDH_3 [24], which is also seen in lymphosarcoma cells from lymph nodes [65]. Bottomley et al. [9] also found lymphocytes from chronic lymphatic leukemia to have high LDH_3. Studies of the LDH isoenzyme changes during transformation of lymphocytes in short-term tissue culture stimulated by phytohemagglutinin (PHA) (extract of *Phaseolus vulgaris*) by Rabinowitz and Dietz (acrylamide electrophoresis) [57] and by Bloom et al. (agarose-gel electrophoresis) [8] revealed an increase of LDH_3. Both sets of authors concluded that PHA stimulates synthesis of M monomers.

GRANULOCYTES

The mature polymorphonucleated leukocyte has an LDH isoenzyme profile distinct from that of the mature lymphocyte with predominating LDH_5. The LDH_5 concentration is high; the LDH_1 concentration is low [26, 65]. The total enzyme per cell is high. In acute granulocytic leukemia, the LDH_3 band becomes prominent and LDH_5 decreases [25, 65]. Maturation is associated with increased levels of LDH_5. In chronic granulocytic leukemia there is only a slight change from the usual pattern [26]. A slight increase of LDH_2 and LDH_3 does occur.

ERYTHROCYTES

The LDH isoenzyme profile of erythrocytes and total LDH per cell are demonstrated in Table 3-3. LDH_1 and LDH_2 are high. The amount of LDH_5 is variable, depending upon the number of nucleated erythrocytes. The more immature the cell, the higher the amount of LDH_5 and the lower the proportion of LDH_1. As a cell matures and the nucleus is lost, LDH_5 decreases and LDH_1 persists. Vessell et al. [73] noted LDH_5 in nucleated erythrocytes. Starkweather et al. [65] showed that LDH_5 is relatively insoluble in nonionic extracting solutions. By increasing the ionic strength of the extracting solution, he revealed that LDH_5 could usually be found in most preparations of erythrocytes even with low reticulocyte counts. Genetic variants of LDH isoenzymes of erythrocytes have been briefly discussed. Beckman [5] reported on several varieties. Genetic variation is always of interest and should be expected to be found in serum as well as cells.

PLATELETS

LDH profiles of platelets show a predominant LDH_3 band (Table 3-3), a pattern called isomorphic by Cohen and Larson [20]. These authors state that the LDH is in an inactive form in plasma and is activated by breakup of platelets during clotting. Platelets have the lowest total LDH of any of the blood cells.

LDH Isoenzymes of Urine

Wacker and Dorfman [74] and Dorfman et al. [27] reported that elevation of total urinary LDH was a useful screening procedure in detection of malignancy of the genitourinary (GU) system. The LDH

TABLE 3-3

PERCENTAGE DISTRIBUTION OF LDH ISOENZYMES OF NORMAL BLOOD CELLS

Cell Type	LDH Bands as Percent of Total				
	1	2	3	4	5
Lymphocytes	25	33	27	9	4
Granulocytes	5	9	17	20	48
Erythrocytes	34	43	14	4	2
Platelets	11	31	42	13	3

of urine probably arises within the GU system since the LDH molecule from serum, because of its large size, does not pass through the glomeruli into the urine [74]. Wacker and Dorfman, and Dorfman et al. [27] reasoned that increased urinary LDH was probably due to malignancy of the urinary system. Although they were aware of other sources of urinary LDH, particularly leukocytes, the point was not adequately explored. Brenner and Gilbert [13] measured urinary LDH, GOT, and catalase in 100 hospitalized patients and 17 healthy adults. In 37 percent of the patients bacilluria was present, but all patients were suspected of having urinary tract infections. Urinary LDH and GOT were not increased unless pyuria was present. Catalase was present under similar situations but also was increased in some sterile urine.

Enzymes in urine may arise from multiple sources, including epithelial cells from pelvis of kidney, renal tubules, urinary bladder, urethra, and ureters. Leukocytes and red cells are by far the most significant source of urinary LDH, GOT, and catalase. Bacteria produce little enzyme, and thus they contribute only a small amount to urinary enzyme values, except possibly for catalase. The pattern of bacilluria, pyuria, and increase of urinary LDH and GOT was also found in rats with experimentally induced acute hematogenous pyelonephritis. Isoenzymes of LDH in urine were studied by Wright and Cawley [83], and a direct correlation between total LDH and leukocyte concentration became evident. Of particular interest is the observation that LDH_5 could be readily demonstrated in urine with high concentration of leukocytes. Unless careful attention is paid to leukocyte concentration of urine, total urinary LDH has little usefulness from the standpoint of detecting malignancy of the GU system. In any event, it seems only logical to perform isoenzyme analysis on urines rather than determining total LDH since the pattern of isoenzymes can aid in predicting the origin of the LDH. Concentration of urine and detection of LDH isoenzymes are discussed in Chap. 15.

Schoenenberger and Wacker [63] have identified two peptides which are responsible for inhibition of urinary total LDH measured from lactate → pyruvate. Peptide I has a molecular weight of 1760 and peptide II has a molecular weight of 2700. Both are removed from urine by dialysis. Even though the inhibitors were removed in test for the isoenzymes by electrophoresis, they would nonetheless have their influence on the total values unless removed by dialysis.

In general, detection of abnormal urinary patterns of LDH isoenzymes is too new for us to be certain that the procedure has little to add for detection of GU pathology. At least two distinct patterns have been found. Elevation of LDH_5 is clearly related to granulocyte breakdown and release of LDH_5 into the urine. A few urines with marked elevation of LDH_1 and LDH_2 have been studied, but thus far no relationship between the pattern and disease has emerged. Also troublesome is the not infrequent elevation of total LDH associated with poor delineation of isoenzymes and vice versa.

References

1. Ambrose, E. J. *Cell Electrophoresis: A Symposium Convened by the British Biophysical Society.* Boston: Little, Brown, 1965.
2. Appella, E., and Markert, C. L. Dissociation of lactate dehydrogenase into subunits with guanidine hydrochloride. *Biochem. Biophys. Res. Commun.* 6:171, 1961.
3. Barnett, H. Electrophoretic separation of lactate dehydrogenase isoenzymes on cellulose acetate. *Biochem. J.* 84:83, 1962.
4. Barth, W. F., Wochner, R. D., Waldmann, T. A., and Fahey, J. L. Metabolism of human gamma macroglobulins. *J. Clin. Invest.* 43:1036, 1964.
5. Beckman, L. *Isozyme Variations in Man. Monographs in Human Genetics.* Philadelphia: Karger, 1966. Pp. 50–58.
6. Blanchaer, M. C. Electrophoresis of serum lactic dehydrogenase. *Clin. Chim. Acta* 6:272, 1961.
7. Blanco, A., and Zinkham, W. H. Lactate dehydrogenase in human testes. *Science* 139:601, 1963.
8. Bloom, A. D., Tsuchivka, M., and Wajinia, T. Lactic dehydrogenase and metabolism of human leukocytes in vitro. *Science* 156:979, 1967.
9. Bottomley, R. H., Locke, S. J., and Ingram, H. C. Lactic dehydrogenase of human chronic lymphocytic leukemic leukocytes. *Blood* 27:85, 1966.
10. Boyd, J. W. The rates of disappearance of L-lactate dehydrogenase isoenzymes from plasma. *Biochim. Biophys. Acta* 132:221, 1967.
11. Boyd, J. W. The disappearance of ^{14}C-labelled isoenzyme 5 of 6-lactate dehydrogenase from plasma. *Biochim. Biophys. Acta* 146:590, 1967.
12. Boyer, S. H., Fainer, D. C., and Watson-Williams, E. J. Lactate dehydrogenase variant from human blood: Evidence for molecular subunits. *Science* 141:642, 1963.
13. Brenner, B. M., and Gilbert, V. E. Elevated levels of lactic dehydrogenase glutamic-oxalacetic transaminase, and catalase in infected urine. *Amer. J. Med. Sci.* 245:31, 1963.

14. Cahn, R. D., Kaplan, N. O., Levine, L., and Zwilling, L. E.   Nature and development of lactic dehydrogenases. *Science* 136:962, 1962.
15. Cawley, L. P.   *Workshop Manual on Electrophoresis and Immunoelectrophoresis* (Rev. ed.).   Commission on Continuing Education, Council on Clinical Chemistry. Chicago: American Society of Clinical Pathologists, 1966.
16. Cawley, L. P.   LDH Isoenzymes—Orchestration. Clinical Chemistry Check Sample No. CC-43, Commission on Continuing Education. Chicago: American Society of Clinical Pathologists, 1967.
17. Cawley, L. P., and Eberhardt, L.   Simplified gel electrophoresis. I. Rapid technic applicable to the clinical laboratory. *Amer. J. Clin. Path.* 38:539, 1962.
18. Coffman, J. R.   Personal communication, 1967.
19. Coffman, J. R., Mussman, H. C., and Cawley, L. P.   Enzymology and Serum Proteins in Disease States of the Major Parenchymatous Organs. An index of current equine research, Morris Animal Foundation, 1967. P. 42.
20. Cohen, L., and Larson, L.   Activation of serum lactic dehydrogenase. *New Eng. J. Med.* 275:465, 1966.
21. Dawson, D. M., Goodfriend, T. L., and Kaplan, N. O.   Lactic dehydrogenases: Function of the two types. *Science* 143:929, 1964.
22. Dewey, M. M., and Conklin, J. L.   Starch gel electrophoresis of lactic dehydrogenase from rat kidney. *Proc. Soc. Exp. Biol. Med.* 105:492, 1960.
23. Dietz, A., and Lubrano, T.   Separation and quantitation of lactic dehydrogenase isoenzymes by disc electrophoresis. *Anal. Biochem.* 20: 246, 1967.
24. Dioguardi, N., and Agostoni, A.   Multiple Molekularformen der Lactatdehydrogenase normaler und leukämischer menschlicher Lymphozyten. *Enzym. Biol. Clin.* (Basel) 5:3, 1965.
25. Dioguardi, N., Agostoni, A., and Fiorelli, G.   Characterization of lactic dehydrogenase in cells of myeloid leukemia. *Enzym. Biol. Clin.* (Basel) 2:116, 1962.
26. Dioguardi, N., Agostoni, A., Fiorelli, G., and Lomanto, B.   Characterization of lactic dehydrogenase of normal human granulocytes. *J. Lab. Clin. Med.* 61:713, 1963.
27. Dorfman, L. E., Amador, E., and Wacker, W. E. C.   Urinary lactic dehydrogenase activity: II. Elevated activities for diagnosis of carcinoma of kidney and bladder. *Biochem. Clin.* 2:41, 1963.
28. Elevitch, F. R., Kelly, D., and Feichtmeir, T. V.   Paper presented at the 113th Annual Meeting of the American Medical Association, June 22–25, 1964.
29. Elliott, B. A., Jepson, E. M., and Wilkinson, J. H.   Serum-hydroxybutyrate dehydrogenase—a new test with improved specificity for myocardial lesions. *Clin. Sci.* 23:305, 1962.
30. Fritz, P. J., and Jacobson, K. B.   Lactic dehydrogenases: Subfractionation of isozymes. *Science* 140:64, 1963.
31. Fritz, P. J., and Jacobson, K. B.   Multiple molecular forms of lactate dehydrogenase. *Biochemistry* (Washington) 4:282, 1965.
32. Gelderman, A. H., and Peacock, A. C.   Increased specific activity

and formation of an inhibitor from the LDH_5 isozyme of lactate dehydrogenase. *Biochemistry* (Washington) 4:1511, 1965.

33. Goldberg, E.   Lactic and malic dehydrogenases in human spermatozoa. *Science* 139:602, 1963.

34. Gonzalez, I. E.   Lactic dehydrogenase isozyme patterns in patients with elevated serum lactic dehydrogenase. *Amer. J. Clin. Path.* 42: 530, 1964.

35. Goodfriend, T. L., and Kaplan, N. O.   Effects of hormone administration on lactic dehydrogenase. *J. Biol. Chem.* 239:130, 1964.

36. Gronvall, C.   On the serum activity of lactic acid dehydrogenase and phosphohexose isomerase in pernicious and hemolytic anemias. *Scand. J. Clin. Lab. Invest.* 13:29, 1961.

37. Hill, B. R.   Further studies of the fractionation of lactic dehydrogenase of blood. *Ann. N.Y. Acad. Sci.* 75:304, 1958.

38. Houssais, J. F.   Transformations moléculaires au niveau des isozymes de la lacticodéhydrogénase de la souris, mises en évidence par électrophorèse en gel d'amidon. *Biochim. Biophys. Acta* 128:239, 1966.

39. Johnson, H. L., and Kampschmidt, R. F.   Lactic dehydrogenase isozymes in rat tissues, tumors, and precancerous livers. *Proc. Soc. Exp. Biol. Med.* 120:557, 1965.

40. Latner, A. L., and Skillen, A. W.   Clinical applications of dehydrogenase isoenzymes. *Lancet* 2:1286, 1961.

41. Loeb, W. F., Nagode, L. A., and Frajola, W. J.   The distribution of four enzymes between canine serum and erythrocytes. *Enzym. Biol. Clin.* (Basel) 7:215, 1966.

42. Lundh, B.   A macromolecular serum lactate dehydrogenase activity in a case of leukemia. *Clin. Chim. Acta* 16:305, 1967.

43. Markert, C. L.   Lactic dehydrogenase isozymes: Dissociation and recombination of subunits. *Science* 140:1329, 1963.

44. Markert, C. L., and Faulhaber, I.   Lactate dehydrogenase isozyme patterns of fish. *J. Exp. Zool.* 159:319, 1965.

45. Markert, C. L., and Moller, F.   Multiple forms of enzymes: Tissue, ontogenetic, and species specific patterns. *Proc. Nat. Acad. Sci. U.S.A.* 45:753, 1959.

46. Markert, C. L., and Ursprung, H.   The ontogeny of isozyme patterns of lactic dehydrogenase in the mouse. *Develop. Biol.* 5:363, 1962.

47. Meister, A.   Reduction of γ_1, γ-diketo and γ-keto acids catalyzed by muscular preparations and by crystalline lactic dehydrogenase. *J. Biol. Chem.* 184:117, 1950.

48. Nachlas, M. M., Margulies, S. I., Goldberg, J. D., and Seligman, A. M.   The determination of lactic dehydrogenase with a tetrazolium salt. *Anal. Biochem.* 1:317, 1960.

49. Nance, W. E., Claflin, A., and Smithies, O.   Lactic dehydrogenase: Genetic control in man. *Science* 142:1075, 1963.

50. Neilands, J. B.   The purity of crystalline lactic dehydrogenase. *Science* 115:143, 1952.

51. Notkins, A. L.   Lactic dehydrogenase virus. *Bact. Rev.* 29:143, 1965.

52. Ohno, S., Klein, J., Poole, J., Harris, C., Destree, A., and Morrison, M.   Genetic control of lactate dehydrogenase formation in the hagfish *Extatutus stoutii*. *Science* 156:96, 1967.

53. Palmer, D. L., and Cawley, L. P.   Factors governing the hemostatic regulation of LDH isoenzymes in human serum. *Amer. J. Clin. Path.* 47:367, 1967.

54. Palmer, D. L., Crow, E. W., and Cawley, L. P.   Subtle changes in isoenzyme LDH$_1$ in early myocardial disease. Presented at the 1966 Annual Meeting of the American College of Physicians, 1967.

55. Plagemann, P. G., Gregory, K. F., and Wroblewski, F.   The electrophoretically distinct forms of mammalian lactic dehydrogenase. Distribution of lactic dehydrogenases in rabbit and human tissue. *J. Biol. Chem.* 235:2282, 1960.

56. Raabo, E.   Colorimetric method for demonstrating lactic dehydrogenase isoenzymes separated by paper electrophoresis. *Scand. J. Clin. Lab. Invest.* 15:405, 1963.

57. Rabinowitz, Y., and Dietz, A. A.   Genetic control of lactate dehydrogenase and malate dehydrogenase isoenzymes in cultures of lymphocytes and granulocytes: Effects of addition of phytohemagglutinin, actinomycin D or puromycin. *Biochim. Biophys. Acta.* 139:254, 1967.

58. Rabinowitz, Y., and Dietz, A. A.   Malic and lactic dehydrogenase isozymes of normal and leukemic leukocytes separated on glass bead columns. *Blood* 29:182, 1967.

59. Ressler, N.   Interpretation of subbands of lactic dehydrogenase isoenzymes. *Nature* (London) 215:284, 1967.

60. Ritzmann, S. E., and Levin, W. C.   II. Polyclonal and monoclonal gammopathies. *Lab. Synopsis* 2:9, 1967.

61. Rosalki, S. B., and Wilkinson, J. H.   Reduction of γ-ketobutyrate by human serum. *Nature* (London) 188:1110, 1960.

62. Sayre, F. W., and Hill, B. R.   Fractionation of serum lactic dehydrogenase by salt concentration gradient elution and paper electrophoresis. *Proc. Soc. Exp. Biol. Med.* 96:695, 1957.

63. Schoenenberger, G. A., and Wacker, W. E. C.   Peptide inhibitors of lactic dehydrogenase (LDH). II. Isolation and characterization of peptide I and II. *J. Biochem.* 5:1375, 1966.

64. Stambough, R., and Buckley, J.   The enzymatic and molecular nature of the lactic dehydrogenase subbands of X$_4$ isozyme. *J. Biol. Chem.* 242:4053, 1967.

65. Starkweather, W. H., Spencer, H. H., and Schoch, H. K.   The lactic dehydrogenases of hemopoietic cells. *Blood* 28:860, 1966.

66. Starkweather, W. H., Spencer, H. H., Schwarz, E. L., and Schoch, H. K.   The electrophoretic separation of lactate dehydrogenase isoenzymes and their evaluation in clinical medicine. *J. Lab. Clin. Med.* 67:329, 1966.

67. Stegink, L. D., and Vestling, C. S.   Rat liver lactate dehydrogenase: Amino-terminal and acetylation status. *J. Biol. Chem.* 241:4923, 1965.

68. Strandjord, P. E., Clayson, K. J., and Freier, E. F.   Heat stable lactate dehydrogenase in the diagnosis of myocardial infarction. *J.A.M.A.* 182:1099, 1962.

69. Van der Helm, H. J.   A simplified method of demonstrating lactic dehydrogenase isoenzymes in serum. *Clin. Chim. Acta* 7:124, 1962.

70. Van der Helm, H. J., Zondag, H. A., Hartog, H. A., and van der Kooi, M. W.   Lactic dehydrogenase isoenzymes in myocardial infarction. *Clin. Chim. Acta* 7:540, 1962.

71. Vesell, E. S.   Genetic Control of Isozyme Patterns in Human Tissue. In Steinberg, A. G., and Bearn, A. G. (Eds.), *Progress in Medical Genetics*, Vol. 4. New York: Grune & Stratton, 1965. Pp. 128–175.
72. Vesell, E. S., and Bearn, A. G.   Localization of lactic acid dehydrogenase activity in serum fractions. *Proc. Soc. Exp. Biol. Med.* 94:96, 1957.
73. Vesell, E. S., and Bearn, A. G.   Isozymes of lactic dehydrogenase in human tissues. *J. Clin. Invest.* 40:586, 1961.
74. Wacker, W. E. C., and Dorfman, L. E.   Urinary lactic dehydrogenase activity: I. Screening method for detection of cancer of kidneys and bladder. *J.A.M.A.* 181:972, 1962.
75. Wacker, W. E. C., Ulmer, D. D., and Vallee, B. L.   Metalloenzymes and myocardial infarction; malic and lactic dehydrogenase activities and zinc concentrations in serum. *New Eng. J. Med.* 255:449, 1956.
76. Warnock, M. L.   Isozymic patterns in organs of mice infected with LDH agent. *Proc. Soc. Exp. Biol. Med.* 115:448, 1964.
77. Wieme, R. J.   Nomenclature of so-called isoenzymes. *Lancet* 1:270, 1962.
78. Wieme, R. J.   Improved agar support for electrophoresis of LDH isoenzymes. *Clin. Chim. Acta* 13:138, 1966.
79. Wieme, R. J., and Demeulenaere, L.   Enzymo-électrophorèse de la déhydrogénase de l'acide lactique, test sélectif d'antégrite parenchymateuse. *Acta Gastroent. Belg.* 22:69, 1959.
80. Wieme, R. J., and Herpol, J. E.   Origin of the lactate dehydrogenase isoenzyme pattern found in the serum of patients having primary muscular dystrophy. *Nature* (London) 194:287, 1962.
81. Wieme, R. J., van Sande, M., Karcher, D., Lowenthal, A., and Van der Helm, H. J.   A modified technique for direct staining with nitroblue tetrazolium of lactate dehydrogenase iso-enzymes upon agar gel electrophoresis. *Clin. Chim. Acta* 7:750, 1962.
82. Wilkinson, J. H.   Isoenzymes, with special reference to new enzyme tests in myocardial infarction. *Proc. Roy. Soc. Med.* 56:177, 1963.
83. Wright, E. J., and Cawley, L. P.   Lactic acid dehydrogenase isoenzyme profile of human urine. *J. Kansas Med. Soc.* 66:499, 1965.
84. Wright, E. J., Cawley, L. P., and Eberhardt, L.   Clinical application and interpretation of the serum lactic dehydrogenase zymogram. *Amer. J. Clin. Path.* 45:737, 1966.
85. Wroblewski, F., and Gregory, K. F.   Lactic dehydrogenase isozymes and their distribution in normal tissues and plasma and in disease states. *Ann. N.Y. Acad. Sci.* 94:912, 1961.
86. Zondag, H. A.   Lactate dehydrogenase isozymes; lability at low temperature. *Science* 142:965, 1963.

4

THE SERUM LIPOPROTEINS

In plasma most small molecules are transported to target cells by specific proteins and therefore are not free to diffuse through the walls of blood capillaries. From a biologic standpoint binding of small molecules to large protein molecules may operate to conserve and transport vital compounds in the blood. Certain serum proteins serve as specific carriers for a variety of substances such as lipids, steroids, metals, and enzymes. This binding property of proteins is relatively specific and involves a particular protein carrier and a bound compound. Examples of the phenomenon are plentiful; some common binding types are haptoglobins (hemoglobin), transferrin (iron), transcortin (cortisone), ceruloplasmin (copper), thyroxine-binding protein (thyroxine), hemopexin (hemoglobin), and glycoproteins (polysaccharides).

Lipoproteins represent a class of proteins capable of transporting various kinds of important lipids throughout the body. With few exceptions the lipids in the blood are attached to protein. Like haptoglobins, lipoproteins are heterogeneous, and they may be under similar genetic control. They also differ in lipid composition. Besides immunologic specific protein, blood lipoproteins contain phospholipids, triglycerides, free cholesterol, cholesterol esters, small quantities of free fatty acids, and monoglycerides and diglycerides [20, 21].

Lipoproteins have been classified on the basis of density, ultracentrifugation flotation units, and electrophoretic mobility. A combination of these classifications often is helpful, although a more useful approach, based on the structures of lipoproteins, has been suggested [41]. The purpose of this chapter is to review briefly what is currently known about lipoproteins and to discuss technics for evaluating them in disease.

Classification of Serum Lipoproteins

Figure 4-1 is a summary of the classification of lipoproteins [20, 37, 40, 41, 51, 54]. Lipoproteins with a density greater than 1.063 are classified as high-density lipoproteins (HDL) and migrate electrophoretically near the α_1-globulins, just behind albumin. Lipoproteins with densities between 1.006 and 1.063 are classified as low-

CLASSIFICATION OF SERUM LIPOPROTEINS

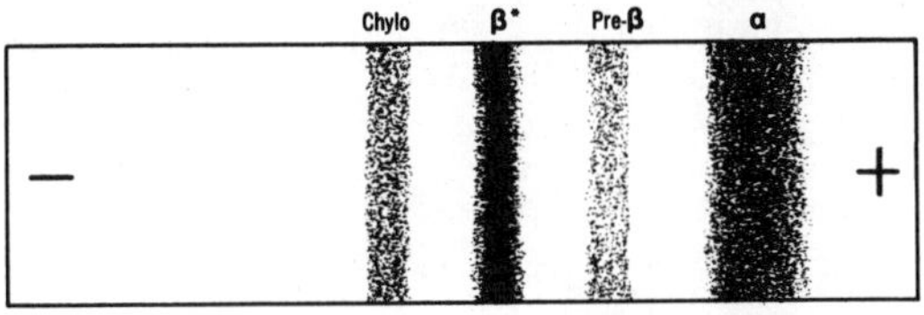

	LDL	VLDL	HDL
DENSITY	(0.000) (1.006)	(1.006) (1.063)	(1.006) (0.93)	(1.063) (1.21)
S_f**	400+	0-20	20-400	—
LIPID CONTENT CHOLESTEROL			
FREE	+	+	+	+
ESTER	+	+++	++	++
TRIGLYCERIDES	++++	++	+++	+
PHOSPHOLIPIDS	+	++	++	+++
FATTY ACIDS	—	—	—	++++
TURBIDITY	++++	—	+	—
PEPTIDES	α+β	β	α+β	α

*β- Lipoproteins can be divided into two groups on basis of S_f and density. S_f 0-12 (density 1.019-1.063) and S_f 12-20 (density 1.006-1.019).

**S_f = Svedberg units of flotation in 10^{-13} cm/sec/dyne/Gm.

FIGURE 4-1

Lipoprotein zones in electrophoretogram correlated with HDL, LDL, VLDL, density, Svedberg flotation units, lipid content, turbidity, and peptide composition. (See text for specific details.)

density lipoproteins (LDL) and travel electrophoretically with the mobility of $α_2$-globulin in the usual type of agar-gel electrophoresis (Fig. 4-2) [56] and with the mobility of β-globulin by paper electrophoresis and agarose or Reinagar agar-gel electrophoresis [31, 41, 54]. Two types of lipoproteins are recognized in the low-density class: a *very*-low-density pre-β-lipoprotein (VLDL), with a high content of triglyceride [21], and a low-density β-lipoprotein (LDL) with a high content of cholesterol [21]. Chylomicrons are characteristically high in triglyceride content, have a density of less than 1.006, and do not migrate in an electrical field but remain at the point of application [19]. Free fatty acids, in addition to being a constituent of lipoproteins, are associated with albumin.

Although four areas on a serum electrophoretogram stain for lipid (Fig. 4-1), a combination of only two lipoproteins is involved. These are designated α- and β-polypeptides, in keeping with the electrophoretic mobility of α- and β-lipoproteins [21, 40]. Different genes control the synthesis of the two polypeptides. Absence of β-polypeptide is found in abetalipoproteinemia, and absence of normal α-polypeptide is associated with Tangier disease (absence of HDL). Antiserum against α- or β-polypeptides discloses them to be separate proteins. It

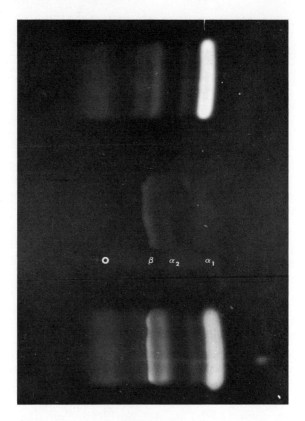

FIGURE 4-2

Agar-gel electrophoresis of human serum. *Upper,* protein stain. *Middle,* lipoprotein stain. *Lower,* lipoprotein counterstain of protein-stained electrophoretogram. A lipoprotein zone is depicted in the α_2-globulin. The pre-β is seen just ahead of the dense β-lipoprotein band, which extends from the α_2-globulin area. The position of the α- and β-lipoproteins and the pre-β in relation to serum proteins is shown by counterstain. Sudan Black stain for fat. Thiazine Red R stain for protein. Anode is on the right.

has also been shown that both the pre-β zone and the chylomicron zone are hybrids of α- and β-polypeptides. These hybrids can be disassociated if the lipid is removed by ether extraction, or if heparin is given to the patient [41].

Changes in Lipoprotein Patterns Relative to Disease

CORONARY DISEASE, NEPHROSIS, AND DIABETES

Increase in pre-β-lipid is found in patients with coronary artery disease [21]. This band becomes more intense following myocardial infarction and is heavily laden with triglycerides, as noted in Figure 4-1. There is also an increase in β-lipoproteins, but rarely in α-lipoproteins. β-lipoproteins are found in atherosclerotic arteries and appear to be derived from the plasma as well as being synthesized locally [47]. In nephrosis the serum electrophoretic pattern shows marked decrease of albumin, slight to moderate reduction of γ-globulin, and an increase of β-globulin. The increase in β-globulin is principally due to the increase of β-lipoproteins [51]. In patients with hypothyroidism, the β-lipoproteins are also elevated. The pre-β band is usually not significantly increased. Diabetes associated with ketosis produces marked increase of pre-β-lipid and complete disorganization of the pattern; this has been revealed by positive fat staining extending from the β-lipoproteins to include the point of application [37].

LIVER DISEASE

Elevation of serum β-lipoproteins follows hepatic parenchymal damage [34, 49]. Persistent elevation of β-lipoproteins is indicative of posthepatitis, if other causes of hyperlipoproteinemia discussed above have been excluded [34]. Dangerfield and Hurworth have noted an unstable β-lipoprotein in a case of biliary cirrhosis [17]. They found an increase of serum turbidity following freezing and thawing, and electrophoretic examination of serum after freezing disclosed that a large portion of the β-lipoprotein remained at the point of origin. We have examined a serum from a patient with jaundice which had similar characteristics. After freezing the turbidity became less, and a fatty layer appeared at the surface of the serum on thawing. Electrophoretically the top layer contained lipoprotein that remained at the point of application.

ABETALIPOPROTEINEMIA

In abetalipoproteinemia there is almost complete absence of LDL, the β-polypeptide synthesis is defective, and lipoproteins with a density of less than 1.063 are absent from the serum [18]. Patients with this disease have a neuromuscular disorder which presumably is based on the degeneration of posterolateral columns and the cerebellar tract. Findings include acanthotic erythrocytes, pigmentary degeneration of the retina, and engorgement of the upper intestinal absorptive cells with triglycerides [18]. The plasma cholesterol and phospholipid levels are low, probably because of a defect involving the HDL in addition to absence of LDL. Detailed studies of HDL from three patients with abetalipoproteinemia show that (1) high-density lipoprotein HDL_3 is reduced and HDL_2 is normal and (2) the phospholipid distribution in both HDL fractions is abnormal, with low content of lecithin [32]. Levy and Fredrickson (see Molecular Heterogeneity of Lipoproteins—α-Lipoproteins, below) have shown that normal α-lipoprotein undergoes changes induced by ultracentrifugation with conversion of HDL_2 into HDL_3 [40].

TANGIER DISEASE

Tangier disease, named after Tangier Island in Chesapeake Bay where the first two cases in siblings were found, is characterized by complete or nearly complete absence of plasma HDL and is associated with the storage of cholesterol esters in reticuloendothelial tissue, particularly in the tonsils [19]. Fredrickson et al. have examined three additional pairs of siblings in families unrelated to the Tangier group [22]. A single pair of autosomal genes exert major control over plasma HDL. The heterozygous mutant may be recognized by decreased HDL concentration, whereas the homozygous abnormal has no significant plasma HDL, and the tissue storage of cholesterol esters is typical of the complete syndrome. Concentration of the serum discloses a small amount of α-lipoprotein which has the same electrophoretic mobility as normal α-lipoprotein but is not antigenetically the same [22]. Tangier α-lipoprotein, as it has been designated, is the only α-lipoprotein found in Tangier disease. In the heterozygous both the Tangier and normal α-lipoprotein are found [22].

The enzymatic disorder responsible for the expression of this disease is not clear. It is postulated that the defect may involve the synthesis of HDL, and presumably the absence of such carrier proteins permits the storage of lipid in tissues. Serum cholesterol and phospholipids are low and similar to values found in abetalipoproteinemia [22].

SKIN DISEASE, MYELOMA, AND MONOCLONAL GAMMOPATHY

Xanthomatosis is divided into two chemical groups. In xanthomatosis associated with hypercholesterolemia, the serum is clear and there is increase of β-lipoprotein on electrophoresis [21, 51]. In xanthomatosis associated with hyperlipemia, the serum is turbid and lipid stain of electrophoretograms shows heavy deposition of lipid from the point of application to the α_2-globulin area [21, 51]. Neufeld et al. reported an unusual case of hyper-β_2-lipoproteinemia in a patient with myeloma; xanthomas were present over the trunk and buttocks [45]. The immunoglobulin abnormality was related to γA immunoglobulin. A serum studied in our laboratory was characteristic of γA monoclonal gammopathy; extensive skin xanthomas were present over the patient's upper trunk. Reticuloendotheliosis without skeletal lesions was disclosed at autopsy, and no evidence of myeloma was detected. Lewis and Page reported on a patient with hyperlipemia associated with high levels of γA immunoglobulin of 11 years' duration [42]. The authors considered that the condition may represent molecular interaction of abnormal immunoglobulins and β-lipoproteins. There was no evidence of myeloma, xanthomatosis, or atherosclerosis in this patient. Kayden et al. described a 40-year-old patient with hyperlipemia (β-lipoprotein), xanthomas, monoclonal peak (7S γ), increased plasma cells in marrow, and hypercholesterolemia. They showed that the abnormal protein formed a complex with β-lipoprotein [33]. Levin

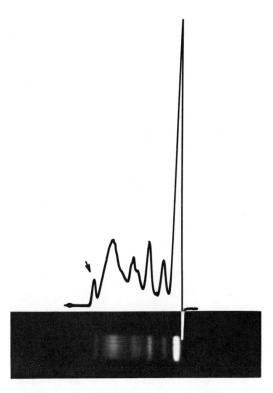

FIGURE 4-3

Electrophoretogram and densitometric scan showing small post-γ monoclonal peak. Anode is on the right.

et al. presented a case of multiple myeloma associated with nonfamilial xanthomatosis, splenomegaly, and absence of osteolytic lesions [39]. Several reports on multiple-myeloma-type proteins with
lichen myxedematosus are indicative of the association of this disease
with dysproteinemia and malignant plasma cell proliferation [42, 43,
48]. The association of γA immunoglobulin increase and skin disease
appears to be real and not based on chance association. If it were on
the basis of chance alone, there should be a higher proportion of cases
involving γG immunoglobulin since in myeloma γG is involved in
about 60 percent of cases and γA in about 16 percent of cases.

It is also unclear whether γA increase and skin disease should be
considered a part of plasma cell malignancy or not. The literature on
the subject is incomplete at this time. It is tempting to reason that
the increase of γA is on the basis of stimulation by the skin pathology.
However, one may also reason that in γA monoclonal gammopathy,
from whatever cause, a toxic reaction between γA and the skin
develops. The latter reaction possibly involves lipoproteins. The skin

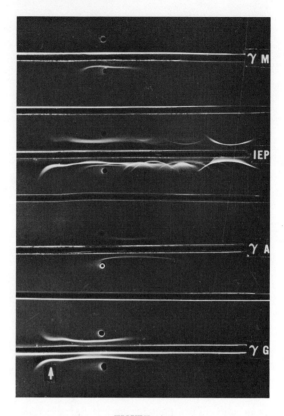

FIGURE 4-4

Immunoelectrophoretogram of serum of Figure 4-3 showing distortion of γG both in polyvalent antiserum system and in specific anti-γG. Anode is on the right.

lesion often contains high concentration of lipid and in our case a high concentration of γA immunoglobulin.

In our laboratory a most unusual lipid band was found by electrophoresis in a patient with long-standing rheumatoid arthritis who was admitted because of rectal abscesses associated with agranulocytosis. He had a small post-γ monoclonal peak on his serum electrophoretogram (Fig. 4-3) which had γG specificity by immunoelectrophoresis (Fig. 4-4) and gave a positive reaction for fat with Sudan Black. Fixation in isopropyl alcohol showed the monoclonal peak to have properties causing precipitation similar to lipoproteins (Fig. 4-5) [53]. The photograph in Figure 4-5 was taken with 35-mm. color film using indirect lighting. Lysozymes of urine and serum were not related to the post-γ peak. Following therapeutic course of steroids, the monoclonal peak disappeared. A Bence Jones protein was present in the urine, which also disappeared after therapy. This example suggests

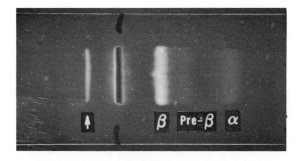

FIGURE 4-5

Photograph of dark-field illuminated agarose-gel electrophoretogram of same serum as in Figure 4-3 after fixation with isopropyl alcohol. Note the sharp appearance of the post-γ-globulin monoclonal peak. Note also the position of α-, β-, and pre-β-lipoproteins. Anode is on the right.

binding of lipids to a slow monoclonal γG immunoglobulin produced in a patient with rheumatoid arthritis during a period of acute stress. Tests by immunoelectrophoresis with anti-β-lipoprotein did not show that β-lipoprotein was part of the monoclonal peak (Fig. 4-6).

MALIGNANCY

Serum β-lipoproteins are increased and α-lipoproteins (HDL$_2$ fraction) are decreased in the presence of cancer in experimental animals

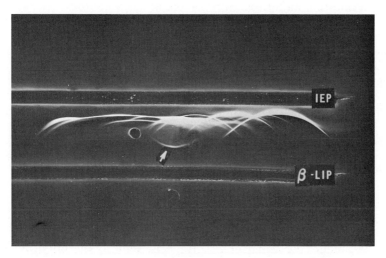

FIGURE 4-6

Immunoelectrophoretic study on serum of Figure 4-3 carried out with anti-β-lipoprotein showing that the position of the lipoprotein is independent of the post-γ monoclonal peak. Anode is on the right.

and man. In a study of breast cancer among young women Barclay et al. found a significant elevation of serum β-lipoproteins (S_f 0-20; D < 1.006) compared to normal young women by ultracentrifugation studies [4]. Also the serum α-lipoproteins (HDL_2 fraction) (S_f 0-4; D < 1.125) were below normal. In experimental animals modification of serum lipoproteins and lipids has been observed to be induced by cancer. Barclay et al. presented information on the effects of two different tumors on serum lipoproteins [5]. Transplanted Walker carcinosarcoma 256 and tumors produced by 9,10-dimethyl-1,2-benzanthracene both caused increase of serum β-lipoprotein (S_f 0-20; D < 1.006) and decrease of HDL_2 fraction of α-lipoprotein. The Moloney rhabdomyosarcoma produces marked turbidity of the serum. Electrophoretically the chylomicrons are increased, and there is a reduction of α-lipoprotein [44].

No satisfactory explanation is known for the influence that cancer exerts on the metabolism of serum lipoproteins. Production, release, and catabolism of lipoproteins are areas to be studied to pinpoint the site of action of cancer. Subtle liver damage may also be a factor in disturbance of lipoproteins. The tumor does not produce lipoproteins, at least in the experimental animal.

Classification of Hyperlipoproteinemias on the Basis of Electrophoresis

Fredrickson and Lees have proposed an electrophoretic system for the classification of hyperlipoproteinemic serum [20]. Their approach appears to have excellent possibilities as a useful laboratory procedure. The electrophoretic position of the various lipoproteins, as indicated in Figure 4-1, is relatively stable when dietary intake is consistent. Chylomicrons represent dietary fat, whereas β-, pre-β-, and α-lipoprotein represents endogenous fat. Fredrickson and Lees have described five types of lipoprotein phenotypes which are characterized by lipoprotein electrophoresis, the concentration of plasma cholesterol and triglyceride, the response of serum lipids to heparin (postheparin lipolytic activity), to the induction of a lipoprotein electrophoretic abnormality by fat or carbohydrate feedings [21, 22].

TYPE I—HYPERLIPOPROTEINEMIA

FAMILIAL FAT-INDUCED HYPERLIPEMIA. This group is characterized by elevated plasma chylomicrons. The lipoprotein electrophoretic pattern is unique, consisting of positive lipid stain at the origin, which differs from the normal patterns as noted in Figures 4-1 and 4-2. Figure 4-7 shows an example of type I compared to normal and other patterns. The postheparin lipolytic activity is low, and the glucose tolerance is normal. This is a rare disorder often associated with

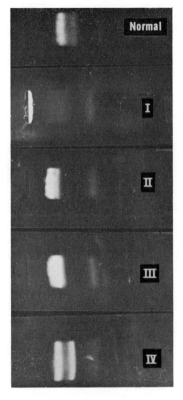

FIGURE 4-7

Agarose electrophoretic lipoprotein patterns of normal, type I, type II, type III, and type IV. Note the sharp band formation and the distinctiveness of the pre-β zone except for "floating beta" in type III. Sudan Black stain. Anode is on the right.

hepatosplenomegaly and may be based on defect of lipoprotein lipase, which normally clears glyceride from serum. Xanthomas may occur. About 35 acceptable cases have been described [22]. The disease is usually manifested at an early age and often is detected before the age of 10. Low fat intake reduces the turbidity of the serum, and the liver and spleen decrease in size. Following a few days of low fat intake, the lipoprotein electrophoretic pattern will change in a predictable pattern. Chylomicrons will disappear, the α-lipoprotein and β-lipoprotein bands will increase, and the pre-β-lipoproteins will increase substantially.

TYPE II—HYPERLIPOPROTEINEMIA

FAMILIAL HYPERBETALIPOPROTEINEMIA. Also referred to as familial hypercholesterolemia, familial hyperbetalipoproteinemia is associated with increased plasma cholesterol, decreased plasma triglyceride, clear plasma, xanthomatosis and atheromatosis, low pre-β-lipoprotein, and increased β-lipoprotein (Fig. 4-7). The genetic transmission is autosomal dominant. As noted, the serum pre-β-lipoprotein may be elevated, and the elevation may be associated with a slight increase of serum triglycerides.

DISEASE-DEPENDENT TYPE II. The type II pattern is seen in hypothyroidism, obstructive jaundice, and nephrotic syndrome. As noted under disease-related lipoprotein patterns, a few cases of myeloma have been reported with a lipoprotein pattern of the type II class.

TYPE III—HYPERLIPOPROTEINEMIA

FAMILIAL XANTHOMATOSIS AND HYPERLIPOPROTEINEMIA. This is often referred to as familial hypercholesterolemia with hyperlipemia. The plasma shows increased amounts of both cholesterol and triglycerides; however, the plasma is clear. Tendon and palpebral xanthomatosis and advanced atheromatosis are common findings in patients of type III. Postheparin lipolytic activity is normal, and glucose tolerance is abnormal. Both β- and pre-β-lipoproteins are increased. This pattern can be induced in the patients by a high-carbohydrate meal. A diabetic family history and coronary artery disease frequently are associated with this phenotype. The β and pre-β zones are merged, and Fredrickson et al. refer to the "floating beta" of type III to describe the pattern (Fig. 4-7) [22]. This feature separates type III from type II. Clinically, the two are also distinct.

TYPE IV—HYPERLIPOPROTEINEMIA

FAMILIAL HYPERPREBETALIPOPROTEINEMIA. The plasma values for cholesterol and triglycerides are slightly elevated; the plasma is usually clear, but may be cloudy or milky. As in type I, xanthomas and hepatosplenomegaly may occur. The postheparin lipolytic activity is low, and an abnormal glucose tolerance is found. The lipoprotein pattern is carbohydrate inducible. Pre-β-lipoproteins are increased. This phenotype is frequently seen in young men who have had myocardial infarction. It may be related to a predisposition to coronary artery disease, and there may be a family history of diabetes (Fig. 4-7).

DISEASE-DEPENDENT TYPE IV. Type IV pattern is frequently observed in familial and nonfamilial disorders, as noted under Changes in Lipoprotein Patterns Relative to Disease, above. Diabetes mellitus, pancreatitis, and alcoholism often show type IV pattern [22]. It is also seen in hypothyroidism, nephrotic syndrome, idiopathic hypercalcemia, and rare examples of dysgammaglobulinemia [22]. In diabetes mellitus, the pattern is most frequently observed in those individuals in poor control. Mechanism for the hyperlipemia in pancreatitis is unknown. In alcoholism the evidence suggests hepatocellular damage as the mechanism leading to accumulation of glycerides in the serum [22]. In nephrotic syndrome there is considerable variation in the lipoprotein pattern. β-Lipoprotein is often increased, although there are occasional cases in which pre-β is quite elevated. Multiple myeloma, as noted under disease-related patterns and type II, is sometimes associated with an abnormal lipoprotein pattern. Type IV patterns in association with myeloma have been reported. γA immunoglobulin is the usual type of immunoglobulin involved.

TYPE V—HYPERLIPOPROTEINEMIA

FAMILIAL HYPERCHYLOMICRONEMIA AND HYPERPREBETALIPOPROTEINEMIA.   Cholesterol and triglycerides are increased, and the plasma is turbid. The plasma chylomicrons are elevated. The electrophoretic lipoprotein pattern is similar to that of type I except for elevation of pre-β-lipoprotein. Postheparin activity is normal, but the glucose tolerance is abnormal. The pattern is subject to fat and carbohydrate inducibility. This phenotype is rare and seems to be due to a defect in both the exogenous and endogenous metabolism of glyceride.

Hyperlipoproteinemia in Coronary Artery Disease

From the foregoing discussion of electrophoretic classification of hyperlipoproteinemias it is logical to assume that persons who are susceptible to coronary artery disease could be detected prior to clinical disease by lipoprotein phenotyping. Fredrickson et al. have successfully shown that a high incidence of coronary artery disease is associated with type III and type IV [22]. Both patterns are inducible by a carbohydrate load. Patients with type III or IV phenotype usually have elevated cholesterol or triglycerides and a high incidence of diabetes or positive family history of diabetes. Kuo studied 286 patients with atherosclerosis to determine their serum lipoprotein phenotypes and serum lipids following an *ad libitum* carbohydrate (35–40 percent) diet [36]. This diet does not cause an abnormal response in normal subjects. In patients who are susceptible or sensitive to carbohydrates, e.g., persons of types III and IV, hyperprebetalipoproteinemia, hyperglyceridemia, and hypercholesterolemia develop. Over 90 percent of the 286 patients responded abnormally to the carbohydrate diet by showing type III or IV lipoprotein phenotype and elevation of serum triglycerides, cholesterol, and phospholipids. Type III phenotype was the most prevalent (81.8 percent). Kuo also found he could reverse the abnormalities and maintain normal lipid values in his patients by placing them on a restricted carbohydrate diet.

The clinical studies carried out thus far indicate that a suitable tolerance test for detection of coronary artery disease should include a diet of carbohydrate designed to cause the susceptible person to show serum lipoprotein and serum lipid abnormalities. The procedure outlined by Kuo meets these criteria.

Molecular Heterogeneity of Lipoproteins

Heterogeneity is used here to refer not only to antigenic variation but also to differences of electrophoretic mobility. The two are altogether separate. The importance of the genetics of the β-lipoprotein antigenic composition cannot be overestimated. The relationship of serum β-lipoproteins to cell membrane lipoprotein is connected with the antigenic composition of normal lipoproteins and tumor lipoproteins.

BETA LIPOPROTEINS

Like haptoglobins, β-lipoproteins have been studied from the standpoint of their molecular heterogeneity and genetic control. Allison and Blumberg found by immunodiffusion that an antibody reacting with normal human serum developed in some persons who had received multiple transfusions [1]. The antigen was designated Ag. One paper reported precipitins in sera of 18 patients who had received multiple blood transfusions [9]. These precipitins reacted with a lipid-containing serum protein present in some, but not all, individuals. Using two of the antisera, the investigators showed the Ag antigen to be a β-lipoprotein (LDL). Frequency of precipitin formation in patients who had been given multiple transfusions appears higher in those with thalassemia than in those without history of the disease. However, this point has not been clearly elucidated. Other examples of polymorphism of β-lipoprotein have been found (Table 4-1), and there are several antigens in the system [2].

Berg, using an antibody made in rabbits against normal human serum, and later against isolated fractions of β-lipoprotein, found evidence of genetic difference of β-lipoprotein in the human population [7]. Employing immunodiffusion, he was able to divide the population into two classes: Lp (a) positive and Lp (a) negative [7]. The precipitin bands were further characterized by lipid staining. The relationship between the antigens of Allison and Blumberg and those found with rabbit anti-β-lipoprotein was studied by Berg [6] with immunodiffusion in gel, and he was able to show that no antigenic relationship or linkage was present between the two. The two antigens appeared to be on different parts of the same β-lipoprotein (LDL) molecule. Thus, it appears that the Lp and Ag systems of human serum β-lipoprotein are independent. The fact that these two antigenic markers have not been separated electrophoretically lends support to the possibility of the involvement of two separate parts of the β-lipoprotein molecule. A new human β-lipoprotein antigen was demon-

TABLE 4-1

POLYMORPHISM OF β-LIPOPROTEIN

Reagents	Antigens	Authors
C. de B.	Ag (a_1)	Allison and Blumberg [1, 2]
	Ag (x)	Adsorbed with L. L.—Hirschfeld, Blumberg, and Allison [26], and Hirschfeld and Okochi [29]
	Ag (z)	Hirschfeld, Blumberg, and Allison [26]
L. L.	Ag (x)	Hirschfeld and Blomback [25]
I. M.	Ag (b)	Blumberg, Alter, Riddell, and Erlandson [9]
B. N.	Ag (t)	Hirschfeld et al. [30]
		Hirschfeld and Bundschuh (one and the same antigen) [27]
	Ag (b)	Gesrick et al. [23]
C. P.	Ag (y)	Contu (formerly Lp-Nuoro) [16]
		Hirschfeld et al. [28, 29]
Australian aborigine	Australia Au (1)	Blumberg et al. [10]

strated by Contu with double-gel diffusion and immunoelectrophoresis in 95 percent of Caucasians. The antigen Lp-Nuoro is not related to the Lp (a) and is genetically transmitted by autosomal chromosomes [16]. As noted in Table 4-1, this antigen has been shown to be identical to Ag (y) [28, 29].

An additional β-lipoprotein antigen characterized by poor lipid staining of immunoprecipitin bands has been described by Blumberg et al. [10] and Alter et al. [3]. It has been designated Australia antigen —Au (1)—since it was first found in an Australian aborigine. An isoprecipitin against the antigen is present in the sera of patients with hemophilia who have received many transfusions. The antigen is found in sera of some normal individuals from subtropical areas (Filipinos, Indians, etc.) but is absent in sera from the population of the United States. Recently Sutnick et al. found the antigen in 10 percent of patients with viral hepatitis and in 28 of 100 patients with Down's syndrome [53]. Serum of some patients with chronic lymphatic and acute myelogenous leukemia also have the antigen [53].

Table 4-1 makes it clear that the genetic variations of β-lipoproteins are considerable and will no doubt grow. The reagents listed on the left are the names used by the original investigator and usually stand for the initials of the persons whose serum was found to contain anti-β-lipoprotein antibodies. Hirschfeld and his associates have done a considerable amount of work on these antigens, of which eight are listed in Table 4-1. Ag (x) and Ag (y) are believed to be products of alleles. High Ag (x) and low Ag (y) frequencies are found in individuals of Japan, India, and Thailand [29].

Seegers et al. have reported the presence of a double β-lipoprotein in a single family [50]. This abnormality was found by paper electrophoresis. One zone had the same mobility as normal β-lipoprotein whereas the abnormal component traveled farther on paper. The reverse was noted by starch electrophoresis. Immunologically, both components react with anti-β-lipoprotein.

PRE-BETA LIPOPROTEINS

This area of the lipoprotein electrophoretogram has received little attention from the standpoint of variation in electrophoretic heterogeneity. Figure 4-8 shows three types of variations noted in our labo-

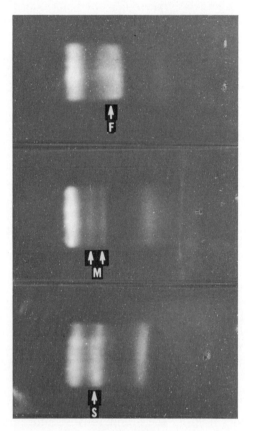

FIGURE 4-8

Agarose electrophoretic lipoprotein patterns from three patients with variability of pre-β-lipoprotein. Note two pre-β zones in middle pattern (M) and variation in mobility of pre-β in upper (F) and lower (S) patterns. Sudan Black stain. Anode is on the right.

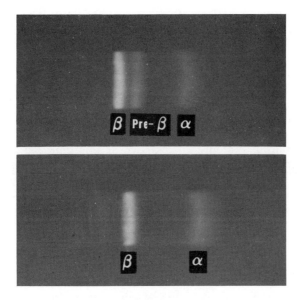

FIGURE 4-9

Close-up photograph of two agarose electrophoretic lipoprotein patterns showing sharp separation of pre-β from β (*upper*) in normal patient. The lower pattern is also from a normal person who has no detectable pre-β. Anode is on the right.

ratories. We have designated them as fast (F), slow (S), and mixed (M). The variation in mobility as noted in the upper (F) and lower (S) patterns may represent differential interaction between agarose and lipoproteins, even though albumin is incorporated in the gel. The upper and lower patterns show homogeneity of the bands. The middle pattern has two separate components which make up the pre-β-lipoprotein. This heterogeneity may represent allotypes, with most individuals being of the S type and the M type representing the heterozygous. Also, the two zones of the M type could be the α- and β-peptide components of pre-β-lipoprotein. Family studies have not been carried out to determine whether this pattern is familial. Occasionally a pattern without a trace of pre-β is found, as shown in the lower portion of Figure 4-9. In paper electrophoretic studies of serum lipoproteins a pre-β zone is often not demonstrable; however, the agarose system rarely fails to show a pre-β zone, which is usually as seen in the upper part of Figure 4-9. The lipoprotein zones stand out as white opaque zones after fixation of strips in isopropyl alcohol and can be photographed at that time (Fig. 4-5).

ALPHA LIPOPROTEINS

Levy and Fredrickson have noted electrophoretic and immunoelectrophoretic heterogeneity of α-lipoproteins (HDL) which corresponds to ultracentrifugal fractions HDL$_2$ and HDL$_3$ [40]. These two components were demonstrated electrophoretically after delipidization and designated LpA (fast fraction) and LpB (slow fraction). The amino acid composition of the two fractions is identical. Immunologically, there is partial cross-reactivity. LpA appears to be the natural form in fresh plasma and undergoes conversion to LpB during storage or ultracentrifugation.

How these findings relate to those of Kranz and Heide, who demonstrated existence of at least two immunologically distinct components of the α-lipoproteins, is unknown at this time [35]. It appears that both groups have shown electrophoretic and immunologic heterogeneity of the α-lipoproteins. Figure 4-10 is an agarose-gel electrophoretogram and scan of serum lipoproteins showing two relatively separate areas of α-lipoprotein. The patient had no unusual abnormalities. This pattern has been seen in a small number of healthy persons. The position of the fast α-lipoprotein corresponds to that of albumin. Berg has shown that in gel diffusion albumin reacts with lipoproteins [7]. The explanation for electrophoretic heterogeneity of α-lipoprotein is not clear. The possibility that increase of certain lipids may saturate the available α-lipoprotein with the excess lipid being bound to albumin is attractive since this mechanism is seen in hemolysis where the available haptoglobin is saturated with hemoglobin and albumin picks up the excess hemoglobin.

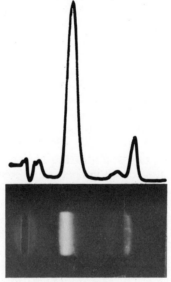

FIGURE 4-10

Electrophoretic heterogeneity of α-lipoprotein in agarose electrophoretogram. Note two α-lipoprotein bands in stained patterns and accompanying scan. The fast component of α-lipoprotein corresponds to position of albumin. Sudan Black stain. Anode is on the right.

Methods of Separation of Serum Lipoproteins

Of the technics currently available for the study of serum lipoproteins in the clinical laboratory, two seem worthy of consideration: electrophoresis and immunoelectrophoresis.

ELECTROPHORESIS

Electrophoresis by paper with albumin-containing buffer, proposed by Lees and Hatch [38], appears to be a satisfactory system for sharp electrophoretic separation of lipoproteins. Separation is followed by staining with Oil Red O or Sudan Black. The staining of lipoproteins following ordinary paper electrophoresis has been performed by Smith for a number of years [51]. The addition of albumin to the buffer overcomes some of the adsorption which normally occurs between β-lipoprotein and the paper.

Lipoproteins have a great affinity for a number of substances other than paper. In agar gel, the sulfonated polysaccharide strongly influences the position of β-lipoproteins [31, 56]. Electrophoretic mobility of β-lipoprotein in agar gel is concentration dependent; that is, the lower the concentration, the farther the migration (Fig. 4-2). The less sulfonated material in the agar, the less variability in the mobility [31, 56]. Agarose, a purified material, has only traces of sulfur and is ideal for lipoprotein electrophoresis; little variability of lipoprotein mobility has been noted. In gelatin electrophoresis, adsorption is negligible, and mobility of various dilutions of α-lipoprotein is essentially identical [54]. When 1 percent albumin is incorporated in the agarose and buffer system of the electrophoretic and immunoelectrophoretic system developed in our laboratory, adsorption of lipoprotein in the agarose is reduced, apparently by the same mechanism which reduces the adsorption of lipoproteins to paper in paper electrophoresis [12, 13, 14, 15]. However, Berg has noted that albumin interacts with lipoprotein at room temperature in gel diffusion [6]. The phenotyping of hyperlipoproteinemias by agarose electrophoresis which incorporates albumin appears to be as useful as paper electrophoresis and requires less time (Figs. 4-7, 4-8, 4-9, and 4-10). The technic is described in detail in Chap. 15.

IMMUNOELECTROPHORESIS

Immunoelectrophoretic analysis of serum lipoproteins offers limited practical application in the clinical laboratory at this time. Lipid staining outlines the position of lipoprotein immunoprecipitin bands and defines their relationship to the other immunoprecipitin bands. The question of whether or not the heterogeneity of lipoproteins manifested electrophoretically is matched by antigenic differences is the subject of several papers [11, 35, 40, 41, 42, 52, 55]. The α- and β-polypeptides are separate proteins, but whether the β-lipoprotein is a single component is controversial [11, 55, 56]. Walton and Darke have shown by a series of immunochemical studies with isolated β-lipoproteins that only one component is present [55]. Storiko and Fisher isolated low-density lipoprotein fractions ($α_2$- and β-lipoprotein) from pooled human serum by ultracentrifugal flotation which immunologically reacted as a single antigen [52]. These findings do not exclude multiple molecular forms of β-lipoproteins, as noted in genetic polymorphism, or in the rare familial example of double β-lipoprotein pattern found by Seegers et al. [50].

Levy et al. have offered excellent evidence that the α-polypeptides are found in conjunction with β-polypeptides in the pre-β-lipoprotein zone [41]. Some of the reported heterogeneity of α-lipoprotein could be related to technic. For example, Levy and his co-workers demonstrated that antisera to α- and β-polypeptide give similar results on untreated serum. However, following ether extraction of serum before electrophoresis to partially remove lipids, polypeptides could be separated. Immunoelectrophoresis performed on agarose with ether-extracted serum may prove useful in study of α- and β-polypeptides in relationship to disease and genetic control. Study of the polymorphism of β-lipoproteins, as depicted by the work of Allison and Blumberg [2] and of Berg [7], unquestionably will be expanded and correlated with the primary structure of the protein.

Summary

Lipids are carried in the blood bound to molecules which have been designated lipoproteins. The bonding is of the hydrophobic type and can be broken by detergents. Localization in lipoprotein electrophoretograms is characterized by fat stains. A combination of the classifications of lipoproteins, based upon density, ultracentrifugation, and electrophoretic mobility, has clarified to some extent their heterogeneity. Apparently only two proteins, α- and β-polypeptides, are involved in the formation of the four fat-positive zones depicted by electrophoresis. The pre-β-lipoprotein zone seems to be a hybrid of α- and β-polypeptides. Abnormalities of the electrophoretic phenotyping of lipoproteins appears to represent a method whereby standardization of the classification of hyperlipoproteinemic serum may be possible.

Genetic polymorphism of β-lipoprotein has been discussed. The β-lipoproteins seem to be one component by electrophoresis and immunoelectrophoresis, but selective immunodiffusion and adsorption studies reveal that these molecules are composed of at least two separate antigens, one designated Ag and the other Lp (a). The "Australia antigen" also belongs with the β-lipoproteins; its relationship to the other antigens is unknown. At this time seven antigens belonging to the Ag system have been described. The usefulness of immunodiffusion for genetic studies is emphasized. Electrophoretic phenotyping appears to be the method of choice for study of serum lipoprotein in pathologic disorders. Immunoelectrophoresis with ether-extracted serum and specific anti-α- and anti-β-polypeptide antiserum should further elucidate the molecular structure and genetic control of the β-lipoproteins. The electrophoretic heterogeneity of the α-lipoproteins is also an area of interest.

References

1. Allison, A. C., and Blumberg, B. S.   An isoprecipitation reaction distinguishing human serum-protein types. *Lancet* 1:634, 1961.
2. Allison, A. C., and Blumberg, B. S.   Serum lipoprotein allotypes in man. *Progr. Med. Genet.* 5:176, 1965.
3. Alter, H. J., and Blumberg, B. S.   Further studies on a "new" human isoprecipitin system (Australia antigen). *Blood* 27:297, 1966.
4. Barclay, M., Escher, G. C., Kaufman, R. J., Terebus-Kekish, O., Greene, E. M., and Skipski, V. P.   Serum lipoproteins and human neoplastic disease. *Clin. Chim. Acta* 10:39, 1964.
5. Barclay, M., Skipski, V. P., Terebus-Kekish, O., Merker, P. L., and Cappucino, J. S.   Serum lipoproteins in rats with tumors induced by 9,10-dimethyl-1,2-benzanthracene and with transplanted Walker carcinosarcoma 256. *Cancer Res.* 27:1158, 1967.
6. Berg, K.   A new serum type system in man—the Lp system. *Acta Path. Microbiol. Scand.* 59:369, 1963.
7. Berg, K.   Comparative studies on the Lp and Ag serum type systems. *Acta Path. Microbiol. Scand.* 62:276, 1964.
8. Berg, K.   Precipitation reactions in agar gel between albumin and beta-lipoprotein of human serum. *Acta Path. Microbiol. Scand.* 62:287, 1964.
9. Blumberg, B. S., Alter, H. J., Riddell, N. M., and Erlandson, M.   Multiple antigenic specificities of serum lipoproteins detected with sera of transfused patients. *Vox Sang.* 9:128, 1964.
10. Blumberg, B. S., Alter, H. J., and Visnich, S.   A "new" antigen in leukemia sera. *J.A.M.A.* 191:541, 1965.
11. Burtin, P.   The Proteins of Normal Human Plasma. In Grabar, P., and Burtin, P. (Eds.), *Immuno-electrophoretic Analysis.* New York: American Elsevier, 1964. P. 94.
12. Cawley, L. P.   *Workshop Manual on Electrophoresis and Immunoelectrophoresis* (Rev. ed.). Commission on Continuing Education, Council on Clinical Chemistry. Chicago: American Society of Clinical Pathologists, 1966.
13. Cawley, L. P., and Eberhardt, L.   Simplified gel electrophoresis: I. Rapid technic applicable to the clinical laboratory. *Amer. J. Clin. Path.* 38:539, 1962.
14. Cawley, L. P., Eberhardt, L., and Schneider, D. R.   Simplified gel electrophoresis: II. Application of immunoelectrophoresis. *J. Lab. Clin. Med.* 65:342, 1965.
15. Cawley, L. P., Schneider, D., Eberhardt, L., Harrouch, J., and Millsap, G.   A simple semi-automated method of immunoelectrophoresis. *Clin. Chim. Acta* 12:105, 1965.

16. Contu, L.   A new human beta-lipoprotein antigen. *Nouv. Rev. Franc. Hémat.* 6:671, 1966.

17. Dangerfield, W. G., and Hurworth, E.   Unstable serum lipoprotein in biliary cirrhosis. *Clin. Chim. Acta* 7:292, 1962.

18. Farguhar, J. W., and Ways, P.   Abetalipoproteinemia. In Stanbury, J. B., Wyngaarden, J. B., and Fredrickson, D. S. (Eds.), *The Metabolic Basis of Inherited Diseases* (2d ed.). New York: McGraw-Hill, 1966. Chap. 24, pp. 509–522.

19. Fredrickson, D. S.   Familial High-Density Lipoprotein Deficiency: Tangier Disease. In Stanbury, J. B., Wyngaarden, J. B., and Fredrickson, D. S. (Eds.), *The Metabolic Basis of Inherited Diseases* (2d ed.). New York: McGraw-Hill, 1966. Chap. 23, pp. 486–508.

20. Fredrickson, D. S., and Lees, R. S.   A system for phenotyping hyperlipoproteinemia. *Circulation* 31:321, 1965.

21. Fredrickson, D. S., and Lees, R. S.   Familial Hyperlipoproteinemia. In Stanbury, J. B., Wyngaarden, J. B., and Fredrickson, D. S. (Eds.), *The Metabolic Basis of Inherited Diseases* (2d ed.). New York: McGraw-Hill, 1966, Chap. 22, pp. 429–485.

22. Fredrickson, D. S., Levy, R. I., and Lees, R. S.   Fat transport in lipoproteins—an integrated approach. *New Eng. J. Med.* 276:34, 44, 94, 148, 215, 273, 1967.

23. Gesrick, G., Bundschuh, G., and Grieger, E.   Uber einen präzipitierenden Isoantikorpes nach über 200 Transfusionen [anti-Ag (b)]. *Z. Immunitaetsforsch.* 131:272, 1966.

24. Hirschfeld, J.   Investigations of a new anti-Ag antiserum with particular reference to the reliability of Ag typing by micro-immunodiffusion tests in agar gel. *Sci. Tools* (the LKB-Journal) 10:45, 1963.

25. Hirschfeld, J., and Blomback, M.   A new anti-Ag serum (L. L.) *Nature* (London) 201:1337, 1964.

26. Hirschfeld, J., Blumberg, B. S., and Allison, A. C.   Relationship of human anti-lipoprotein allotypic sera. *Nature* (London) 202:706, 1964.

27. Hirschfeld, J., and Bundschuh, G.   Relation between anti-Ag(b) and anti-Ag(t) reagents. *Vox Sang.* 13:21, 1967.

28. Hirschfeld, J., Contu, L., and Blumberg, B. S.   Antilipoprotein nuoro serum (C. P.) and its relation to sera C. de B. and L. L. *Nature* (London) 214:495, 1967.

29. Hirschfeld, J., and Okochi, K.   Distribution of Ag (x) and Ag (y) antigens in some populations. *Vox Sang.* 13:1, 1967.

30. Hirschfeld, J., Unger, P., and Ramgren, O.   A new anti-Ag serum. Serum B. N. *Nature* (London) 212:206, 1966.

31. Houtsmuller, A. J., Huysson-Haasdijk, A., Huysman, A., and Rinkel-Van Driel, E.   The application of Reinagar for the quantitative separation of alpha- and beta-lipoproteins. *Clin. Chim. Acta* 9:497, 1964.

32. Jones, J. W., and Ways, P.   Abnormalities of high density lipoproteins in abetalipoproteinemia. *J. Clin. Invest.* 46:1151, 1967.

33. Kayden, H. J., Franklin, E. C., and Rosenberg, B.   Interaction of myeloma gamma globulin with human beta-lipoprotein (P). *Circulation* 26:659, 1962.

34. Kellen, J. A.   The estimation of beta-lipoproteins as a screening test for infectious hepatitis. *Clin. Chim. Acta* 9:138, 1964.

35. Kranz, T., and Heide, K. Zur Struktur der Lipoproteine. In Peeters, H. (Ed.), *Protides of the Biological Fluids*. Proceedings of the Thirteenth Colloquium, Bruges, 1965. Amsterdam: Elsevier, 1966. Sec. B, Lipoproteins, p. 289.

36. Kuo, P. T. Hyperglyceridemia in coronary artery disease and its management. *J.A.M.A.* 201:87, 1967.

37. Lees, R. S., and Fredrickson, D. S. The differentiation of exogenous and endogenous hyperlipemia by paper electrophoresis. *J. Clin Invest.* 44:1968, 1965.

38. Lees, R. S., and Hatch, F. T. Sharper separation of lipoprotein species by paper electrophoresis in albumin-containing buffer. *J. Lab. Clin. Med.* 61:518, 1963.

39. Levin, W. C., Aboumrad, M. H., Ritzmann, S. E., and Braulty, C. Gamma-type 1 myeloma and xanthomatosis. *Arch. Intern. Med.* (Chicago) 114:688, 1964.

40. Levy, R. I., and Fredrickson, D. S. Heterogeneity of plasma high density lipoproteins. *J. Clin. Invest.* 44:426, 1965.

41. Levy, R. I., Lees, R. S., and Fredrickson, D. S. The nature of prebeta (very low density) lipoproteins. *J. Clin. Invest.* 45:63, 1966.

42. Lewis, L. A., and Page, I. H. An unusual serum lipoprotein-globulin complex in a patient with hyperlipemia. *Amer. J. Med.* 38:286, 1965.

43. McCarthy, J., Osserman, E., Lombardo, J. C., and Takatsuki, K. An abnormal serum globulin in lichen myxedematosus. *Arch. Derm.* (Chicago) 89:446, 1964.

44. Moloney, J. B. Personal communication, 1967.

45. Neufeld, A. H., Halpenny, G. W., and Morton, H. S. Beta-2 lipoprotein myelomatosis *Canad. J. Biochem.* 42:1499, 1964.

46. Osserman, E. F., and Takatsuki, K. Role of an abnormal myeloma-type serum gamma globulin in the pathogenesis of the skin lesions of papular mucinosis (lichen myxedematosus). *J. Clin. Invest.* 42:962, 1963.

47. Papenberg, J., and Hollander, W. Immunological and chemical characteristics of lipoproteins in human atherosclerotic arteries. *J. Clin. Invest.* 46:1102, 1967.

48. Perry, H. O., Montgomery, H., and Stickney, J. M. Further observations on lichen myxedematosus. *Ann. Intern. Med.* 53:955, 1960.

49. Ribeiro, L. P., and McDonald, H. J. Serum lipoprotein levels in rats with liver damage induced by carbon tetrachloride. *Clin. Chim. Acta* 8:727, 1963.

50. Seegers, W., Hirschhorn, K., Burnett, L., Robson, E., and Harris, H. Double beta-lipoprotein: A new genetic variant in man. *Science* 147:303, 1965.

51. Smith, I. Paper Electrophoresis at Low Voltage. In Smith, I. (Ed.), *Chromatographic and Electrophoretic Technique*, Vol. II. New York: Interscience, 1960. Chap. 1, pp. 23–25.

52. Storiko, K., and Fisher, G. B. Immunochemical Investigations on Low Density Lipoproteins. In Peeters, H. (Ed.), *Protides of the Biological Fluids*. Proceedings of the Twelfth Colloquium, Bruges, 1964. Amsterdam: Elsevier, 1965. Sec. B, Lipoproteins, p. 296.

53. Sutnick, A., London, T., and Blumberg, B. S. Australia antigens, Down's syndrome and hepatitis. *J. Clin. Invest.* 46:1122, 1967.
54. Walton, K. W. Apparent variation in the characteristics in agar. *Immunochemistry* 1:279, 1964.
55. Walton, K. W., and Darke, S. J. Immunological characteristics of human low-density lipoproteins. *Immunochemistry* 1:267, 1964.
56. Wieme, R. J. *Agar Gel Electrophoresis.* Amsterdam: Elsevier, 1965. P. 108.

5

URINARY PROTEIN ELECTROPHORESIS AND IMMUNOELECTROPHORESIS

The purpose of this chapter is to present a guide for the interpretation of protein electrophoretic patterns of normal urine and of urine from various diseases. After concentration of the specimen by ultrafiltration, electrophoresis is done in agarose gel, cellulose acetate, or paper with barbital buffer at pH 8.6 followed by staining for proteins [10, 33]. Starch-gel and acrylamide-gel electrophoresis may also be used. Details on methods are found in Chap. 15.

Normal urine has a protein content of 40 to 80 mg. per day [28, 29]. In our laboratory this is roughly equivalent to 0 to 10 mg. per 100 ml. for random specimens. Any value above this is considered to be abnormal proteinuria. All urine studied by electrophoresis should have a total protein determination performed before interpretation is attempted.

Normal urine electrophoretograms are variable, but in general a certain basic, although poorly resolved, pattern is recognizable [33]. Toward the anode, the most rapidly migrating protein observed is albumin. This is followed by α_1-globulin which is actually α_1-antitrypsin [15]. At least two α_2-globulins may be present. A β_1-globulin, which is transferrin, may appear in trace amounts. The γ-globulins cover the γ_1 area only and are of low molecular weight [3, 12] (Fig. 5-1). Admittedly, the globulin pattern is generally rather diffuse, and definite bands may not be recognizable as such in a normal urine specimen with the agarose electrophoretic system described in Chap. 15. Always absent from normal urine electrophoretograms, as compared to serum protein electrophoretograms, are β-lipoproteins and β_1C-globulin [29, 33], which is thought to be related to one of the components of complement [21, 33].

Immunoelectrophoresis of normal urine, on the other hand, reveals a large number of components identical to their serum protein counterparts [1, 2, 4, 6, 13, 14, 18, 29, 33]. Regularly occurring are albumin, α_1-acid-glycoprotein (uromucoid of Tamm and Horsfall, which is thought to be of renal origin), α_1-antitrypsin, α_1-lipoprotein, α_2HS-

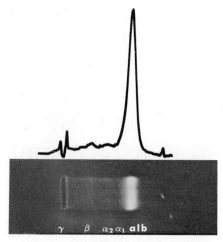

FIGURE 5-1

Normal urinary protein electrophoretogram

glycoprotein, Zn-α_2-glycoprotein, α_2Gc-globulin, transferrin, hemo-pexin, β_1E-globulin (the fourth component of complement), γA and γG immunoglobulins. Prealbumin, ceruloplasmin, haptoglobin, and traces of α_2-macroglobulin may be present. Fibrinogen, β_1-lipoprotein, and γM immunoglobulins do not occur in normal urine.

Pathologic Urine

Pathologic protein excretion involves low-molecular-weight pro-teins (under 200,000) and occurs in conditions which may be grouped into four main types [17, 29]: (1) proteinuria due to increased glo-merular permeability, (2) proteinuria of prerenal origin with normal glomerular permeability, (3) proteinuria due to renal tubular disease, and (4) chyluria. Postural or orthostatic and exercise proteinuria must also be mentioned since they are normal variations which must be differentiated from pathologic states. Proteins found in these condi-tions are shown in Table 5-1.

GLOMERULAR PERMEABILITY

Proteinuria associated with increased glomerular permeability ex-hibits an electrophoretic pattern dominated by albumin, with a markedly elevated total urinary protein [29] (see Fig. 5-2). The classic clinical syndrome of nephrosis is a primary example of this type of proteinuria, in which the main problem is some form of damage to the basement membrane of the glomerular capillary. The damage may be caused by inflammation, amyloidosis, or increased venous pressure, such as occurs with renal vein thrombosis or congestive heart failure.

TABLE 5-1

URINARY PROTEIN ELECTROPHORESIS

Pattern	Total Protein	Albumin	α_1-Globulin	α_2-Globulin	β-Globulin	γ_1-Globulin	Post-γ
1. Normal	40–80 mg./day	++	+	+	±	±	—
2. Glomerular	Markedly elevated	++++	++	+	+++	+	—
3. Prerenal							
a. Paraproteinemias	Elevated	++	+	+	± to +++	+++ to ±	—
b. Hemoglobinuria	Elevated	++	+	+	++++	±	—
c. Myoglobinuria	Elevated	++	+	+	±	± to ++++	—
d. Inflammatory syndrome	Elevated	++	+++	++	±	±	—
e. Monocytic leukemia	Slightly elevated	++	+	+	±	±	++++ (cationic)
4. Combined inflammatory and glomerular	Elevated	++++	+++	++	+	±	—
5. Tubular	Slightly elevated	+	+	++	++++	++	+
6. Chyluria	Variable—normal serum pattern	++++	+	++	++	+++	—
7. Postural	Elevated upright	++++	++	+	+++	+	—
8. Exercise	Slightly elevated—accentuated normal urine pattern	+++	++	++	+	+	—
Normal serum	6.0–7.5 gm.	++++	+	++	++ + β_2	+++	—

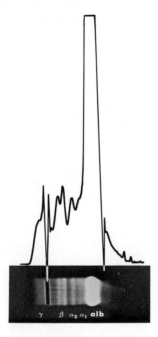

FIGURE 5-2

Urinary protein electrophoretogram with a pattern of increased glomerular permeability. Note the large albumin peak and the increase in α_1- and β-globulins.

In addition to a large albumin peak, the electrophoretogram displays a large β peak corresponding to transferrin [29]. The α_1 peak is also elevated and corresponds to α_1-antitrypsin and α_1-acid-glycoprotein [29]. These are all low-molecular-weight urinary proteins, which indicates that the molecular sieve effect is still largely functioning in nephrotic kidneys. Haptoglobin 1-1, the smallest molecular form, may be excreted in increased amounts in nephrotic syndrome and travels in the α_2-globulin range [5, 20]. The amount of γG excreted in the urine in nephrotic syndrome depends upon the plasma level [30].

PRERENAL

Proteinuria of prerenal origin with normal capillary permeability occurs with normal kidneys when there are increased plasma levels of low-molecular-weight excretable proteins, which may be either normal or abnormal proteins. The total urinary protein excretion is increased even in the presence of normal glomerular permeability [29]. The electrophoretic patterns are variable, depending upon the particular low-molecular-weight protein present in excess in the serum. For example, in paraproteinemias with Bence Jones proteinuria, there

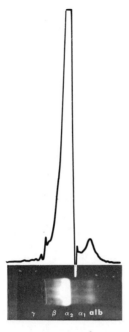

FIGURE 5-3

Prerenal pattern produced by a paraproteinemia (multiple myeloma) with positive Bence Jones proteinuria. Note the large monoclonal peak in the α_2 zone.

is a monoclonal peak in the electrophoretogram (Figs. 5-3, 5-4). Immunoelectrophoretic analysis or immunodiffusion studies with specific antisera to $\varkappa$ and λ chains are required to determine the type of L chain [11, 16]. For more detailed discussion of urinary proteins in dysproteinemias, see Chap. 2.

With acute intravascular hemolysis and hemoglobinuria, the β_1-globulin is increased with hemoglobin A_1. Hemoglobin usually appears in the urine after the plasma level exceeds 135 mg. per 100 ml. Myoglobin, occurring in urine following crush injuries or electrocution, travels electrophoretically slower than hemoglobin A_1; it is less negatively charged than hemoglobin and moves toward the cathode in the approximate position of hemoglobin A_2 [10, 27, 33] (see Chap. 15, under Myoglobin Electrophoresis). The pattern associated with generalized destructive and proliferative processes, which are grouped together as "inflammatory syndrome" and include cancer, burns, acute infection, collagen diseases, hyperthyroidism, and pregnancy, is that of elevated α_1-acid-glycoprotein and α_1-antitrypsin [29] (Fig. 5-5). Many of these inflammatory syndrome cases will also have associated increased glomerular permeability, and the pattern shows both elevated albumin and α_1 peaks [29] (Fig. 5-6).

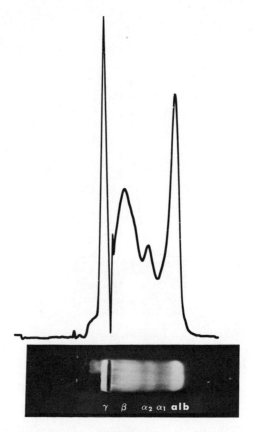

FIGURE 5-4

Prerenal pattern produced by a paraproteinemia (multiple myeloma) with positive Bence Jones proteinuria. Note the large monoclonal peak in the γ zone.

In cases of monocytic and monomyelocytic leukemia a cationic protein peak corresponding to lysozyme is evident in the electrophoretic pattern [22] (Fig. 5-7). Quantitation of the lysozyme by gel diffusion is shown in Figure 5-8. The enzyme originates from lysed leukocytes. Lesser amounts of cationic protein may also appear in nephrotic urines, but they are probably due to a different mechanism.

TUBULAR DISEASE

Proteinuria associated with renal tubular disorders is typified by chronic cadmium poisoning [23, 24, 25] phenacetin and vitamin D intoxication, and severe chronic hypokalemia [7, 9]. Chronic pyelonephritis, acute renal tubular failure, and polycystic renal disease also often display predominantly tubular patterns [31, 32]. Hereditary dis-

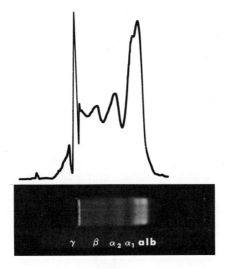

FIGURE 5-5

Urinary protein electrophoretogram with a prerenal inflammatory syndrome pattern. Note the reasonably normal albumin peak with a large α_1 peak. The sharply spiked peak in the pre-γ zone is an artifact from the trench in the agarose.

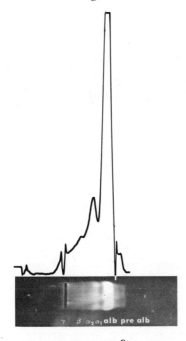

FIGURE 5-6

Urinary protein electrophoretogram with a combined inflammatory syndrome and glomerular permeability pattern. Note the large albumin peak with a prominent α_1 peak.

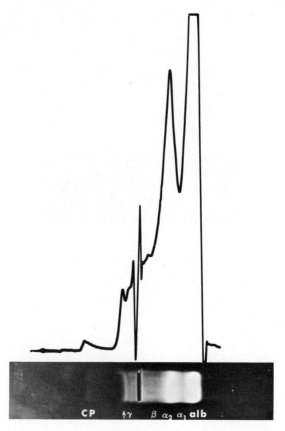

FIGURE 5-7

Urinary protein electrophoretogram with a combined pattern of increased glomerular permeability plus prerenal proteinuria due to the presence of lysozyme. Note the post-γ cationic protein peak (CP).

orders affecting renal tubules include Wilson's disease, galactosemia, oculocerebrorenal syndrome, and Fanconi syndrome [7, 9]. Total proteinuria in these cases is above normal but is rarely above 1 gm. per day [29]. Electrophoretic patterns show markedly elevated α_2 and β_2 peaks with a proportionate decrease in albumin. A post-γ zone may also be present [8, 29] (Fig. 5-9).

CHYLURIA

Chyluria is a reflection of inflammation, obstruction, or trauma to the retroperitoneal lymphatics with establishment of abnormal communications between intestinal chyle ducts, retroperitoneal lymph vessels, and the urinary tract [29]. The urinary protein content is extremely variable. The urine may appear milky. The electrophoretic

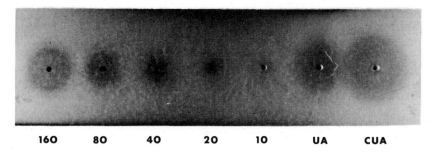

160   80   40   20   10   UA   CUA

μg/ml

FIGURE 5-8

Urinary lysozyme assay study on patient whose urinary electrophoretogram is shown in Figure 5-7. Patient's concentrated urine (CUA) is at far right with the unconcentrated urine (UA) immediately to the left of the CUA. Standard dilutions (160 to 10 mg. per milliliter) are from the left. Note that the patient's unconcentrated specimen (UA) contains approximately 160 mg. per milliliter of lysozyme. Normal urinary lysozyme concentration is 4 mg. per milliliter.

pattern is markedly similar to the serum protein pattern with the additional presence of fibrinogen and the high lipid content [29].

Nonpathologic Variants

Postural proteinuria must be distinguished from nephrotic syndrome in that albumin, α_1-globulin, and transferrin (β_1-globulin) are increased along with total protein excretion in the upright position [29]. However, in the recumbent position the electrophoretic pattern of orthostatic proteinuria reverts to normal or nearly normal [29, 32]. The phenomenon of orthostatic or postural proteinuria is thought to be due to reduced renal blood flow with greater exposure time and subsequent increased protein diffusion per unit time across the glomerular filter in the upright position [19, 29].

Exercise proteinuria must also be distinguished from pathologic states. Total protein excretion is elevated following exercise, but not as markedly as in postural proteinuria. The electrophoretic pattern is more nearly that of an accentuated normal pattern [26]. The process is possibly merely an accentuation of normal proteinuria [29].

Summary

In summary, three main types of pathologic urinary protein electrophoretic patterns should be kept in mind. First is the increased glomerular permeability pattern, with a markedly elevated albumin

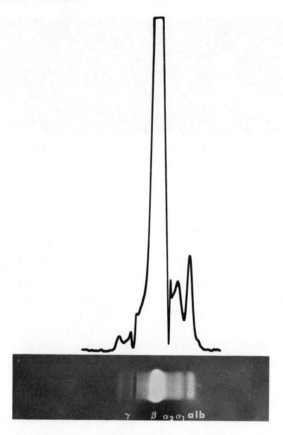

FIGURE 5-9
Urinary protein electrophoretogram used to illustrate a tubular pattern. Although this pattern is actually of a monoclonal gammopathy in the β zone, the markedly elevated β peak, elevated α_2, correspondingly decreased albumin, and the presence of a post-γ peak are similar to the pattern of a renal tubular disorder.

fraction and prominent α_1- and β_1-globulin fractions. Second is the prerenal pattern, which is variable and depends upon the particular low-molecular-weight pattern that is increased in the serum. Third is the tubular pattern, characterized by decidedly elevated α_2 and β_2 peaks and the proportionate decrease in albumin. A fourth pathologic pattern which should be mentioned is that of chyluria; it is similar to the normal serum electrophoretic pattern.

Two important normal variants occur which must be distinguished from the pathologic urines. Orthostatic proteinuria produces a glomerular permeability pattern in the upright position, which reverts to normal in recumbency. Exercise proteinuria is an overall accentuation of the normal pattern.

References

1. Berggard, I. The plasma proteins in normal urine. *Nature* (London) 187:776, 1960.
2. Berggard, I. Studies on the plasma proteins in normal human urine. *Clin. Chim. Acta* 6:413, 1961.
3. Berggard, I. On a γ-globulin at low molecular weight in normal human plasma and urine. *Clin. Chim. Acta* 6:545, 1961.
4. Berggard, I. Identification and Isolation of Urinary Proteins. In Peeters, H. (Ed.), *Protides of the Biological Fluids.* Proceedings of the Twelfth Colloquium, Bruges, 1964. Amsterdam: Elsevier, 1965. Pp. 285–291.
5. Berggard, I., and Bearn, A. G. Excretion of haptoglobin in normal urine. *Nature* (London) 195:1311, 1960.
6. Berggard, I., Cleve, H., and Bearn, A. G. The identification of five plasma proteins previously unidentified in normal human urine. *Clin. Chim. Acta* 10:1, 1964.
7. Butler, E. A., and Flynn, F. V. The proteinuria of renal tubular disorders. *Lancet* 2:978, 1958.
8. Butler, E. A., and Flynn, F. V. The occurrence of post-gamma protein in urine: A new protein abnormality. *J. Clin. Path.* 14:172, 1961.
9. Butler, E. A., Flynn, F. V., Harris, H., and Robson, E. B. A study of urine proteins by two-dimensional electrophoresis with special reference to the proteinuria of renal tubular disorders. *Clin. Chim. Acta* 7:34, 1962.
10. Cawley, L. P. *Workshop Manual on Electrophoresis and Immunoelectrophoresis* (Rev. ed.). Commission on Continuing Education. Chicago: American Society of Clinical Pathologists, 1966. Pp. 115–118.
11. Edelman, G. M., and Gally, J. A. The nature of Bence Jones proteins. *J. Exp. Med.* 116:207, 1962.
12. Franklin, E. C. Physicochemical and immunologic studies of gamma globulins of normal human urine. *J. Clin. Invest.* 38:2159, 1959.
13. Grant, G. H. The proteins of normal urine. *J. Clin. Path.* 10:360, 1957.
14. Grieble, H. G., Courcon, J., and Grabar, P. The immunochemical heterogeneity of proteins and glycoproteins in normal human urine. *J. Lab. Clin. Med.* 66:216, 1965.
15. Hardwicke, J., and de Vaux St. Cyr, C. Human α_1-glycoproteins, separation from pathological urine and examination by immunological and electrophoretic methods. *Clin. Chim. Acta* 6:503, 1961.
16. Harrison, J. F., Blainey, J. D., Hardwicke, J., Rowe, D. S., and Soothill, J. F. Proteinuria in multiple myeloma. *Clin. Sci.* 31:95, 1966.
17. Heremans, J. F. Résultats de l'électrophorèse des colloïdes urinaires. *Clin. Chim. Acta* 3:34, 1958.
18. Keutel, H. J., Hermans, G., and Licht, W. Immuno-electrophoretic studies on the urinary colloids identical with serum colloids and their significance in the formation of urinary calculi. *Clin. Chim. Acta* 4:665, 1959.
19. Lathem, W., Roof, B. S., Nickel, J. F., and Bradley, S. E. Urinary protein excretion and renal hemodynamic adjustments during ortho-

stasis in patients with acute and chronic renal diseases. *J. Clin. Invest.* 33:1457, 1954.

20. Marnay, A.   Haptoglobinuria in nephrotic syndrome. *Nature* (London) 191:74, 1961.

21. Müller-Eberhard, N. J., and Biro, C.   Isolation and description of the fourth component of complement. *J. Exp. Med.* 118:447, 1963.

22. Osserman, E. F., and Lawton, D. P.   Serum and urinary lysozyme (muromidase) in monocytic and monomyelocytic leukemia. *J. Exp. Med.* 124:921, 1966.

23. Piscator, M.   Proteinuria in chronic cadmium poisoning 1. *Arch. Environ. Health* (Chicago) 4:607, 1962.

24. Piscator, M.   Proteinuria in chronic cadmium poisoning 2. *Arch. Environ. Health* (Chicago) 5:325, 1962.

25. Piscator, M.   Proteinuria in chronic cadmium poisoning 3. *Arch. Environ. Health* (Chicago) 12:335, 1966.

26. Poortmans, J., and Van Kerchove, E.   La protéinurie d'effort. *Clin. Chim. Acta* 7:229, 1962.

27. Prankerd, T. A. J.   Electrophoretic properties of myoglobin and its character in sickle-cell diseases and paroxysmal myoglobinuria. *Brit. J. Haemat.* 2:80, 1956.

28. Rigas, D. A., and Heller, G. G.   The amount and nature of urinary proteins in normal human subjects. *J. Clin. Invest.* 30:853, 1951.

29. Schultze, H. E., and Heremans, J. F.   *Molecular Biology of Human Proteins.* Vol. I, *Nature and Metabolism of Extracellular Proteins.* New York: American Elsevier, 1966. Sect. IV, Chap. 2, The Urinary Proteins, pp. 670–731.

30. Slater, R. J., and Kunkel, H. G.   Filter paper electrophoresis with special reference to urinary proteins. *J. Lab. Clin. Med.* 41:619, 1953.

31. Tidstrom, B.   Urinary protein in glomerulonephritis and pyelonephritis. Electrophoretic analysis. *Acta Med. Scand.* 174:385, 1963.

32. Tidstrom, B.   Paper Electrophoresis in the Diagnosis of Renal Disease. In Peeters, H. (Ed.), *Protides of the Biological Fluids.* Proceedings of the Twelfth Colloquium, Bruges, 1964. Amsterdam: Elsevier, 1965. Pp. 311–319.

33. Wieme, R. J.   *Agar Gel Electrophoresis.* Amsterdam: Elsevier, 1965. Pp. 292–298.

6

CEREBROSPINAL FLUID PROTEIN ELECTROPHORESIS AND IMMUNOELECTROPHORESIS

T he origin of the proteins in the cerebrospinal fluid (CSF) appears to be better understood today than at any time in the past. In large measure these proteins originate from serum protein and pass into the cerebrospinal fluid via the choroid plexus and meninges. Apparently there are also proteins which arise within the central nervous system and are not found in the peripheral blood [20]. In the main, however, the bulk of the proteins originate from serum protein. Pathologic alterations of the cerebrospinal fluid protein, therefore, can be understood by grasping the significance of the origin of the proteins and perceiving how the origin may be influenced by diseases both within and without the central nervous system.

Protein Components of Spinal Fluid

Technical difficulties plague the interpretation of results derived from electrophoretic technics and immunoelectrophoretic technics [26, 59]. Unconcentrated cerebrospinal fluid can be used on acrylamide electrophoretic columns [15, 21, 22]. However, for paper, agar gel, agarose, cellulose acetate [33, 34, 50], starch gel [36, 47], or immuno-electrophoresis satisfactory results are obtained only with concentrated material [8]. The method of concentration found successful in our laboratory is that of ultrafiltration. This is followed by agarose electrophoresis and immunoelectrophoresis. The exact procedures are described in Chap. 15.

The most extensive work done on cerebrospinal fluids, involving somewhat over 400 cases, was reported by Dencker [16]. He was able to show by immunoelectrophoresis 36 different protein fractions, 13 to 14 of which were found regularly [16, 19]. Table 6-1 lists the proteins identified by immunoelectrophoresis and their suggested origin.

Under normal circumstances the electrophoretic pattern shows five well-defined zones (see Figs. 6-1 and 6-2). From anode to cathode these are: prealbumin, albumin, α-globulin, β-globulin, and γ-globulin.

TABLE 6–1

CEREBROSPINAL FLUID PROTEINS

I. Serum Proteins	II. CSF-Specific Proteins
a. Albumin	a. Tryptophan-rich pre-albumin?
b. α_1-Glycoprotein	b. α_1-Lipoprotein$_B$?
c. α_1-Lipoprotein$_A$	c. $\alpha_{1:3}$ globulin
d. α_1-Seromucoid	d. $\alpha_{2:4}$ globulin
e. α_2-Haptoglobin	e. $\alpha_{2:5}$ globulin
f. α_2-Ceruloplasmin	f. $\alpha_{2:6}$ globulin
g. α_2-Lipoprotein	g. $\alpha_{2:7}$ globulin
h. α_2-Macroglobulin	h. $\alpha_{2:8}$ globulin
i. $\alpha_2$gc-Globulin	i. Transferrin II?
j. Transferrin I	j. $\beta_{1:1}$ globulin
k. β_1C	k. $\beta_{1:2}$ globulin
l. β_1-Lipoprotein	l. $\beta_{1:3}$ globulin
m. Fibrinogen	m. $\beta_{1:4}$ globulin
n. β_2B	n. $\beta_{1:5}$ globulin
o. β_2M (or γM)[a]	o. $\beta_{1:6}$ globulin
p. γB	p. β_2X
q. γX (? C-reactive protein)	q. γR (or γC)
r. γA	
s. γG	

[a]Two immunologically identical components.
? May be present in other body fluids.

The prealbumin protein is present in much higher concentration in spinal fluid than in serum and is found in the highest concentration in the ventricular system. In the β region a protein is present that is immunochemically identical with transferrin, but slower in mobility. It is probably transferrin minus sialic acid [12, 45, 46, 53]. The γ region contains a protein that migrates slower than any comparable component in serum [11, 53]. This fraction changes its mobility on freezing and thawing and retreats into the mass of other γ-globulins [30, 31]. This protein has been designated as γ_c by MacPherson [39, 40, 42, 43]. Several other proteins have been defined by anti–cerebrospinal fluid antibodies [29].

Protein Patterns in Disease

Rarely can a specific diagnosis be made from the electrophoretic pattern, but several generalizations as to the category of the neurologic disease can be advanced. In the absence of hemorrhage, four basic electrophoretic patterns can be seen (Table 6-2).

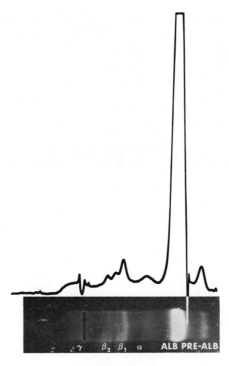

FIGURE 6-1

Electrophoretogram of normal spinal fluid. Prealbumin is present in normal cerebrospinal fluid in a concentration of 4–6 percent. Albumin is the predominate protein in normal cerebrospinal fluid. The α region is characteristically diffuse and lower than the prealbumin. The β_1 protein consists primarily of transferrin I or fast transferrin. Note particularly the comparative peak heights of prealbumin and β_1. In normal spinal fluids using our technic the β_1 is slightly higher than the prealbumin. The β_2 region is lower in peak height than β_1 and prealbumin. The γ region is diffuse and lower than the other components.

The first pattern in Table 6-2 indicates that specific plasma proteins have accumulated in the cerebrospinal fluid primarily because of their elevated levels in the serum. A γG paraprotein, for example, may be found in the cerebrospinal fluid in cases of multiple myeloma. The findings of an abnormal immunoglobulin in the spinal fluid in the presence of an obstructive spinal mass and absence of serum protein changes may occasionally be indicative of a central nervous system plasmacytoma [60, 61, 62]. Therefore, it may be of interpretative value to analyze the serum proteins if the cerebrospinal fluid shows this type of pattern.

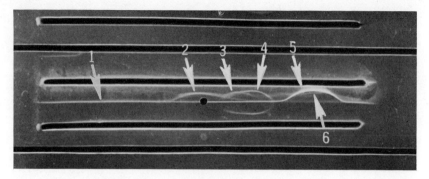

Immunoelectrophoretogram of normal spinal fluid. In the upper trench an antiserum to cerebrospinal fluid is present and in the lower trench antiserum to transferrin is present. Arrow 1 points to a line of precipitation between antisera made in different species which have been placed in the two trenches and diffused toward each other. Arrow 2 points to the diffuse γ-globulin precipitin line. Arrow 3 points to transferrin II, or slow transferrin, which is the major component of the β_2 region. Arrow 4 points to transferrin I or fast transferrin, the major component of the β_1 region. Arrow 5 points to albumin, which is the major component of cerebrospinal fluid. Arrow 6 points to one of the prealbumin proteins. Anode is on the right.

TABLE 6–2

BASIC ELECTROPHORETIC PATTERNS IN DISEASE

    I. Extra cerebrospinal fluid pattern
   II. Capillary permeability pattern
 III. Degenerative pattern
 IV. γ-Globulin pattern

The second pattern is an increase in normal plasma proteins in the cerebrospinal fluid (see Figs. 6-3, 6-4, 6-5, and 6-6). As might be imagined, any inflammatory process would increase the flow of plasma proteins across the choroid plexus and meninges. This pattern is one of increased γ-globulin, α-globulin, β_1-globulin, and albumin with decreases in the prealbumin and slow transferrin portion of the electrophoretogram. The presence of an unusually high α-globulin in the pattern is occasionally related to a serum protein electrophoretic pattern showing acute-phase-reactant protein elevations [53]. Bisalbuminemia may be manifested in the cerebrospinal fluid as a split albumin peak and thus may be misleading; however, it may be separated from the capillary permeability pattern by the lack of the other electrophoretic changes present in the capillary permeability pattern [2].

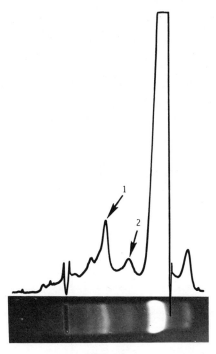

FIGURE 6-3

A capillary permeability pattern. Arrow 1 points to a β_1 peak that is higher than normal. It has been our experience that an elevated β_1 peak height in the presence of a normal total cerebrospinal fluid protein is probably not abnormal. This elevated β_1 fraction is primarily transferrin I, or fast transferrin, which has a molecular weight of about 90,000. The α-globulin at arrow 2 appears more discrete and slightly elevated. This may be seen in capillary permeability patterns since several proteins in the area have molecular weights below 100,000, but it is not consistently seen. Isolated α increases may be seen when the serum α-globulins are elevated. Occasionally other proteins with higher molecular weight such as γ-globulin are increased, but this finding is not consistent. Anode is on the right.

The capillary permeability pattern may be found in acute and chronic inflammatory processes [58], neoplasms [53], diabetes with neurologic symptoms [35, 53], and cerebroarterial disease [6]. With rare exceptions this pattern has also been seen in Guillain–Barré syndrome [49, 53].

The third alteration is referred to as a degenerative pattern and consists of an increase of proteins in the β_2-globulin area on the electrophoretic pattern. Lowenthal feels that the proteins in this area are distinct from those in the plasma β_2-globulin region [41]. One of them may be increased slow transferrin. β-lipoprotein has also been described but it may be due to diffusion from the blood [1, 10, 17, 55,

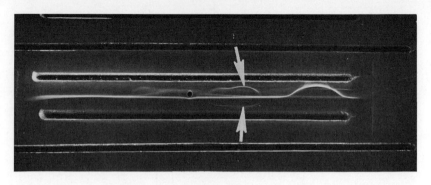

FIGURE 6-4

Immunoelectrophoresis of a simple capillary permeability pattern such as the one shown in Figure 6-3. Heavier precipitation arc is due to an increased transferrin I, or fast transferrin. The arrows point to the transferrin arc derived from a cerebrospinal fluid antiserum at the top and specific anti-transferrin at the bottom. Anode is on the right.

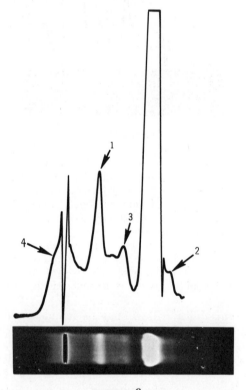

FIGURE 6-5

Capillary permeability pattern alluded to in Figure 6-3 with elevated fractions other than β_1. Arrow 1 points to a marked β_1 elevation in comparison to the prealbumin at arrow 2. Arrows 3 and 4 point respectively to elevated α- and γ-globulins, both of which have a peak height greater than that of prealbumin. Anode is on the right.

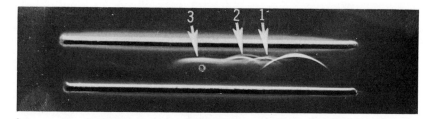

FIGURE 6-6

Immunoelectrophoretogram of Figure 6-5 showing a plethora of proteins present in the α-globulin region denoted by arrow 1. Arrow 2 shows the increased transferrin I arc. Arrow 3 shows the increased γ-globulin arc. This pattern is typical of Guillain-Barré syndrome and other disease states.

56]. The mobilities of hemopexin, α_1-antitrypsin, and α_1-acid glyco-protein are also decreased in cerebrospinal fluid, and these may constitute a portion of the slow β_2 fraction [53]. The role played by minor cerebrospinal fluid proteins or as yet unidentified proteins is unknown. Even though the complete explanation for the degenerative pattern is not available, it is felt to be due to either specific central nervous system protein production or protein alteration. This type of pattern has been seen in cerebral atrophy, syringomyelia, occasional cases of amyotrophic lateral sclerosis, and several other conditions where chronic or recurrent anoxia may play a part [53]. We have not seen this pattern in over 200 cerebrospinal fluid electrophoretic patterns.

The fourth pattern consists of either a generalized or specific increase in one of the immunoglobulins without significant serum protein changes (see Figs. 6-7 and 6-8). The implication is that either specific antigens are stimulating the reticuloendothelial components of the central nervous system or the reticuloendothelial components are responding abnormally to otherwise innocuous substances [52]. Electrophoretic patterns show a diffuse increase in the γ region in multiple sclerosis [4, 7, 9, 13, 14, 23, 25, 32, 48, 51, 54, 63]. Immunoelectrophoresis shows the increased γ-globulin in multiple sclerosis to be of the γG type, which is clearly differentiated from a γG paraprotein of the type I pattern. The recent finding that the γG proteins of multiple sclerosis do not react to brain tissue when γG can be shown in the plaques has led to the interpretation that the γG in the cerebrospinal fluid is a spillover of nonspecific γG proteins [57]. If true, this could explain the clinical observation that multiple sclerosis may be present in the absence of increased cerebrospinal fluid γ-globulin [24]. A correlative observation is that $\beta_1 C$ is decreased in the cerebrospinal fluid of patients with multiple sclerosis [37]. When γG is elevated in multiple sclerosis, the concentrations may parallel closely the clinical state of the disease [5].

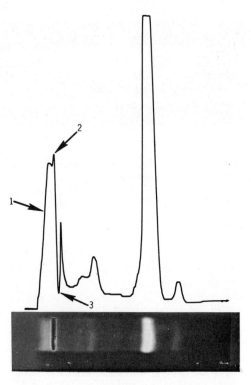

FIGURE 6-7

Example of an increased γ-globulin pattern. Arrow 1 points to the obvious elevation of γ-globulin. Arrow 2 points to an artifact resulting from protein precipitation on the edge of the inoculation trench. Arrow 3 points to an artifact resulting from the inoculation trench alone. This pattern is seen most commonly in multiple sclerosis. Anode is on the right.

FIGURE 6-8

Immunoelectrophoresis of an increased γ-globulin pattern. The arrow points to a γ-globulin precipitation arc that is heavier than any other arc except that of albumin. Anode is on the right.

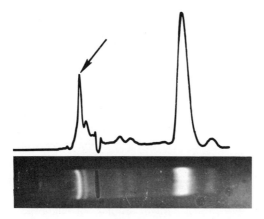

FIGURE 6-9

Variant of the increased γ-globulin pattern. The arrow points to a thin spike in the γ area which is similar to the monoclonal γ peaks seen in multiple myeloma. Patients with multiple myeloma may have such spikes in the γ region even in the presence of a normal total cerebrospinal fluid protein. This particular pattern was obtained from a proved case of subacute sclerosing leukoencephalitis. Anode is on the right.

Optic neuritis has been shown to have increased γ-globulin levels by paper electrophoresis [27] but this has not been studied further. γM has been shown to be increased in neurosyphilis and trypanosomiasis [44, 53]. Other chronic afflictions have been shown to have elevated γ-globulin levels [8, 53]. In subacute sclerosing leukoencephalitis a "monoclonal" γ peak may be seen [3, 38]. This peak is present in the slow γ region. In the one case we have seen the monoclonal peak

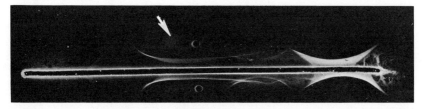

FIGURE 6-10

Immunoelectrophoresis of the cerebrospinal fluid in Figure 6-9. The bottom well contains patient's serum and the top well contains patient's cerebrospinal fluid. The trench contains polyvalent antiserum. Arrow points to a precipitate in the spinal fluid that is not seen in the serum. This kind of phenomenon is commonly seen in serum of patients with monoclonal γ peaks due to myeloma. The precipitate corresponds to the monoclonal peak. The γG band likewise is broadened, indicating an abnormal protein. Anode is on the right.

reacted with anti-γG, which corresponds with reports in the literature [18]. The peak was also present in the same position in the serum. The serum peak decreased on therapy with ACTH (see Figs. 6-9 and 6-10).

The γ region has been divided by routine electrophoresis into several components and analyzed extensively [41]. However, for current clinical applications, electrophoretic analysis and delineation of the specific type of immunoglobulin with quantitation [28] may prove to be more useful.

References

1. Aird, R. B., Gofman, J. W., Jones, H. B., Campbell, M. B., and Garoutte, B. Ultracentrifuge studies of lipoproteins in multiple sclerosis. *Neurology* (Minneapolis) 3:22, 1953.
2. Bohls, S. W., Lara, F. Y., Thurman, N., and Keyes, J. Manifestation of bisalbuminemia in serum and cerebrospinal fluid. *Dis. Nerv. Syst.* 27:727, 1966.
3. Booÿ, J. The C.S.F. aspects in leuco-encephalitis. *Folia Psychiat. Neerl.* 61:352, 1958.
4. Brackenridge, C. J. Cerebrospinal fluid protein fractions in health and disease. *J. Clin. Path.* 15:206, 1962.
5. Bradshaw, P. The relation between clinical activity and the level of gamma globulin in the cerebrospinal fluid in patients with multiple sclerosis. *J. Neurol. Sci.* 1:374, 1964.
6. Bronsky, D., Kaplitz, S. E., Ade, R. D., and Dubin, A. Proteins in cerebrovascular disease. *Amer. J. Med. Sci.* 244:54, 1962.
7. Bronsky, D., Kaplitz, S. E., Muci, J., Dubin, A., and Chesrow, E. J. Electrophoretic partition of cerebrospinal fluid and serum proteins in multiple sclerosis. *J. Lab. Clin. Med.* 56:382, 1960.
8. Burrows, S. Simple method for concentration of cerebrospinal fluid for protein electrophoresis. *Clin. Chem.* 11:1068, 1965.
9. Chodirker, W. B., and Tomasi, T. B. Gamma globulins: Quantitative relationships in human serum and nonvascular fluids. *Science* 142: 1080, 1963.
10. Clausen, J. The beta lipoprotein of serum and cerebrospinal fluid. *Acta Neurol. Scand.* 42:153, 1966.
11. Clausen, J., Matzke, J., and Gerhardt, W. Agar-gel micro-electrophoresis of proteins in the cerebrospinal fluid. Normal and pathological findings. *Acta Neurol. Scand.* 40 (Suppl. 10):49, 1964.
12. Clausen, J., and Munkner, T. Transferrin in normal cerebrospinal fluid. *Nature* (London) 189:60, 1961.
13. Cosgrove, J. B. R., and Aguis, P. Studies in multiple sclerosis: I. Normal values for paper electrophoresis of serum and cerebrospinal fluid proteins; description of the method with special reference to a practical concentration technique for cerebrospinal fluid proteins. *Neurology* (Minneapolis) 15:1155, 1965.
14. Cosgrove, J. B. R., and Aguis, P. Studies in multiple sclerosis: II. Comparison of the beta-gamma globulin ratio, gamma globulin

elevation, and first-zone colloidal'gold curve in the cerebrospinal fluid. *Neurology* (Minneapolis) 16:197, 1966.

15. Cunningham, V. R.   Analysis of "native" cerebrospinal fluid by the polyacrylamide disc electrophoresis technique. *J. Clin. Path.* 17:143, 1964.

16. Dencker, S. J.   Immuno-electrophoretic investigation of cerebrospinal fluid gamma gobulins in multiple sclerosis. *Acta Neurol. Scand.* 40 (Suppl. 10):57, 1964.

17. Dencker, S. J., Bronnestam, R., and Swahn, B.   Demonstration of large blood proteins in cerebrospinal fluid. *Neurology* (Minneapolis) 11:441, 1961.

18. Dencker, S. J., and Kolar, O.   The cerebrospinal fluid gamma G profile in subacute sclerosing leukoencephalitis. *Acta Neurol. Scand.* 41 (Suppl. 13):563, 1965.

19. Dencker, S. J., and Swahn, B.   Clinical value of protein analysis in cerebrospinal fluid. A micro-immuno-electrophoretic study. From the Departments of Neurology and Medicine, Lund University. Lund, Sweden: Gleerup, 1961.

20. Dencker, S. J., and Swahn, B.   Proteins of central nervous system origin present in cerebrospinal fluid. *Nature* (London) 194:288, 1962.

21. Evans, J. H., and Quick, D. T.   Polyacrylamide gel electrophoresis of spinal fluid proteins. Neurological Disorders. *Arch. Neurol.* (Chicago) 14:64, 1966.

22. Evans, J. H., and Quick, D. T.   Polyacrylamide gel electrophoresis of spinal-fluid proteins. *Clin. Chem.* 12:28, 1966.

23. Foster, J. B., and Horn, D. B.   Multiple sclerosis and spinal fluid gamma globulin. *Brit. Med. J.* 1:1527, 1962.

24. Gilland, O.   Multiple sclerosis classification scheme integrating clinical and CSF findings. *Acta Neurol. Scand.* 41 (Suppl. 13):563, 1965.

25. Goa, J., and Tveten, L.   Electrophoresis of cerebrospinal fluid proteins in certain neurological diseases. *Scand. J. Clin. Lab. Invest.* 15:152, 1963.

26. Greenhouse, A. H., and Speck, L. B.   The electrophoresis of spinal fluid proteins. *Amer. J. Med. Sci.* 248:333, 1964.

27. Hallett, J. W., Wolkowicz, M. I., Leopold, I. H., and Wijewski, E.   Cerebrospinal fluid electrophoretograms in optic neuritis. *Arch. Ophthal.* (Chicago) 72:68, 1964.

28. Hartley, T. F., Merrill, D. A., and Claman, H. N.   Quantitation of immunoglobulins in cerebrospinal fluid. *Arch. Neurol.* (Chicago) 15:472, 1966.

29. Hockwald, G. M., and Thorbecke, G. J.   Use of antiserum against cerebrospinal fluid in demonstration of trace proteins in biological fluids. *Proc. Soc. Exp. Biol. Med.* 109:91, 1962.

30. Hockwald, G. M., and Thorbecke, G. J.   Trace proteins in cerebrospinal fluid and other biological fluids: I. Effect of various fractionation procedures on beta-trace and gamma-trace proteins and methods for isolation of both proteins. *Arch. Biochem.* 101:325, 1963.

31. Hockwald, G. M., and Thorbecke, G. J.   Trace proteins in cerebrospinal fluid and other biological fluids: II. Effect of storage and enzymes on the electrophoretic mobility of gamma-trace and beta-trace proteins in cerebrospinal fluid. *Clin. Chim. Acta* 8:678, 1963.

32. Ivers, R. R., McKenzie, B. F., McGuckin, W. F., and Goldstein, N. P. Spinal fluid gamma globulin in multiple sclerosis and other neurologic diseases. Electrophoretic patterns in 606 patients. *J.A.M.A.* 176:515, 1961.

33. Kaplan, A. Electrophoresis of cerebrospinal fluid proteins. *Amer. J. Med. Sci.* 253:549, 1967.

34. Kaplan, A., and Johnstone, M. Concentration of cerebrospinal fluid proteins and their fractionation by cellulose acetate electrophoresis. *Clin. Chem.* 12:717, 1966.

35. Kutt, H., Hurwitz, L. J., Ginsburg, S. M., and McDowell, F. Cerebrospinal fluid protein in diabetes mellitus. *Arch. Neurol.* (Chicago) 4:31, 1961.

36. Kutt, H., McDowell, F., Chapman, L., Pert, J. H., and Hurwitz, L. J. Abnormal protein fractions of cerebrospinal fluid demonstrated by starch gel electrophoresis. *Neurology* (Minneapolis) 10:1064, 1960.

37. Kuwert, E., and Pette, E. Demonstration of complement in spinal fluid in multiple sclerosis. *Ann. N.Y. Acad. Sci.* 122:429, 1965.

38. Laterre, E. C. *Les Protéines du Liquide Céphalo-rachidien à l'État Normal et Pathologique.* Paris: Maloine, 1965.

39. Laterre, E. C., and Heremans, J. F. A note on proteins apparently "specific" for cerebrospinal fluid. *Clin. Chim. Acta* 8:220, 1963.

40. Laterre, E. C., Heremans, J. F., and Carbonara, A. Immunological comparison of some proteins found in cerebrospinal fluid, urine and extracts from brain and kidney. *Clin. Chim. Acta* 10:197, 1964.

41. Lowenthal, A. *Agar-Gel Electrophoresis in Neurology.* New York: American Elsevier, 1964.

42. MacPherson, C. F. C. Quantitative estimation of the gamma$_c$-globulin in normal and pathological cerebrospinal fluid by an immunochemical method. *Clin. Chim. Acta* 11:298, 1965.

43. MacPherson, C. F. C., and Saffran, M. An albumin and gamma-globulin characteristic of bovine cerebrospinal fluid. *J. Immun.* 95:629, 1965.

44. Mattern, P., Sandor, G., and Pillot, J. Contribution to the study of immunoglobulins of the cerebrospinal fluid in the course of the neurosyphilis human by the techniques of analysis by immunoelectrophoresis and by immunofluorescence. *Ann. Inst. Pasteur* (Paris) 109 (Suppl. 5):120, 1965.

45. Parker, W. C., and Bearn, A. G. Studies on the transferrins of adult serum, cord serum, and cerebrospinal fluid. The effect of neuraminidase. *J. Exp. Med.* 115:83, 1962.

46. Parker, W. C., Hagstrom, J. W. C., and Bearn, A. G. Additional studies on the transferrins of cord serum and cerebrospinal fluid. Variation in carbohydrate prosthetic groups. *J. Exp. Med.* 118:975, 1963.

47. Pert, J. H., and Kutt, H. Zone electrophoresis of cerebrospinal fluid proteins in starch gel. *Proc. Soc. Exp. Biol. Med.* 99:181, 1958.

48. Prineas, J., Teasdale, G., Latner, A. L., and Miller, H. Spinal fluid gamma globulin and multiple sclerosis. *Brit. Med. J.* 2:922, 1966.

49. Ravin, H., and Jensen, K. Cerebrospinal fluid proteins in the Guillain-Barré syndrome. *Acta Path. Microbiol. Scand.* 65:93, 1965.

50. Rice, J. D., Jr., and Bleakney, B.   Electrophoresis of unconcentrated cerebrospinal fluid using cellulose acetate strips and the dye nigrosin. *Clin. Chim. Acta* 12:343, 1965.
51. Roboz, E., Hess, W. C., and Forster, F. M.   Quantitative determination of gamma globulin in cerebrospinal fluid. Its application in multiple sclerosis. *Neurology* (Minneapolis) 3:410, 1953.
52. Scheinberg, L. C., Kotsilimbas, D. G., Karpf, R., and Mayer, N.   Is the brain an immunologically privileged site? III. Studies based on homologous skin grafts to the brain and subcutaneous tissues. *Arch. Neurol.* (Chicago) 15:62, 1966.
53. Schultze, H. E., and Heremans, J. F.   The Proteins of Cerebrospinal Fluid. In Schultze, H. E., and Heremans, J. F. (Eds.), *Molecular Biology of Human Proteins.* New York: American Elsevier, 1966. Vol. 1, pp. 732–761.
54. Sibley, W. A., and Wurz, L.   Immunoassay of cerebrospinal fluid gamma globulin. *Arch. Neurol.* (Chicago) 9:386, 1963.
55. Someda, K., and Kagevama, N.   Lipoproteins in the cerebrospinal fluid: Immunoelectrophoretic analysis. *Klin. Wschr.* 43:230, 1965.
56. Swahn, B., Bronnestam, R., and Dencker, S. J.   On the origin of the lipoproteins in the cerebrospinal fluid. *Neurology* (Minneapolis) 11: 437, 1961.
57. Tourtellotte, W. W., and Parker, J. A.   Multiple sclerosis: Brain immunoglobulin-G and albumin. *Nature* (London) 214:683, 1967.
58. Ursing, B.   Clinical and immunoelectrophoretic studies on cerebrospinal fluid in virus meningoencephalitis and bacterial meningitis. *Acta Med. Scand.* 177 (Suppl. 429):1, 1965.
59. Watson, D.   Modern methods for determining cerebrospinal fluid protein. *Clin. Chem.* 10:412, 1964.
60. Weiner, L. P., Anderson, P. N., and Allen, J. C.   Cerebral plasmacytoma with myeloma protein in the cerebrospinal fluid. *Neurology* (Minneapolis) 16:615, 1966.
61. Weiss, A. H., and Christoff, N.   The effect of cerebrospinal fluid paraproteins on the colloidal gold test. *Arch. Neurol.* (Chicago) 14:100, 1966.
62. Weiss, A. H., Smith, E., Christoff, N., and Kochwa, S.   Cerebrospinal fluid paraproteins in multiple myeloma. *J. Lab. Clin. Med.* 66:280, 1965.
63. Yokoyama, M., and Einstein, E. R.   Immunochemical analysis of cerebrospinal fluid. *Ann. N.Y. Acad. Sci.* 122:439, 1965.

7

SERUM ALKALINE PHOSPHATASE
ISOENZYMES

Serum alkaline phosphatase is not a single enzyme. It is composed of a mixture of enzymes which originate from several tissues. The main sources are liver, intestine, and osteoblasts of bone [15]. During pregnancy another alkine phosphatase, originating in the placenta, is found in the serum [7, 12, 33]. Genetic factors play a definite role in the observed variation of phosphatase activity [7, 29].

Alkaline Phosphatase Isoenzymes and Blood Groups

Two main zones of activity of alkaline phosphatase, a fast (A) and a slow (B) band, are depicted by starch-gel electrophoresis. The slowly migrating phosphatase component was observed by Hodson et al. in 1962 [17]. It was thought to have originated from the intestinal mucosa, and its intensity and presence were variable from one individual to another. Similar variations of the slowly migrating component were observed by Cunningham and Rimer by starch-gel electrophoresis in 1963 [12]. By the same means Arfors et al. [2] classified alkaline phosphatase into types I and II on the basis of presence or absence of band B (slowly migrating band) and called attention to an association between the alkaline phosphatase type and the ABO blood group system. They later presented evidence to show an association with Lewis blood groups and suggested that the alkaline phosphatase component might be a complex of enzyme and blood group substance. Shreffler [29] believes that the expression of alkaline phosphatase is masked or inhibited in individuals of blood group A and in nonsecretors of ABH blood group substances. The nonsecretor effect is apparently not directly related to the Lewis blood group system. Shreffler [29] attempted to show closer genetic correlation by grading the intensity of the slow-moving alkaline phosphatase band, also found to vary from one individual to another. Rather than indicating just type I or type II alkaline phosphatase isoenzyme pattern, which indicated presence or absence of the slow-moving band (B), he graded the intensity of band B from no appearance (0) to intense staining (4).

Band A was always present, and although there was some variation in intensity it was not considered significant. Shreffler also described a slow band (C) but attached no significance to it. In some serum a very slow band (D) close to the origin was found which stained very intensely in some cases, and in each such instance there was an intense staining band (A). The position of band D correlates with β-lipoprotein [7]. Bamford et al. [3], also using gradation of stain intensity of the B band, found higher serum alkaline phosphatase levels in people of blood group O or B whose red cells were Le(a⁻)—that is, in persons who were ABH secretors—than in those who were nonsecretors, probably on account of increased amounts of intestinal alkaline phosphatase in the serum rather than the presence of blood group substances complexing with serum alkaline phosphatase, as proposed by Arfors et al. [2]. Beckman [4] confirmed the association of the extra alkaline phosphatase band in secretors of ABH substance in a large percentage of the population. Rendel et al. [26] have reported that the level of alkaline phosphatase activity is higher in sheep of blood group O, all of which have an extra phosphatase band.

Origin of Alkaline Phosphatase Isoenzymes

The above studies and others have suggested specific tissue origin of individual alkaline phosphatases which may give rise to variable electrophoretic patterns. The polymorphism, therefore, is an expression of multiple molecular forms. From a clinical pathology standpoint, zymograms of alkaline phosphatase might play an important role in distinguishing diseases in which alkaline phosphatase elevation is present. In an alkaline phosphatase zymogram of serum, three zones of activity are usually disclosed in agarose or agar-gel electrophoretograms. The fast component has the electrophoretic mobility of α_2-globulin and the slow components are electrophoretically equivalent to β-globulin. The most rapidly migrating band has been shown in the agar-gel system to be of liver origin (Fig. 7-1) [33]. The second band is from bone, and the slowest is from intestine. Placental alkaline phosphatase travels slightly slower than liver alkaline phosphatase and just anodic to bone alkaline phosphatase and overlaps with it to a considerable degree.

Identification of alkaline phosphatase isoenzymes can be augmented by additional physical-chemical tests. Besides the electrophoretic mobility which is the primary consideration for identification, heat lability, sensitivity to L-phenylalanine, and ethylenediamine tetra-acetic acid (EDTA) inhibition have been found useful tests in distinguishing the various alkaline phosphatase isoenzymes. Immunologic differences have also been found between the various alkaline phosphatases. In one study antibody to liver phosphatase and to placental phosphatase showed that the antisera were specific for their homologous antigen

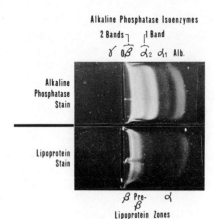

FIGURE 7-1

Alkaline phosphatase zymogram (*upper*) compared to lipoprotein stain (*lower*). Technic of electrophoresis is segmental low ionic strength with high voltage drop. This is done by separating agar or agarose strip into two parts by application trench placed completely across the strip. Sample is introduced after dilution with low ionic substance. This method gives no endosmosis and results in fine separation just anodic to the point of application. Note that β and pre-β lipoprotein zones correspond to slow and mid-zone alkaline phosphatase band. Anode is on the right. (Modified from L. P. Cawley [8].)

and did not react with enzyme from bone, neutrophils, kidney, or intestine [30]. As noted above, individuals of blood group O normally show by starch-gel electrophoresis a more intense, slow alkaline phosphatase (band B), which apparently originates from the intestine and corresponds to the intestinal band seen in agar-gel electrophoretograms. This additional band in group O individuals is thought to be responsible for their increased concentration of total serum alkaline phosphatase as measured by the usual method.

LIVER ALKALINE PHOSPHATASE

Liver alkaline phosphatase has the most rapid mobility and is the predominant alkaline phosphatase in serum. In young children, however, the bone enzyme is the most prevalent; it is sensitive to heat while liver and intestinal enzymes are not [23]. The usual procedure for heat inactivation of alkaline phosphatase enzymes is to place serum in a water bath at 56° C. for 30 minutes. The increase of serum alkaline phosphatase in infectious hepatitis, obstructive biliary disease, and hepatitis from drugs such as chlorpromazine is due to liver alkaline phosphatase. Histochemically, alkaline phosphatase originating from the liver is probably identical to that from endothelial cells of blood vessels in other organs. There is probably a second alkaline phosphatase originating in the biliary system. Increased alkaline phosphatase in hepatobiliary disease, therefore, could be produced by two mechanisms: (1) release of blood-vessel-type enzyme originating within the liver itself due to cirrhosis, hepatitis, or a metastatic disease, and (2) increased pressure from the biliary system. Hill and Sammons [16], using starch-gel electrophoresis, studied alkaline phosphatases from bile and serum from seven patients who had undergone a cholecystectomy for cholelithiasis and did not have liver disease. In the serum the major band was in the β-globulin position, whereas in the bile specimen the major band was in the γ-globulin region. These observations confirm those by Pope and Cooperband [24], who also showed that by paper electrophoresis the alkaline phosphatases of liver and biliary origin maintained their same electrophoretic position, thus indicating that the difference between the two isoenzymes is one of charge rather than size. Also, ultrafiltration demonstrated that the two types of molecules were essentially the same size [24].

As pointed out by Hill et al. [16] and Posen et al. [25] and as discussed in Chap. 3, elevation of serum enzymes must be considered in the light of production, transport, and clearance. Production of liver alkaline phosphatase may be from endothelial cells of blood vessels or the surface of hepatic cells. High production results in a high serum level because the rate of clearance is not able to keep up with the rate of production. Eventually, as production ceases, homeostasis is reestablished as alkaline phosphatase is removed directly from the serum. This finding is contrary to the usual belief that alkaline phosphatase is cleared by the liver and excreted in the intestinal tract via the biliary system. The molecular size of alkaline phosphatase is over 100,000, and it is thus large enough to fall in the same category as plasma proteins where clearance is concerned. The clearance of serum proteins has been established by radioisotope studies. Alkaline phosphatase has essentially the same clearance rate as do plasma proteins. Clubb, Neale, and Posen [10] have infused placental alkaline phosphatase into normal individuals and into those with biliary obstruction and found no difference in the clearance rate. One difference between lactate dehydrogenase (LDH) isoenzymes and alkaline phosphatase is that in the Riley virus studies the virus obstructs certain lymphoid cells or lymph nodes and clearance of LDH is impaired, but this does not influence the clearance for alkaline phosphatase. Thus alkaline phosphatase is removed by a different system from that operating for LDH.

BONE ALKALINE PHOSPHATASE

The alkaline phosphatase originating from bone has electrophoretic mobility corresponding to the β_1 position by agarose electrophoresis. In children bone alkaline phosphatase is the prominent alkaline phosphatase, as shown by its electrophoretic mobility, its heat inactivation, and its inactivation by EDTA. Placental alkaline phosphatase is not inactivated by either of these two methods. The bone enzyme is elevated in bone disease. Yong [33] found the bone band increased in 93 percent of 42 patients with osteoblastic lesions. Histochemically the enzyme originates in osteoblasts of the bone, and osteoblastic lesions give rise to detectable bone alkaline phosphatase in the serum even when the total activity is normal [33]. The same isoenzyme is present in the serum of growing children. The isoenzyme disappears or becomes faint by 15 years of age [33].

INTESTINAL ALKALINE PHOSPHATASE

The alkaline phosphatase isoenzyme has the electrophoretic mobility of β_2-globulin. This enzyme is inhibited by L-phenylalanine. Liver, bile, bone, lung, kidney, and spleen isoenzymes are not inhibited by L-phenylalanine, as reported by Fishman et al. [14], whereas alkaline phosphatase from the intestinal tract is inhibited. The electrophoretic mobility of intestinal alkaline phosphatase remains unchanged after treatment with neuraminidase, in contrast to liver and bone isoenzymes, which are affected [27]. Posen et al. [25] characterized the slow serum alkaline phosphatase by using anti–human intestinal alkaline phosphatase. They found good correlation in individual sera between L-phenylalanine sensitivity and antibody reaction to slow-moving alkaline phosphatase. Their findings constitute further evidence that slow-moving akaline phosphatase accounts for approximately 20 percent of the total alkaline phosphatase activity in some sera.

The intestinal component of serum alkaline phosphatase was shown by Langman et al. [22] to be influenced by diet. Successive blood samples taken during the day disclosed differences in concentration of intestinal alkaline phosphatase, particularly in individuals of blood group O or B who were secretors and to a lesser degree in blood group A individuals who were secretors. Nonsecretors, irrespective of ABO group, showed little change. Shreffler [29] also noted some quantitative differences in the amount of weak or slow-moving alkaline phosphatase band in individuals of blood group O. Fluctuation was not circadian but was related to diet. The type of diet was important. Those who received an isocaloric meal containing 60 gm. of corn oil demonstrated increased intestinal alkaline phosphatase isoenzyme, in contrast to the individuals who did not receive oil in their meal. A rise in thoracic duct lymph alkaline phosphatase has been noted by Blomstrand and Werner [6], but this did not occur if the diet contained only protein and carbohydrate. The relationship of the diet and serum alkaline phosphatase in the rat was reported by Jackson as early as 1952 [18]. He studied the effect of food ingestion on intestinal and serum alkaline phosphatases in the rat and found that fasting reduced them. Glucose had no effect. Ligation of the bile duct or introduction of an external draining biliary drain did not change the rise in intestinal alkaline phosphatase. Intestinal phosphatase was not increased two hours after ingestion of fat by tube but was increased at six hours. Feeding of a mixed ration resulted in elevation of both serum and intestinal phosphatase. Cortner et al. [11] and Wills [32] had previously noted increase in alkaline phosphatase following a diet of fat.

The alkaline phosphatase content of intestinal cells may vary according to ABH blood group secretor status, and there may be some relationship between fat absorption in some individuals and blood group secretor status. The enzyme from the intestinal mucosa may be solubilized by lipid in the diet of persons of blood group O, and to a lesser extent B, who are secretors and transported via the lymphocytes to the blood. Considerable interest in the association of ABO blood groups and secretor status in various diseases has resulted in extensive studies in recent years. A number of disorders involving stomach and duodenum such as gastric ulcer, gastric carcinoma, and pernicious anemia show relationship with ABH blood group, particularly blood group O and blood group A.

Keiding [19] studied the intestinal alkaline phosphatase in human urine and serum to determine its substrate specificity. The intestinal fluid phosphatase had a lower substrate specificity for *p*-nitrophenol phosphate than did serum alkaline phosphatase. He concluded that since the difference between serum reactivity toward the two substrates phenol phosphate and *p*-nitrophenol phosphate is very small it excludes the presence of quantities of lymph or intestinal alkaline phosphatase in normal serum. He felt that in unusual circumstances intestinal alkaline phosphatase may be found in the serum, but in general its concentration is too low to make much difference. This is in contrast to findings that in normal serum intestinal alkaline phosphatase accounts for approximately 20 percent of the total alkaline phosphatase activity [25] in some sera.

PLACENTAL ALKALINE PHOSPHATASE

A number of investigators have noted elevated serum alkaline phosphatase in pregnancy. There is also an increase of cytoplasmic leukocyte alkaline phosphatase and of the enzyme in the urine during pregnancy. In agar-gel electrophoretograms the placental enzyme is electrophoretically comparable to that of bone alkaline phosphatase, and from that standpoint the two cannot be distinguished. Since bone alkaline phosphatase is heat labile, however, a method for distinguishing the two isoenzymes is available. Furthermore, the enzymes can be distinguished by substrate specificity. Sadosky and Zuckerman [28] depended on migration in starch gel. Birkett et al. [5] found alkaline phosphatase at term to be partially inactivated by anti–human placental alkaline phosphatase. Kitchener et al. [20], by heat inactivation, EDTA inactivation, and electrophoretic mobility on cellulose acetate, found that approximately half of the enzyme in the maternal circulation resembled placental alkaline phosphatase. The placental alkaline phosphatase disappeared from the maternal circulation at a rate similar to that observed in nonpregnant subjects infused with placental alkaline phosphatase.

Alkaline Phosphatase Isoenzymes and Lipoprotein

As noted above, alkaline phosphatase isoenzymes in agarose zymograms disclosed three zones which correspond to α_2-globulin and β-globulin. The two zones in the β-globulin area are often close together and in some instances cannot be clearly distinguished. The localization of the two alkaline phosphatase zones in the β-globulin area suggested to us that there may be binding between the enzyme and β-lipoprotein [8, 9]. Lawrence and Melnick [23] showed by immunoelectrophoresis that a large number of enzymes including alkaline phosphatase are bound to the β-lipoprotein immunoprecipitin band. Boyer [7], in starch-gel electrophoretic studies of serum alkaline phosphatase and placental alkaline phosphatase, noted a zone of enzyme activity which corresponded to the β-lipoprotein. Taswell and Jeffers [31] found eight zones of alkaline phosphatase activity by starch-gel electrophoresis in a study of 202 sera. One zone corresponded to the β-lipoproteins and was present in almost all normal sera; it was increased in both bone and liver diseases. In Figure 7-1, the position of the alkaline phosphatase isoenzyme bands in agar-gel electrophoretograms is depicted and compared to the same serum stained for lipoproteins [8]. The α_2 alkaline phosphatase isoenzyme band does not appear to be related to the lipoprotein. However, the two alkaline phosphatase bands in the β-globulin area seem to correspond to the position of the pre-β- and β-lipoproteins. Figure 7-2 shows a series of four patients whose total serum alkaline phosphatase values by King-Armstrong method are, from top to bottom, 13, 28, 85, and 168. In review of Figure 7-2, it appears that, as the serum alkaline phosphatase increases, the α_2 alkaline phosphatase band becomes more intense. This suggested to us that after the β-lipoprotein is saturated with enzyme the excess enzyme spills over and travels at its own electrophoretic mobility, which corresponds to that of the α_2-globulin. The four patients, from top to bottom, had the following conditions: gout, child with Wilms's tumor, jaundice with clinical diagnosis of infectious mononucleosis (in this patient additional laboratory findings disclosed LDH 865 I.U., glutamic oxalacetic transaminase (GOT) 127 units, test for mononucleosis was negative, and LDH zymogram showed LDH$_3$ increased), and cancer of the head of the pancreas with obstruction of the common duct with no evidence of liver damage (bromsulphalein elevated, normal LDH, slight increase of GOT, and normal cholesterol).

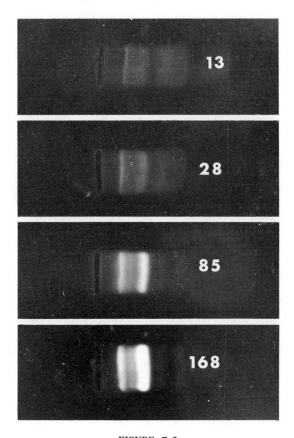

FIGURE 7-2

Four alkaline phosphatase zymograms. Observe increase in fast band as concentration increases (*top to bottom*). Anode is on the right (From L. P. Cawley [8].)

To test the idea that binding between lipoprotein and alkaline phosphatase isoenzymes was responsible for the isoenzyme patterns of the type seen in Figure 7-2, serum was delipidized by several different organic solvents. The delipidized serum was then tested for total alkaline phosphatase activity and alkaline phosphatase isoenzymes (see Chap. 15, under Alkaline Phosphatase Isoenzymes). Ether incompletely removed lipids and did not prove satisfactory, as will be shown later. A mixture of equal parts of ethanol and ether was very effective in removing serum lipids. Butanol was likewise effective. Serum delipidized by butanol extraction showed all the enzyme activity to be confined to the zone having mobility of α_2-globulin [9]. However, the electrophoretic mobility was increased in some instances (Fig.

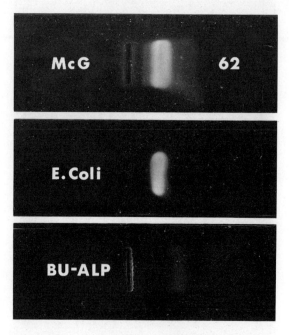

FIGURE 7-3

Patient with ascending cholangitis. Total serum alkaline phosphatase value was 62 King-Armstrong units. Alkaline phosphatase stain is shown (*upper*) compared to alkaline phosphatase stain of *E. coli* and butanol extract of serum (*lower*). Anode is on the right. (From L. P. Cawley [8].)

7-3, lower). The patient had ascending cholangitis, and three zones of alkaline phosphatase were evident (Fig. 7-3, upper). The pattern at the bottom shows one band which has a faster electrophoretic mobility than the normal α_2-globulin alkaline phosphatase. At the top of Figure 7-3 the raw serum shows two well-defined bands and a slower small faint band. After delipidization, one band with slightly greater mobility to *Escherichia coli* (middle) remained. The α-naphthol gave a purple color in both the *E. coli* and the delipidized serum, suggesting some interference with the color reaction by serum lipids. Triton X-100, a nonionic detergent, also altered the color of the enzymatically produced end product (α-naphthol) [1]. Allen et al. [1] considered Triton X-100 to be an activator of certain esterases. As will be noted later, butanol extraction of serum also resulted in enhancement of alkaline phosphatase activity.

Additional studies on the same patient are shown in Figure 7-4. In the top zymogram the three bands are well delineated. The lipoprotein stain shows (middle zymogram) relationship of the three

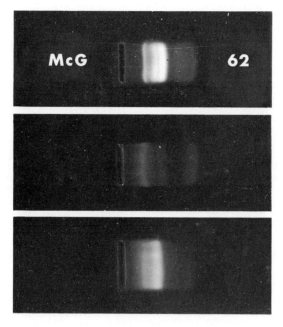

FIGURE 7-4

Same serum as in Figure 7-3 compared to lipoprotein stain (*middle*) and
alkaline phosphatase stain after ether extraction of serum. Slow-moving
zone of enzyme activity in raw serum (*upper*) is absent in ether-extracted
serum. Anode is on the right.

zones to lipoproteins. The ether-extracted serum shown at the bottom
discloses two remaining bands; the cathodic band is gone.

Figure 7-5 is a study of a patient with chronic ulcerative colitis
who had resection of the colon in 1957. Hepatomegaly was present,
and the total serum LDH was 66 I.U. The LDH_5 was elevated, bilirubin was 2.0 mg. per 100 ml., and GOT was 40 units. The alkaline phosphatase bands were reduced from three to two (lower pattern) after
extraction of the serum with ether.

In Figure 7-6 three patients selected because of some interesting
relationship to alkaline phosphatase are shown, and their alkaline
phosphatase zymograms are compared to their serum protein electrophoretograms. The serum protein electrophoretogram is mounted
above the corresponding alkaline phosphatase zymogram. In all instances, three isoenzyme fractions of alkaline phosphatase are noted.
In patient A, who had metastatic carcinoma of the liver, four bands
of alkaline phosphatase are identified. He was jaundiced, the β-lipoprotein serum band was increased, and the slow-moving alkaline phos-

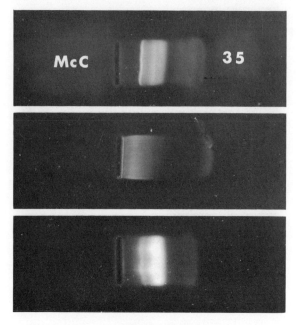

FIGURE 7-5

Another example of slow-moving alkaline phosphatase band after ether extraction of serum (*lower*) compared to lipoprotein stain (*middle*) and raw serum (upper). Anode is on the right.

phatase band was relatively prominent. In patient B the electrophoretic pattern shows an increase and broadening of β_2 (β-lipoprotein). The slow-moving alkaline phosphatase band matches these features. In addition, there is a very faint band just cathodic to the origin. In patient C the middle band (bone alkaline phosphatase) is elevated. The serum pattern is not unusual. The patient had multiple fractures and elevated serum alkaline phosphatase.

In any instance, the alkaline phosphatase isoenzyme fractionation from the standpoint of usefulness in the clinical laboratory still must be proved. In some cases we have been able to correlate the isoenzyme fractions with pathologic conditions, particularly where the bone or liver alkaline phosphatase band is concerned. The slow-moving band seems to be dependent on β-lipoprotein content.

In the study to determine the effect of the lipid extraction on the total serum alkaline phosphatase, it was found that, rather than a decrease of activity, there was an increase. Forty patients selected at random for total serum alkaline phosphatase performed on raw serum and butanol-extracted serum disclosed that in almost every instance

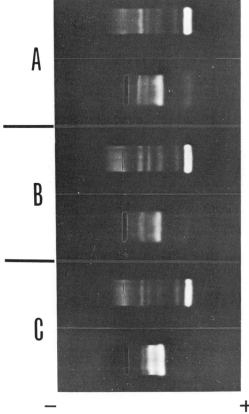

FIGURE 7-6

Paired serum electrophoretograms (*upper*) and alkaline phosphatase zymo-grams (*lower*) on three patients. The top pair is from a patient (A) who had metastatic carcinoma of liver with jaundice and serum alkaline phos-phatase of 27.1 King-Armstrong units. The B pair is from a patient with infectious hepatitis with slight jaundice increase of serum β-lipoprotein and serum alkaline phosphatase of 23.2 King-Armstrong units. Note broad β₂ serum band (β-lipoprotein) and also slow-moving alkaline phosphatase band. A very faint slow band is shown just anodic to point of application. The C pair of patterns is from a patient with multiple fractures. Serum alkaline phosphatase value was 69.8 King-Armstrong units. Observe the very dense middle band. Anode is on the right.

there was an increase in the enzyme activity after extraction (Table 7-1). In two cases a decrease was noted, and in two others no change was found. The overall increase amounted to 18.8 percent. There ap-pears at this point to be no difference in the blood group of the patient and the alkaline values before and after extraction.

TABLE 7–1

DIFFERENCE IN TOTAL ALKALINE PHOSPHATASE IN NATURAL
AND BUTANOL-EXTRACTED SERUM

Patient	Blood Group	Serum	Butanol Ext.	Differential
1	B	6.0	7.5	+ 1.5
2		3.2	4.5	+ 1.3
3	B	12.0	13.9	+ 1.9
4		18.9	20.0	+ 1.1
5		8.3	9.0	+ 0.7
6		11.5	9.0	− 2.5
7	O	10.5	12.2	+ 1.7
8	A	6.5	7.4	+ 0.9
9	A	23.0	27.0	+ 4.0
10	B	7.4	9.7	+ 2.3
11	O	6.5	5.6	− 0.9
12	A	3.5	5.6	+ 2.1
13	A	8.0	11.4	+ 3.4
14	O	9.0	9.5	+ 0.5
15	A	8.0	8.5	+ 0.5
16	O	9.0	10.0	+ 1.0
17	A	9.5	10.0	+ 0.5
18	A	5.8	5.8	0.0
19	A	8.5	9.5	+ 1.0
20	A	7.4	8.5	+ 1.1
21	B	9.0	9.0	0.0
22	B	10.0	12.2	+ 2.2
23	A	4.7	6.2	+ 1.5
24	O	4.0	5.2	+ 1.2
25	A	8.5	10.0	+ 1.5
26	A	6.2	7.4	+ 1.2
27	O	10.5	12.2	+ 1.7
28	A	14.3	16.0	+ 1.7
29	A	35.5	45.8	+ 10.3
30	O	5.8	8.0	+ 2.2
31	A	10.0	13.2	+ 3.2
32	O	9.5	12.2	+ 2.7
33	O	3.5	5.2	+ 1.7
34	O	26.5	32.0	+ 5.5
35	B	12.2	16.0	+ 3.8
36	A	8.5	10.0	+ 1.5
37	B	6.8	8.0	+ 1.2
38	O	8.5	11.7	+ 3.2
39	B	6.8	8.5	+ 1.7
40	A	11.2	15.3	+ 4.1

References

1. Allen, S. L., Allen, J. M., and Licht, B. M. Effects of Triton X-100 upon the activity of some electrophoretically separated acid phosphatases and esterases. *J. Histochem. Cytochem.* 13:434, 1965.
2. Arfors, K. E., Beckman, L., and Lundin, L. G. Genetic variations of human serum phosphatases. *Acta Genet.* (Basel) 13:89, 1963.
3. Bamford, K. F., Harris, H., Luffman, J. E., Robson, E. B., and Cleghorn, T. E. Serum-alkaline-phosphatase and the ABO blood groups. *Lancet* 1:530, 1965.
4. Beckman, L. Associations between human serum alkaline phosphatases and blood groups. *Acta Genet.* (Basel) 14:286, 1964.
5. Birkett, D. J., Done, J., Neale, F. C., and Psen, S. Serum alkaline phosphatase in pregnancy; an immunological study. *Brit. Med. J.* 1:1210, 1966.
6. Blomstrand, R., and Werner, W. B. Alkaline phosphatase activity in human thoracic duct lymph. *Acta Chir. Scand.* 129:177, 1965.
7. Boyer, S. H. Alkaline phosphatase in human sera and placentae. *Science* 134:1002, 1961.
8. Cawley, L. P. *Workshop Manual on Electrophoresis and Immunoelectrophoresis* (Rev. ed.). Commission on Continuing Education, Council on Clinical Chemistry. Chicago: American Society of Clinical Pathologists, 1966.
9. Cawley, L. P., and Eberhardt, L. Association of alkaline phosphatase isoenzymes with serum lipoproteins. *Amer. J. Clin. Path.* 47:364, 1967.
10. Clubb, J. S., Neale, F. C., and Posen, S. The behavior of infused human placental alkaline phosphatase in human subjects. *J. Lab. Clin. Med.* 66:493, 1965.
11. Cortner, J. A., and Schnatz, J. D. Electrophoretic behavior of alkaline lipolytic activity in human adipose tissue. *Biochim. Biophys. Acta* 139:107, 1967.
12. Cunningham, V. R., and Rimer, J. G. Isoenzymes of alkaline phosphatase of human serum. *Biochem. J.* 89:50P, 1963.
13. Evans, D. A. P. Confirmation of association between ABO blood groups and salivary ABH secretor phenotypes and electrophoretic patterns of serum alkaline phosphatase. *J. Med. Genet.* 2:126, 1965.
14. Fishman, W. H., Ingilis, N. I., and Krant, M. J. Serum alkaline phosphatase of intestinal origin in patients with cancer and with cirrhosis of the liver. *Clin. Chim. Acta* 12:298, 1965.
15. Haije, W. G., and De Jong, M. Iso-enzyme patterns of serum alkaline phosphatase in agar-gel electrophoresis and their clinical significance. *Clin. Chim. Acta* 8:620, 1963.
16. Hill, P. G., and Sammons, H. G. An interpretation of the elevation of serum alkaline phosphatase in disease. *J. Clin. Path.* 20:654, 1967.
17. Hodson, A. W., Latner, A. L., and Raine, L. Iso-enzymes of alkaline phosphatase. *Clin. Chim. Acta* 7:255 1962.
18. Jackson S. H. The effect of food ingestion of intestinal and serum alkaline phosphatase in rats. *J. Biol. Chem.* 198:553, 1952.
19. Keiding, N. R. Intestinal alkaline phosphatase in human lymph and serum. *Scand. J. Clin. Lab. Invest.* 18:134, 1966.

20. Kitchener, P. N., Neale, F. C., Posen, S., and Brudenell-Woods, J. Alkaline phosphatase in maternal and fetal sera at term and during the puerperium. *Amer. J. Clin. Path.* 44:654, 1965.

21. Kreisher, J. H., Close, V. A., and Fishman, W. H. Identification by means of L-phenylalanine inhibition of intestinal components separated by starch gel electrophoresis of serum. *Clin. Chim. Acta* 11:122, 1965.

22. Langman, M. J. S., and Leuthold, E. Influence of diet on the "intestinal" component of serum alkaline phosphatase in people of different ABO blood groups and secretor status. *Nature* (London) 212:41, 1966.

23. Lawrence, S. H., and Melnick, P. J. Enzymatic activity related to human serum beta-lipoprotein: Histochemical, immunoelectrophoretic and quantitative studies. *Proc. Soc. Exp. Biol. Med.* 107:998, 1961.

24. Pope, C. E., II, and Cooperband, S. R. Protein characteristics of serum and bile alkaline phosphatase. *Gastroenterology* 50:631, 1966.

25. Posen, S., Neale, F. C., Birkett, D. J., and Brudenell-Woods, J. Intestinal alkaline phosphatase in human serum. *Amer. J. Clin. Path.* 48:81, 1967.

26. Rendel, J., and Stormont, C. Variants of ovine alkaline serum phosphatases and their association with the R-O blood groups. *Proc. Soc. Exp. Biol. Med.* 115:853, 1964.

27. Robinson, J. C., and Pierce, J. E. Differential action of neuraminidase on human alkaline phosphatases. *Nature* (London) 204:472, 1964.

28. Sadovsky, E., and Zuckerman, H. An alkaline phosphatase specific to normal pregnancy. *Obstet. Gynec.* 26:211, 1965.

29. Shreffler, D. C. Genetic studies of blood group-associated variations in a human serum alkaline phosphatase. *Amer. J. Hum. Genet.* 17:71, 1965.

30. Sussman, H. H., Smoll, P. A., and Cotlove, C. Human alkaline phosphatase. *J. Biol. Chem.* 243:160, 1968.

31. Taswell, H. F., and Jeffers, D. M. Isoenzymes of serum alkaline phosphatase in hepatobiliary and skeletal disease. *Amer. J. Clin. Path.* 40:349, 1963.

32. Wills, E. D. The effect of surface active agents on pancreatic lipase. *Biochem. J.* 60:529, 1955.

33. Yong, J. M. Origins of serum alkaline phosphatase. *J. Clin. Path.* 20:647, 1967.

8

AMYLASE

Multiple molecular forms of human amylase were conclusively demonstrated by Norby [23] in 1964 by an application of agar-gel electrophoresis in which the completed electrophoretogram was sandwiched to "starch slides" for a suitable period of time. Localization of amylase activity was depicted on the starch slides by staining with iodine. The zones of enzyme activity were clearly seen against a background of blue. Saliva, duodenal fluid, salivary gland extract, and pancreatic extract were studied. Norby showed that pancreas and salivary glands contain unique amylases all in the γ-globulin zone with different electrophoretic mobility. The amylase from saliva and that from salivary gland were identical and consisted of two components, a major and a minor, both migrating cathodically. Pancreatic extract contained one amylase band that migrated more cathodically than saliva or salivary gland amylase (SA). Duodenal fluid showed in some instances (7 of 34 cases) two bands of amylase activity, one corresponding to the major SA band and one corresponding to pancreatic amylase (PA). In the other cases only one band was present and it corresponded to PA.

Subsequent papers utilizing comparable technics established that both PA and SA existed in serum and urine (Fig. 8-1). Oger and Bischops [24] used agar-gel electrophoresis with starch substrate overlay and iodine stain to demonstrate two zones of amylase activity in serum, one corresponding to SA and the other to PA. They also demonstrated two bands in duodenal fluid, a fast and a slow band corresponding to SA and PA respectively. Poort et al. [25] used agarose electrophoretograms sandwiched to starch-agar slides to detect amylase in extract of rat pancreas. When placed in ethanol-acetic acid, the starch slides became white and opaque except where starch had been digested by amylase. Aw [2] employed cellulose-acetate electrophoresis for separation and, making a sandwich with a thin layer of starch, demonstrated the presence of multiple amylase zones comparable to those found by the above authors. Alfonso [1] likewise demonstrated the presence of multiple amylase bands in serum using an agar-gel electrophoretic stage with histochemical localization by a starch overlay. Of interest is the fact that by immunoelectrophoresis he could not show precipitin bands corresponding to zones of amylase

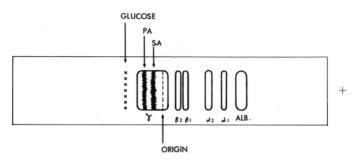

FIGURE 8-1
Drawing of serum protein electrophoretogram depicting position of PA and
SA isoamylases in reference to serum proteins and glucose

activity. Joseph et al. [15], with a technic of agar-gel electrophoresis
followed by a histochemical overlay consisting of starch, glucose oxidase, peroxidase, o-dianisidine, and maltase, described differences in
the electrophoretic mobility of amylases from pancreas and salivary
gland but found only one broad zone in serum. In addition, an extract
of liver showed a zone of activity cathodic to the PA which did not
display activity by the amyloclastic method. The appearance of an
additional zone of amylase activity anodic to SA and in the α_2-globulin
area after incubation of human sera at 45° C. for 16 hours was reported
by Fric and Cincibuchova [13]. Berk [4], by analysis of acrylamide
serum electrophoretogram segments by a saccharogenic method, demonstrated two amylase zones corresponding to PA and SA with the
same relative mobilities as described in agar, cellulose acetate, and
agarose electrophoretograms.

Of the aforementioned investigators only Norby described two
amylase zones in salivary gland extract or saliva. At least three chro-
matographically distinct amylases were found in human saliva by
Millin and Smith [21] by gel filtration, calcium phosphate gel, and
DEAE-cellulose chromatography. In contrast, a simple, pure, homogeneous, highly active α-amylase was isolated from human saliva by
Sephadex G-25 followed by DEAE-cellulose acetate column chromatography of active component [26]. By paper and acrylamide electrophoresis only a single band was present. Quite to the contrary, Muus
et al. [22] in 1964 reported additional studies on separation of crys-
tallin salivary amylase into four zones by disc electrophoresis. Amylase
activity was detected from segments of the acrylamide gel tested
against starch and iodine. The enzyme activity corresponded to protein zones. Wolf and Taylor [34] in 1967 extended the study by using
paper electrophoretically pure salivary amylase in disc electrophoresis.
Detection was by starch-slide iodine reaction. These authors found
from five to seven bands of activity. The influence of protein on the

amyloclastic reaction, as will be pointed out later, must be excluded as the cause of false amylase activity in disc electrophoretograms. In any event, more studies are needed to determine whether these conflicting findings can be reconciled. Chromatographic and electrophoretic heterogeneity are not necessarily one and the same. In the latter case net electrical charge differences determine electrophoretic mobility. In reference to the demonstrable electrophoretic differences between PA and SA, gel filtration [31] and immunologic tests with hog anti–pancreatic amylase [18, 19, 20] did not show a difference between salivary and pancreatic amylase.

Historical Review

It is of some interest to review historically the developments leading to the concept of isoenzymes of amylase [16] and to point out some of the pitfalls that caused false interpretation of the presence or the absence of isoenzymes in the protein zones of an electrophoretogram. Analysis for amylase of extracts of fractions removed from paper electrophoretograms corresponding to protein zones leads to the false conclusion that amylase was present in each protein zone. The major amylases were found to occupy positions corresponding to albumin and γ-globulin. These early reports engendered considerable enthusiasm, and the report of isoenzymes of amylase in various serum proteins by McGeachin and Lewis [17] was followed shortly by the work of Dreiling et al. [12], who also found high amylase activity in the albumin and γ-globulin fraction with lesser activity in the other protein zones. In pancreatitis these observers noted marked increase in the amylase corresponding to the γ-globulin with relatively no change in the albumin amylase. Berk et al. [5] reported that amylase activity in rabbit serum was predominantly in the γ-globulin zone but noted activity in the albumin fraction with very little activity in α_2- and β-globulin areas.

In 1954 Baker and Pellegrino [3] determined amylase activity in paper electrophoretograms using starch-impregnated agar plates for the detection of enzyme followed by iodine stain. The amylase, they showed, was located in the γ-globulin fraction. Later Wilding [32] and Ujihira et al. [30] reaffirmed that all the amylase of serum was located in the γ-globulin area, but no evidence for isoamylases was presented. Wilding [31] noted that the blue color of the starch-iodine reaction can be discolored by protein having no enzyme activity. He also found that although serum amylase had the same electrophoretic mobility as portions of the γ-globulin it was unbound to these proteins. A year later Wilding et al. [33] reported a case in which the serum amylase had an electrophoretic mobility distinctly different from normal. In this instance the amylase was bound to a high-molecular-weight protein. In a case reported by Berk [4] the urinary level of amylase was low and the serum of amylase was high. By gel filtration he detected two peaks, one associated with a globulin and the other free, as is the usual finding.

Our interest in the puzzle came accidentally when, in the course of developing a simplified system of gel electrophoresis for the clinical laboratory [9], we incorporated starch into the agar gel hoping to increase resolution. Inadvertently a strip was set aside prior to fixation, and the amylase zone was depicted by clearing of the starch from the electrophoretogram leaving a clear zone in the otherwise "frosted glass" pattern associated with a mixture of starch and agar (see Chap. 15, under Amylase Isoenzymes). This approach was used in delineation of amylase activity in plant proteins [10]. In an attempt to augment the localization of the amylase, a set of agar-starch zymograms were stained with iodine, periodic acid–Schiff (PAS), and Alcian blue stain. The results of the iodine stain pointed up the problem of false localization of amylase by the amyloclastic method; namely, zones with amylase activity appeared to exist in all protein fractions, but only the γ-globulin zone contained true enzyme capable of digesting the starch leaving only transparent agar to indicate that the starch had been digested [10]. Elsewhere there was obvious competition between the protein and starch for the iodine. Final proof was obtained in a study designed to measure inhibition on the slope of rate reaction of a standard glucose solution (200 mg. per 100 ml.) utilizing glucose oxidase, peroxidase, potassium iodide (KI), and starch. In an aqueous environment no difficulty was encountered. However, as soon as protein, either serum or hemoglobin, was added, there was a discrepancy in the rate of reaction, and this was shown to be due to binding of iodine by protein.

Rate reaction measurements utilizing the Analytrol as a continuous monitoring device [8] to obtain the slope of the curve proved to be a very sensitive system to depict inhibition of the reaction (Fig. 8-2). Adulteration of the system with various concentrations of protein, bilirubin, ascorbic acid, etc., permitted easy recognition of inhibition of the reaction. The system was used to demonstrate inhibition of the reaction by A_1 hemoglobin isolated from a hemoglobin electrophoretogram (Fig. 8-3). Searcy et al. [26, 27], Alfonso [1], and Wilding [32] demonstrated clearly the competitive nature of serum proteins, particularly albumin, for the iodine, which gave the impression of enzyme activity in the amyloclastic method of analysis. The saccharogenic method of analysis causes no difficulty. Boiling the fractions removed from paper electrophoretograms removed all the enzyme activity from the γ-globulin zone but caused an increase in the activity in the other protein zones, particularly albumin; thus, as the story unfolds it can be seen why in the earliest efforts toward finding heterogeneity of the amylases, confusion resulted from reports dealing with indirect methods of measuring enzyme activity [7]. In those methods wherein

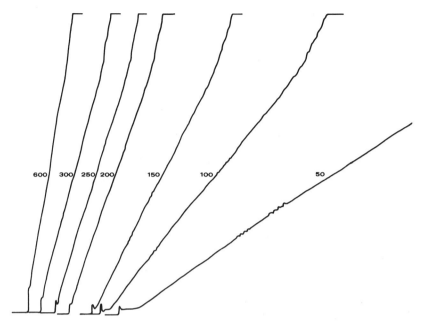

FIGURE 8-2

Analytrol recording of series of glucose standards in rate reaction studies with glucose oxidase, peroxidase, starch, and KI. The slopes are reproducible, and a plot of slope versus concentration is linear.

the electrophoretogram was studied directly with histochemical technics with the proper controls, little confusion developed.

An exception is the report by Joseph et al. [15], who used a histochemical reaction on agar-gel electrophoretograms to localize the amylase. The electrophoretograms were incubated with starch, glucose oxidase, peroxidase, *o*-dianisidine, maltase. The reaction indicated that amylase activity was located in the γ-globulin area. These investigators also demonstrated that a liver amylase was present which was more cathodic in its position than PA. Hoeke et al. [14], however, correctly pointed out that the amylase band found in the liver was in fact glucose. Previously, we had determined the position of glucose in an agar electrophoretogram in an attempt to develop a method for measuring serum amylase in a one-step fashion by utilizing glucose oxidase, peroxidase, *o*-dianisidine, and maltase. Electrophoresis to separate serum amylase from serum glucose was shown to be possible by using glucose oxidase test tape laid on an agar-gel electrophoretogram of serum protein to disclose the position of serum glucose. Glucose is cathodic to amylase and distinct from amylase (Fig. 8-1) and thus does not interfere with the detection of amylase in the glucose oxidase–maltase system although it would react. In this study starch was incor-

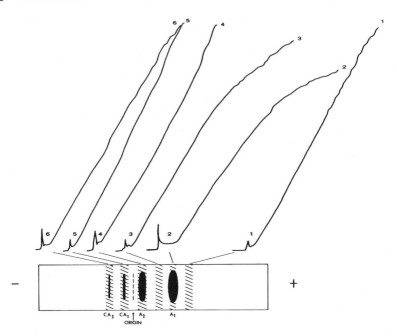

FIGURE 8-3

Six rate reaction curves generated as in Figure 8-2 in combination with extract of six zones of hemoglobin electrophoretogram. A_1 hemoglobin, A_2 hemoglobin, carbonic anhydrase (CA_1 and CA_2) are the only zones detected by protein stain. The glucose standard was 200 mg. per 100 ml. Serum, 1µl. in reaction chamber of 5-ml. volume, completely inhibits the reaction. As can be seen, fraction corresponding to A_1 hemoglobin influences the slope of the reaction. Hemoglobin hemolysate (3µl.) was used for the hemoglobin electrophoretogram.

porated in the agar gel. Detection of substrate specificity of glucose oxidase against starch can be performed by reversing the system and subjecting glucose oxidase to electrophoresis [11]. Glucose is not bound to protein and is a nonionic substance which has been used to measure endosmosis [3]. Dextran travels essentially in the same electrophoretic position as glucose. The fact that glucose is completely separate from protein in the completed electrophoretogram offers a new method for deproteinization. Independently we used essentially the same reasoning as Joseph et al. [15] and arrived at the concept of using glucose oxidase coupled with maltase to determine the presence of isoenzymes of amylase. The method proved effective but not any more effective than the single observation of the presence or the loss of the "frosted glass" pattern by the activity of the enzyme on a substrate which is incorporated into the agarose support (see Chap. 15, under Amylase Isoenzymes).

Clinical Usefulness

Additional work has to be done in the area of interpretation of amylase isoenzymes in serum and urine before useful clinical information becomes available. From the standpoint of establishing guidelines of interpretation it has been shown that in mumps the salivary enzyme is present in the serum and the urine. In pancreatitis the predominant amylase is the pancreatic form. There are not many instances in which a differentiation of this type is warranted. The clinical significance of anomalous serum amylase reported by Wilding et al. [33] and Berk [4] remains to be established.

Histochemically amylase is found in acini of human submaxillary gland, parotid, and pancreas [29]. The enzyme could not be demonstrated in human heart, kidney, or liver by Tremblay [29], who employed a very sensitive starch-film sandwich technic to frozen sections. The area of enzyme activity was well demarcated by PAS stain of starch films. Serum and urinary amylase, therefore, must originate principally from the salivary and pancreatic tissues. In the rat other tissues than salivary and pancreatic are involved in production of serum amylase since removal of these organs does not cause its disappearance [15]. The same may be true in man. Berk [4] shows that, although patients with pancreatectomy have lower serum and urine total amylase activity, the enzyme is still present.

References

1. Alfonso, E.   On human serum amylase. *Clin. Chim. Acta* 14:195, 1966.
2. Aw. S. E.   Separation of urinary isoamylases on cellulose acetate. *Nature* (London) 209:298, 1966.
3. Baker, R. W. R., and Pellegrino, C.   Separation and detection of serum enzymes by paper electrophoresis. *Scand. J. Clin. Lab. Invest.* 6:94, 1954.
4. Berk, J. E.   Serum amylase and lipase. *J.A.M.A.* 199:99, 1967.
5. Berk, J. E., Kawagiechi, M., Zeineh, R., Ujihira, I., and Searcy, R.   Electrophoretic behavior of rabbit serum amylase. *Nature* (London) 200:572, 1963.
6. Berk, J. E., Searcy, R. L., Hayaski, S., and Ujihira, I.   Distribution of serum amylase in man and animals. *J.A.M.A.* 192:389, 1965.
7. Bussard, A., and Perrin, G.   Electrophoresis in agar plates. *J. Lab. Clin. Med.* 46:689, 1955.
8. Cawley, L. P., and Eberhardt, L.   A permanent record for determination of prothrombin time, using the Spinco Analytrol as a recorder. *Amer. J. Clin. Path.* 37:219, 1962.
9. Cawley, L. P., and Eberhardt, L.   Simplified gel electrophoresis: I. Rapid technic applicable to the clinical laboratory. *Amer. J. Clin. Path.* 38:539, 1962.
10. Cawley, L. P., Eberhardt, L., and Goodwin, W. L.   Agar-gel electro-

phoretic and analytic agar-gel electrophoretic study of bromelin. *Transfusion* 4:441, 1964.

11. Cawley, L. P., Eberhardt, L., and Musser, B. O. Zymographic and gas-liquid chromatographic study of substrate specificity of glucose oxidase. *Amer. J. Clin. Path.* 47:366, 1967.

12. Dreiling, D. A., Janowitz, N. D., and Josephberg, L. J. Serum iso-amylases. *Ann. Intern. Med.* 58:235, 1963.

13. Fric, P., and Cincibuchova, D. Effect of prolonged incubation on the electrophoretic pattern of amylase. *Clin. Chim. Acta* 15:370, 1967.

14. Hoeke, J. O. O., Voorrips, C., and DeWael, J. Isoamylase pitfalls. *Clin. Chim. Acta* 16:171, 1967.

15. Joseph, R. R., Olivero, E., and Ressler, N. Electrophoretic study of human isoamylases. *Gastroenterology* 51:377, 1966.

16. McGeachin, R. L., Gleason, J. R., and Adams, M. R. Amylase distribution in extrasalivary tissues. *Arch. Biochem.* 75:403, 1958.

17. McGeachin, R. L., and Lewis, J. P. Electrophoretic behavior of serum amylase. *J. Biol. Chem.* 234:795, 1959.

18. McGeachin, R. L., and Reynolds, J. M. Difference in mammalian amylases demonstrated by enzyme inhibition with specific antisera. *J. Biol. Chem.* 234:1456, 1959.

19. McGeachin, R. L., and Reynolds, J. M. Inhibition of amylases by rooster antisera to hog pancreatic amylase. *Biochim. Biophys. Acta* 39:531, 1960.

20. McGeachin, R. L., and Reynolds, J. M. Serological differentiation of amylase isoenzymes. *Ann. N.Y. Acad. Sci.* 94:996, 1961.

21. Millin, D. J., and Smith, M. H. Gel filtration and chromatography of human salivary proteins. *Biochim. Biophys. Acta* 62:450, 1962.

22. Muus, J., and Vnenchak, J. M. Isozymes of salivary amylase. *Nature* (London) 204:283, 1964.

23. Norby, S. Preliminary notes. Electrophoretic non-identity of human salivary and pancreatic amylases. *Exp. Cell Res.* 36:663, 1964.

24. Oger, A., and Bischops, L. Les iso-enzymes de l'amylase. *Clin. Chim. Acta* 13:670, 1966.

25. Poort, C., and Van Venrooy, W. J. W. Detection of enzymes in agar electrophoretograms. *Nature* (London) 204:684, 1964.

26. Searcy, R. L., Hayaski, S., Hardy, E. M., and Berk, J. E. The interaction of human serum protein fractions with the starch-iodine complex. *Clin. Chim. Acta* 12:631, 1965.

27. Searcy, R. L., Ujihira, I., Hayashi, S., and Berk, J. E. An intrinsic disparity between amyloclastic and saccharogenic estimation of human serum isoamylase activities. *Clin. Chim. Acta* 9:505, 1964.

28. Shainkin, R., and Birk, Y. Isolation of pure alpha-amylase from human saliva. *Biochim. Biophys. Acta* 122:153, 1966.

29. Tremblay, G. The localization of amylase activity in tissue sections by starch film method. *J. Histochem. Cytochem.* 2:202, 1963.

30. Ujihira, I., Searcy, R. L., Berk, J. E., and Hayashi, S. A saccharogenic method for estimating electrophoretic and chromatographic distribution of human serum amylase. *Clin. Chem.* 11:97, 1965.

31. Wilding, P. Use of gel filtration in the study of human amylase. *Clin. Chim. Acta* 8:918, 1963.

32. Wilding, P. The electrophoretic nature of human amylase and the effect of protein on the starch-iodine reaction. *Clin. Chim. Acta* 12: 97, 1965.
33. Wilding, P., Cooke, W. T., and Nicholson, G. I. Globulin-bound amylase. A cause of persistently elevated levels in serum. *Ann. Intern. Med.* 60:1053, 1964.
34. Wolf, R. O., and Taylor, L. I. Isoamylase of human parotid saliva. *Nature* (London) 213:1128, 1967.

9

ALBUMIN

Albumin appears as a homogeneous symmetrical band in a zone electrophoretogram. For a number of reasons this homogeneity by zone electrophoresis was assumed to imply that the molecular composition of albumin was a single antigen and that antiserum prepared against albumin was specific to the whole molecule and other antigen sites did not exist. The recent biochemical characterization of the tetrameric structure of immunoglobulins and the existence of several hereditary variants of albumin have called attention to the possibility that the structure of albumin may be more complex than was originally contemplated.

Molecular Heterogeneity

Bisalbuminemia, first described by Scheurlen [28], is characterized by double albumin peaks or bands on electrophoresis (Fig. 9-1). In the original studies these two components were labeled A, for the fraction with the same electrophoretic mobility as normal albumin, and B, for the slow component. Both are present in essentially the same concentration, and numerous later studies showed that immunologically they were identical, both carried bilirubin, they stained equally with bromphenol blue, and for the most part they were considered identical [32]. Isolation of the individual components followed by enzymatic hydrolysis and fingerprinting revealed that albumin A had the same peptide pattern as normal albumin. Albumin B, on the other hand, contained a lysine in place of an amino acid with a free carbonyl group [15]. Payne and Dickinson [24], by immunodiffusion and immunoelectrophoresis on separated A and B albumins in a new family with bisalbumineria, found no immunochemical difference between the two albumins. Except for reported low serum cholesterol no significant abnormal laboratory or clinical findings are known. Weitkamp et al. [34] reported close linkage between albumin B and Gc system. It is important to recognize that the electrophoresis should be carried out at pH 8.6 since, as pointed out by Tarnoky [31], diamers and polymeric forms, fetal albumin, and N-F isomerization are formed at acid pH, which points up the heterogeneity of albumin under different physical conditions. Bell et al. [2] described a fast type of bisal-

FIGURE 9-1
Various types of albumins described in bisalbuminemia. (Modified from Melartin et al. [23].)

buminemia occurring in a Cree Indian family. He was able to trace it back to a homozygote who had a single fast band. There are now several electrophoretic types of bisalbuminemia. Albumin Naskapi, described by Melartin et al. [21], has electrophoretic mobility greater than albumin A (Fig. 9-1) and is found in relatively high frequency in North American Indians. The fast albumin reported by Bell et al. [2] in Crees and by Weitkamp et al. [34] in Chippewas is probably the same as albumin Naskapi. Another albumin variant with mobility slower that that of albumin A, also found by Melartin et al. [23] in sera from Mexicans, is called albumin Mexico (Fig. 9-1). The mode of inheritance is simple autosomal codominant, as has been reported for the others. Albumin Naskapi and albumin Mexico are immunologically identical [20]. Efremov and Braend [13] used starch-gel electrophoresis to screen 1015 individuals, mostly Norwegians, for heterogeneity of albumin and proposed a genetic theory of two alleles to explain the phenotypes found. Two albumin phenotypes were found and designated FF (major band preceded by a weak band) and FS (two major bands with a weak band in front). The investigators did note that the FF major band sometimes appeared to be composed of two components although they were not able to obtain separation in each instance. The FS phenotype was found in only one individual. The many studies of bisalbuminemia support the hypothesis that the determinants for the albumins are alleles at a single locus [2, 21, 22]. Laurell and Nilehn [19] described a new type of inherited serum albumin anomaly characterized in agarose-gel electrophoretograms by a broad albumin band appearing to tail on the cathodic side. By starch-gel electrophoresis the anomaly had an abnormal small band between the postalbumin and the fast α_2 globulin. The anomalous albumin seemed to undergo dimerization more easily than normal serum. Out of 1550 tested, five subjects with this abnormality were patients in the department of orthopedic surgery. In contrast, there was only one such example in 3200 control cases screened by agarose-gel electrophoresis. The family study indicated that the biochemical abnormality was the result of heterozygosity of an autosomal gene. Some of these patients had a high incidence of bone and joint complaints and bad hearing.

The nomenclature of albumin variants suggested by Melartin and Blumberg [21, 22] and Melartin [20] is used in Table 9-1, in which

TABLE 9–1

NOMENCLATURE OF ALBUMIN VARIANTS

Phenotype	Gene	Electrophoretic Mobility
Al Gent	Al^{Ge}	+
Al Naskapi	Al^{Na}	Fast Albumins
Al Reading	Al^{Re}	
Al A	Al^{A}	Normal Albumin
Al Mexico	Al^{Me}	Slow Albumins
Al B	Al^{B}	−

the known albumin variants are listed in order of decreasing electrophoretic mobility. The proposed nomenclature distinguishes phenotype from genetic notation. Phenotype is designated by the word *albumin* followed by the specific name. Abbreviated forms are also acceptable, i.e., Al Naskapi or Al Na. The abbreviation *Al* is used for locus at which the genes segregate, and the alleles are described by superscript on the locus symbol, i.e., Al^{Na}. (These symbols are underlined in typescript and italicized in print.) The phenotypic expressions for some albumins other than those already discussed are: albumin Gent for the fast albumin described by Wieme [35] (Fig. 9-1) and albumin Reading for a slower variant described by Tarnoky [31].

According to Melartin [20] starch-gel electrophoresis is capable of detecting all albumin variants thus far described and also alterations induced by drugs in mobility of normal albumin or variants. Cellulose acetate and agar- or agarose-gel electrophoresis (zone electrophoresis) are about equal in their ability to detect the albumin variants, but are not as satisfactory as a molecular sieve system of electrophoresis (starch gel or acrylamide gel) [20]. Melartin also reports that penicillin in large amounts either in vitro or in vivo produces distortion of the starch-gel electrophoretic mobility of normal albumin and albumin variants. In all instances the drug causes appearance of a fast component which appears to arise from the faster albumin component. This suggests to us that penicillin binds to a segment of the albumin and increases its electronegativity. In some cases complete separation of albumin into two zones may occur. However, there is usually only blurring and enlargement of the albumin zone. This also occurs with Keflin; we have noted a comparable effect with γ-globulin and erythrocytes treated with Keflin. In both instances there is an increase of electronegativity and, therefore, an increase of electrophoretic mobility of the γ-globulins (see Appendix) and erythrocytes (see Chap. 13, Fig. 13-2).

We have seen three examples of bisalbuminemia of the slow type (A/B) in many thousands of electrophoretograms. One was in a resident physician from Pakistan. Unfortunately, further studies were not possible. One example is shown in Figure 2-10F in Chap. 2, Monoclonal, Polyclonal, and Dysclonal Gammopathies.

Several investigators have shown bovine and human albumin to consist of multiple units. Foster [1, 14] has designated the units N and F. He has studied the relationship of these units under a number of circumstances and shown that at low pH the dissociation between the N and F forms is more complete. Apparently there are series of N forms and corresponding F forms. In acrylamide-gel electrophoresis bovine albumin has been shown to consist of several units [30]. Human albumin isolated from paper electrophoretograms separates into subunits by starch-gel electrophoresis (Saifer et al. [27]). Raymond et al. [26] have studied the molecular size of the subunits of bovine albumin by a molecular sieve electrophoretic system employing increasing concentrations of acrylamide. Their studies disclose that the subunits vary in molecular weight and also in size. By zone electrophoresis these subunits migrate as a single unit, indicating uniformity in net charge density of the subunits. There is a possibility that interunit bonding of albumin as proposed by Foster is strong enough to hold the albumin units together during zone electrophoresis, but this bonding is broken during molecular sieve electrophoresis.

Antigenic Heterogeneity

Peterson and Sober [25] obtained three separate peaks for albumin by calcium phosphate column chromatography. Lapresle in 1955 [17] reported splitting human albumin into three specific antigens by using proteolytic enzymes from rabbit spleen. His excellent study demonstrated at least three antigenic groups of albumin, which can be characterized by antialbumin antibodies by gel diffusion and immunoelectrophoresis. The subunits of albumin found by molecular sieve electrophoresis are not necessarily related to the antigenic heterogeneity of albumin reported by Lapresle. Kaminski and Tanner [16] and Lapresle et al. [18] used trypsin, chymotrypsin, and pepsin in a study of antigenic structure of albumin. Lapresle et al. [18] found that trypsin does not split albumin but that chymotrypsin splits it into two antigenic components and pepsin-like rabbit spleen proteolytic enzyme splits it into three antigenic components. The splenic enzyme resembles cathepsin A. Smit et al. [29] hydrolyzed bovine albumin with pepsin and isolated two protein fractions by Sephadex G-100 column chromatography. Sedimentation coefficients on the fragments of 4.22 S and 2.84 S for the fast and slow components respectively compared to 4.14 S for whole molecule were obtained. These investigators showed that the fast component was probably an immunologically active fragment of albumin with a molecular weight of 35,000. Enzymatic cleavage of serum bovine albumin by pepsin studied by Weber [33] gave two fractions by column chromatography. He has shown that rotational relaxation time and ultracentrifugation sedimentation constants are different for each fraction.

In the course of an electrophoretic and immunoelectrophoretic investigation of the effect of bromelin on the interaction between human erythrocytes and serum products, degradation of serum proteins was observed [4–10]. The most outstanding change was noted in albumin, which was split into two well-defined fractions and one less well-defined component all disclosed by agar-gel or agarose-gel electrophoresis (Fig. 9-2, AB) [7, 8, 9]. Band 2, the major band, had the same electrophoretic mobility as normal albumin, and band 3, the slow band, varied quantitatively from about 50 percent to 100 percent of the major band. The fast component, band 1, was very weak. The original observation of the splitting of albumin was made while employing polyvalent anti–human serum, which revealed only two components of albumin. In the system employing albumin and anti–human albumin three antigenic components were present. Immunoelectrophoretic studies have shown the antigenic differences to match the electrophoretic differences. Figure 9-2 shows electrophoretic strips of bromelin (B), albumin (A), and albumin-bromelin mixture (AB). Albumin consists of a single substance and the bromelin consists of several components; none of the components of bromelin overlap with degradation products of albumin. In Figure 9-2, AB, three components of albumin are seen and are designated 1, 2, and 3. Band 1 is very faint compared to bands 2 and 3, which appear to be roughly equal. The longer the period of incubation, the greater the quantity of band 3, although the increase after one hour's incubation at 37° C. does not necessarily give a proportionate increase in band 3. Albumin removed from agar-gel electrophoretogram and subjected to acrylamide electrophoresis shows several components (Fig. 9-3, Ac) [9, 11]. In the figure, Ad contains the same material as Ac but has been diluted with phosphate buffer to make its albumin content roughly equivalent with sample AB, which contains the albumin-bromelin mixture. The three components 1, 2, and 3 in Figure 9-3, AB, correspond to the agar-gel electrophoretogram components in Figure 9-2, AB. The immunoelectrophoretogram of the albumin-bromelin mixture against goat anti–

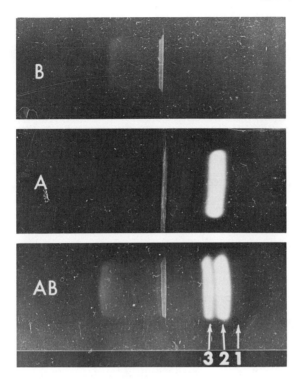

FIGURE 9-2

Agar-gel electrophoretogram disclosing B—bromelin; A—albumin; and AB—effect of bromelin on albumin. Note peaks 1, 2, and 3 of albumin. Band 1 is quite faint. Anode is on the right.

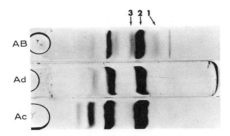

FIGURE 9-3

Disc electrophoretograms. AB—mixture of bromelin and albumin— showing the effect of the enzyme on the albumin. Bands 1, 2, and 3 are new bands, distinct from those present in Ad, which represents dilute albumin to make it comparable in concentration to AB. Ac stands for concentrated albumin and depicts the number of components in albumin which are independent of those produced by bromelin but represent the normal subcomponents of albumin. Anode is on the right.

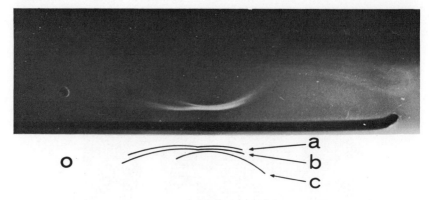

FIGURE 9-4

Immunoelectrophoretogram of albumin after treatment with bromelin against goat anti–human albumin. The diagram below delineates the position of the bands as present in the actual precipitin pattern above. Anode is on the right.

human albumin is shown in Figure 9-4. Three precipitin zones, a, b, and c, are designated in the drawing. Immunodiffusion study of the bromelin-albumin mixture compared to albumin with goat anti–human albumin discloses three precipitin lines against the albumin-bromelin mixture which blend into one against albumin (Fig. 9-5).

At the moment it is unclear whether the individual electrophoretic components are matched by antigen heterogeneity. There are at least three and possibly four separate antigenic components of human

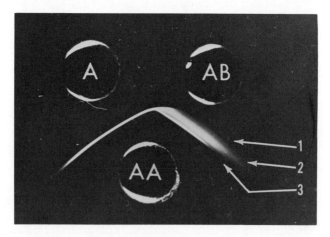

FIGURE 9-5

Immunodiffusion study of albumin (A) and albumin plus bromelin (AB) against goat antialbumin (AA). Note flaring of the band between AB and AA, composed of components 1, 2, and 3.

serum albumin. Two are major components and have electrophoretic mobility comparable to that of A and B (Fig. 9-1). Whether these are related to A and B of bisalbuminemia remains to be seen. Each of the major zones shares two antigens. The major band has an additional antigen, and there is a suggestion of one more antigen corresponding to the fast component. Electron microscopic examination of albumin reveals a ring structure with a control hole [12]. Further electron microscopic studies of this type combined with enzyme hydrolysis and immunochemical studies with ferritin-labeled antiserum will help elucidate the structure of albumin.

Analbuminemia

Analbuminemia, first described by Bennhold, Peters, and Roth [3], is a condition in which there is complete absence of circulating albumin. It is surprising that those with the condition have few symptoms, which raises the question of the biologic value or function of albumin. Albumin is known to be a carrier of a number of important chemicals—estrogen, bilirubin, and fatty acids, to name the most common ones. Some of the individuals with analbuminemia had edema, and biochemically there were some clinical laboratory abnormalities. Erythrocyte sedimentation was rapid; the serum cholesterol and phospholipids were elevated; flocculation tests were abnormal. All of these findings occur in hypoalbuminemia to a lesser degree. The half-life of albumin in analbuminemia in one individual was three times normal, whereas the survival of I^{131} globulin was essentially normal [32].

References

1. Adkins, B. J., and Foster, J. F.   Limited subtilisin hydrolysis of bovine plasma albumin: Evidence for structural subunits. *Fed. Proc.* 23:474, 1964.
2. Bell, H. E., Nicholson, S. F., and Thompson, Z. R.   Bisalbuminemia of the fast type with a homozygote. *Clin. Chim. Acta* 15:247, 1967.
3. Bennhold, H., Peeters, H., and Roth, E.   Über einen Fall von kompletter Analbuminaemie ohne wesentliche klinische Krankheitszeichen. *Verh. Deutsch. Ges. Inn. Med.* 60:630, 1954.
4. Cawley, L. P., and Eberhardt, L.   Agar-gel electrophoretic investigation of bromelin. *Transfusion* 3:421, 1963.
5. Cawley, L. P., and Eberhardt, L.   Immunoelectrophoretic study of plasma proteins absorbed by erythrocytes after treatment with tannic acid or bromelin. *Transfusion* 3:422, 1963.
6. Cawley, L. P., Eberhardt, L., and Goodwin, W. L.   Agar-gel electrophoretic and analytic agar-gel electrophoretic study of bromelin. *Transfusion* 4:441, 1964.
7. Cawley, L. P., Eberhardt, L., Schneider, D., and Goodwin, W.   Im-

munochemical study of fragments of human serum albumin obtained by action of bromelin. *Transfusion* 5:383, 1965.

8. Cawley, L. P., Schneider, D., and Eberhardt, L. Immunoelectrophoretic and serologic study of the immunologically active components of bromelin with rabbit anti-bromelin. *Vox Sang.* 11:81, 1966.

9. Cawley, L. P., Schneider, D., and Goodwin, W. L. Molecular and antigenic heterogeneity of albumin. *Amer. J. Clin. Path.* 42:532, 1964.

10. Cawley, L. P., Wiley, J. L., Schneider, D., and Harrouch, J. Immunoelectrophoretic and immunochemical characterization of antibodies against bromelin. *Transfusion* 4:316, 1964.

11. Davis, B. J., and Ornstein, L. A New High Resolution Electrophoresis Method. Delivered at meeting of the Society for the Study of Blood, at the New York Academy of Medicine, March 24, 1959.

12. Deutsch, K., Segal, J., and Kalaidjiev, A. Electron microscope examination of some globular proteins. *Nature* (London) 195:177, 1962.

13. Efremov, G., and Braend, M. Serum albumin: Polymorphism in man. *Science* 146:1679, 1964.

14. Foster, J. F. Plasma Albumin. In Putnam, F. W. (Ed.), *The Plasma Proteins*. New York: Academic, 1960. Vol. 1, p. 179.

15. Gitlin, D., Schmid, K., Earle, D. P., and Givelber, H. Observation on double albumin: II. A peptide difference between two genetically determined human serum albumins. *J. Clin. Invest.* 40:820, 1961.

16. Kaminski, M., and Tanner, C. E. Étude comparative par électrophorèse en gélose et immunoélectrophorèse, de la dégradation de la sérumalbumine humaine par la pepsine, la trypsine et la chymotrypsine. *Biochim. Biophys. Acta* 33:10, 1959.

17. Lapresle, C. Étude de la dégradation de la sérumalbumine humaine par un extrait de rate de lapin; mise en évidence de trois groupements spécifiques différents dans le motif antigénique de l'albumine humaine et de trois anticorps correspondants dans le sérum de lapin anti-albumine humaine. *Ann. Inst. Pasteur* (Paris) 89:654, 1955.

18. Lapresle, C., Kaminski, M., and Tanner, C. Immunochemical study of the enzymatic degradation of human serum albumin: An analysis of the antigenic structure of a protein molecule. *J. Immun.* 82:94, 1959.

19. Laurell, C. B., and Nilehn, J. E. A new type of inherited serum albumin anomaly. *J. Clin. Invest.* 45:1935, 1966.

20. Melartin, L. Albumin polymorphism in man. *Acta Path. Microbiol. Scand.* Suppl. 191, 1967.

21. Melartin, L., and Blumberg, B. S. Albumin Naskapi: A new variant of serum albumin. *Science* 153:1664, 1966.

22. Melartin, L., and Blumberg, B. S. Inherited variants of human serum albumin. *Clin. Res.* 14:482, 1966.

23. Melartin, L., Blumberg, B. S., and Lisker, R. Albumin Mexico, a new variant of serum albumin. *Nature* (London) 215:1288, 1967.

24. Payne, R. B., and Dickinson, J. P. Immunochemical studies on bis-albuminaemia. *Nature* (London) 215:536, 1967.

25. Peterson, E. A., and Sober, H. Chromatography of the Plasma Proteins. In Putnam, Frank W. (Ed.), *The Plasma Proteins*, Vol. 1, *Isolation Characterization and Function*. New York: Academic, 1960. Pp. 114, 115.

26. Raymond, S., and Nakamichi, M.   Electrophoresis in synthetic gels: II. Effect of gel concentration. *Anal. Biochem.* 7:225, 1964.
27. Saifer, A., Robin, M., and Ventrice, M.   Starch-gel electrophoresis of "purified" albumins. *Arch. Biochem.* 92:409, 1961.
28. Scheurlen, P. G.   Über Serumeiweissveränderungen beim Diabetes mellitus. *Klin. Wschr.* 33:198, 1955.
29. Smit, F., Lontie, R., and Preaux, G.   Immunologically Active Fragment of Bovine Serum Albumin Obtained by Peptic Digestion. In Peeters, H. (Ed.), *Protides of the Biological Fluids.* Proceedings of the 11th Colloquium, Bruges, 1963. Amsterdam: Elsevier, 1964. P. 119.
30. Sogami, M., and Foster, J. F.   Resolution of oligomeric and isomeric forms of plasma albumin by zone electrophoresis on polyacrylamide gel. *J. Biol. Chem.* 237:2514, 1962.
31. Tarnoky, A. L., and Lestas, A. N.   A new type of bisalbuminaemia. *Clin. Chim. Acta* 9:551, 1964.
32. Waldmann, T. A.   Hereditary Disorders of Albumin Synthesis. In Sunderman, F. W., and Sunderman, F. W., Jr. (Eds.), *Serum Proteins and Dysproteinemias.* Philadelphia: Lippincott, 1964. Pp. 406–409.
33. Weber, G.   Nature of the configurational changes of serum albumin in acid. *Biochem. J.* 84:74, 1962.
34. Weitkamp, L. R., Rucknagel, D. L., and Gershowitz, H.   Genetic linkage between structural loci for albumin and group specific component (Gc). *Amer. J. Hum. Genet.* 18:559, 1966.
35. Wieme, R. J.   On the presence of two albumins in certain normal human sera and its genetic determination. *Clin. Chim. Acta* 5:443, 1960.

10
HEMOGLOBINOPATHIES

Hemoglobin variants may be separated in an electrical field at a fixed pH because of differences in electrical charge specific to each type. The hemoglobin variants therefore can be identified by their relative mobilities. A system of agarose-gel electrophoresis provides an effective method of screening for hemoglobin variants [2, 19, 27].

Improvement of resolution by agarose-gel electrophoresis for hemoglobins can be obtained by a variety of alterations, such as different additives to the gel and buffer, and pH and buffer variations [3, 27]. The addition of glycine [4, 18] to the barbital buffer–agarose gel system and a discontinuance system of buffer produces sharp delineation of hemoglobins C, S, E, A, and A_2. A_2 can also be quantitated by densitometry. Under ideal conditions F and A_1 hemoglobins can be separated. The entire procedure is found in Chap. 15.

Biochemistry

Hemoglobin is a conjugated protein composed of four heme groups and a globin molecule. Heme is an iron-porphyrin complex. Globin is the protein portion of hemoglobin and is a tetramer composed of four polypeptide chains, two α and two β [17]. Each chain contains roughly 150 amino acids having a slightly different sequence in the α and β chains. The complete hemoglobin molecule is a compact mass approximately 55 by 55 by 70 Angstrom (Å) units with a molecular weight of 66,000 [20]. The tertiary and quaternary properties of the molecule account for the uniform molecular size and shape. The difference in primary structure is responsible for differences in electrostatic charge of the molecule. Hemoglobin A, which is normal adult hemoglobin, has two α chains and two β chains and is designated $\alpha_2{}^A\beta_2{}^A$ [10]. The human adult erythrocyte also contains minor hemoglobins, fetal (F), A_2, and A_3 accounting for approximately 15 percent of the total hemoglobin content of the cell [20]. Fetal and A_2 together total about 4 to 5 percent. Hemoglobin F, or fetal hemoglobin, has two α chains and two γ chains and is indicated as $\alpha_2{}^A\gamma_2{}^F$ [5]. Specific amino acid differences are responsible for β- and γ-chain differences. There is another type of chain which may replace the normal β chain, and this is the δ chain of hemoglobin A_2, which is $\alpha_2{}^A\delta_2{}^{A2}$ [25].

Hemoglobin variants may be of two types [28]. First, one or both of the normal adult polypeptide chains may be reduced or absent, as in thalassemia, which may have either α- or β-chain defects. Examples of α-chain defects are Hb H ($\beta_4{}^A$) [12] and Hb Barts ($\gamma_4{}^F$) [8]. Examples of β-chain defects are thalassemia major, with a high percentage of Hb F [22], and thalassemia minor, with increased Hb A_2 [13]. Second, the amino acid sequence of one of the polypeptide chains, usually the β chain, may be abnormal. For example, sickle cell hemoglobin (Hb S) is $\alpha_2{}^A\beta_2{}^{6\ valine}$ [6], indicating that the glutamic acid at position 6 in the normal Hb A β chain is replaced by valine in the Hb S β chain. Similarly, Hb E is $\alpha_2{}^A\beta_2{}^{26\ lysine}$ [7], indicating that the glutamic acid at position 26 in the normal Hb A β chain is replaced by lysine in the Hb E β chain.

A number of hemoglobin variants have been detected and named

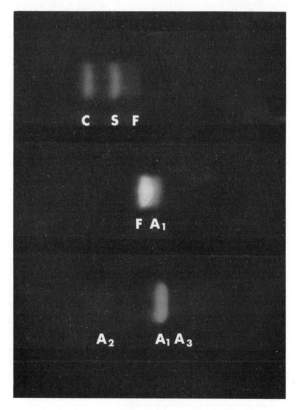

FIGURE 10-1

Agarose electrophoretogram of Hb C, S, F, A_1, A_2, and A_3. Note excellent separation of C and S (glycine added). Anode is on the right. (From L. P. Cawley [2].)

[15]. It is possible to separate many of these by electrophoresis. The amino acid differences in the composition of the hemoglobin peptides are responsible for differences in electrical charge, which produce the differences in electrophoretic mobility [14]. For example, Hb A_2 carries a greater negative charge than Hb C. Therefore, it migrates faster toward the anode at alkaline pH than Hb C. Similarly, Hb C has the smallest negative charge and migrates slowest toward the anode at alkaline pH. In the agarose-gel electrophoretic system the more common hemoglobin variants migrate toward the anode in the following order: Hb A_3 migrates most rapidly, followed by Hb A_1. Hb F is next, after which Hb S and Hb D migrate in the same position. Hb A_2 comes next, followed by Hb E, and Hb C occupies the position nearest the cathode (see Figs. 10-1 and 10-2).

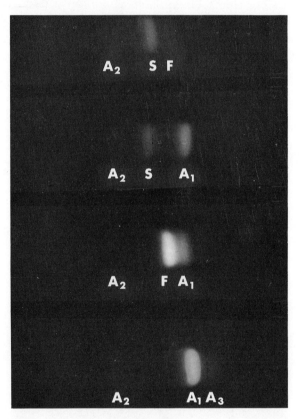

FIGURE 10-2

Four hemoglobin electrophoretograms with glycine showing Hb S from homozygous sickle cell disease above a heterozygous sickle cell disease. Fetal hemoglobin is shown above a normal Hb pattern at the bottom of the photograph. The system results in separation of Hb F from A_1 and Hb A_2 from A_3. Anode is on the right. (From L. P. Cawley [2].)

Discussion

The action of glycine in the electrophoretic phase is not yet clearly defined but is under study. There is apparently some reduction in heat produced during electrophoresis by the addition of glycine. It does not change the pH, but there is evidence that it does change the electrostatic charge of the hemoglobin molecule (see Figs. 10-2 and 10-3).

Studies on fractional precipitation of proteins suggest that dipolar substances such as glycine increase the solubility of proteins and raise the dielectric constant [21]. In a medium with a high dielectric constant, electrolyte disassociation is enhanced while electrostatic attraction between polar groups of dissolved protein molecules is decreased. This would help to account for the more distinct separation of hemoglobin types by the presence of glycine in this electrophoretic system.

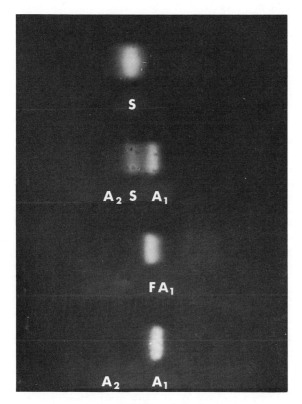

FIGURE 10-3

Four hemoglobin electrophoretograms without glycine showing slower migration and indistinct separation. Compare with Figure 10-2. Anode is on the right.

Agarose-gel electrophoresis is a screening procedure that categorizes the various hemoglobin groups. Ferrosolubility [11] and sickle cell preparation [1] are then used to identify S and D hemoglobins, which migrate together in this system. Hb S shows the sickling property when reduced and has a low ferrohemoglobin solubility. Hb D has a high ferrohemoglobin solubility. The alkali denaturation test [22] is used to quantitate Hb F, which is alkali resistant, while Hb A is 99 percent denatured by alkali. Hemoglobins A_2 and E are separated arbitrarily by the proportion of abnormal hemoglobin traveling in this region. If the proportion is greater than 20 percent, it is probably Hb E; if it is less than 20 percent, it is probably A_2 [26]. For more definitive separation, acrylamide-gel [4, 16] or starch-gel electrophoresis [23, 24] may be employed.

As a screening procedure for the more commonly encountered hemoglobinopathies, this method is easily adapted to any system of electrophoresis, is relatively rapid, and gives sharp separation of the hemoglobin types. Agarose is easily handled, and the stained patterns can be kept indefinitely without deterioration.

Summary

Glycine is incorporated into a barbital buffer–agarose gel system for more distinct separation of hemoglobins as a screening procedure. The different hemoglobin fractions can be quantitated by densitometry or elution.

References

1. Beck, J. S. B., and Hertz, C. S.   Standardizing sickle cell method and evidence of sickle cell trait. *Amer. J. Clin. Path.* 5:325, 1935.
2. Cawley, L. P.   *Workshop Manual on Electrophoresis and Immunoelectrophoresis* (Rev. ed.). Commission on Continuing Education, Council on Clinical Chemistry. Chicago: American Society of Clinical Pathologists, 1966.
3. Cawley, L. P., and Eberhardt, L.   Simplified gel electrophoresis: I. Rapid technique applicable to the clinical laboratory. *Amer. J. Clin. Path.* 38:539, 1962.
4. Davis, B. J.   *Disc Electrophoresis*, Part II. Rochester, N.Y.: Distillation Products Industries, 1961. P. 8.
5. Hunt, J. A.   Identity of the α chains of adult and foetal human haemoglobins. *Nature* (London) 183:1373, 1959.
6. Hunt, J. A., and Ingram, V. M.   Allelomorphism and the chemical differences of the human hemoglobins A, S and C. *Nature* (London) 181:1062, 1958.
7. Hunt, J. A., and Ingram, V. M.   Human haemoglobin E: The chemical effect of gene mutation. *Nature* (London) 184:870, 1959.
8. Hunt, J. A., and Lehmann, H.   Haemoglobin "Bart's": A foetal haemoglobin without α chains. *Nature* (London) 184:871, 1959.

9. Ingram, V. M.   A specific chemical difference between the globins of normal human and sickle cell anemia hemoglobin. *Nature* (London) 178:792, 1956.
10. Ingram, V. M.   Gene mutations in human haemoglobin: The chemical difference between normal and sickle cell haemoglobin. *Nature* (London) 180:326, 1957.
11. Itano, H. A.   Solubilities of naturally occurring mixtures of human hemoglobins. *Arch. Biochem.* 47:148, 1953.
12. Jones, R. T., Schroeder, W. A., Balog, J. E., and Vinogrod, J. R.   Gross structure of hemoglobin H. *J. Amer. Chem. Soc.* 81:3161, 1959.
13. Kunkel, H. G., Appellini, R., Muller-Eberhard, U., and Wolf, J.   Observation on the minor basic hemoglobin component in the blood of normal individuals and patients with thalassemia. *J. Clin. Invest.* 36:1615, 1957.
14. Miale, J. B.   *Laboratory Medicine—Hematology* (2d ed.). St. Louis: Mosby, 1962. Pp. 330–346.
15. Nomenclature of abnormal hemoglobins: Report. *Blood* 17:125, 1961.
16. Ornstein, L.   *Disc Electrophoresis,* Part I. Rochester, N.Y.: Distillation Products Industries, 1961.
17. Perutz, M. F., Rossman, M. G., Cullis, A. F., Muirhead, H., Will, G., and North A. C. T.   Structure of haemoglobin. A three-dimensional Fourier synthesis at 5·5-Å. resolution, obtained by X-ray analysis. *Nature* (London) 185:416, 1960.
18. Porath, J., and Ui, N.   Chemical studies on immunoglobulins: I. A new preparative procedure for gamma globulins employing glycine-rich solvent systems. *Biochim. Biophys. Acta* 90:324, 1964.
19. Robinson, A. R., Robson, M., Harrison, A. P., and Zuelges, W. W.   A new technique for differentiation of hemoglobin. *J. Lab. Clin. Med.* 50:745, 1957.
20. Rucknagel, D. L., and Neel, J. V.   The hemoglobinopathies. *Progr. Med. Genet.* 1:158, 1961.
21. Schultze, N. E., and Heremans, J. A.   *Molecular Biology of Human Proteins,* Vol. I. New York: American Elsevier, 1966. P. 241.
22. Singer, K., Chernoff, A. I., and Singer, L.   Studies on abnormal hemoglobins: I. Their demonstration in sickle cell anemia and other hematologic disorders by means of alkali denaturation. *Blood* 6:413, 1951.
23. Smithies, O.   Zone electrophoresis in starch gels: Group variations in the serum proteins of normal human adults. *Biochem. J.* 61:629, 1955.
24. Smithies, O.   An improved procedure for starch gel electrophoresis: Further variations in the serum proteins of normal individuals. *Biochem. J.* 71:535, 1959.
25. Stretton, A. O. W., and Ingram, V. M.   An amino acid difference between human hemoglobins A and A_2. *Fed. Proc.* 19:343, 1960.
26. Sunderman, F. W., and Sunderman, F. W., Jr.   *Hemoglobin. Its Precursors and Metabolites.* Philadelphia: Lippincott, 1964. Pp. 94–112.
27. Wieme, R. J.   *Agar Gel Electrophoresis.* Amsterdam: Elsevier, 1965. Pp. 192–203.
28. Wintrobe, M. M.   *Clinical Hematology* (5th ed.). Philadelphia: Lea & Febiger, 1962. Pp. 150–163.

11

BLOOD CELL PROTEINS

E lectrophoresis and immunoelectrophoresis applied to cellular components, particularly blood cells, is a new field of investigation which shortly may have practical clinical usefulness. It is only natural that erythrocytes should come under first examination since they are readily obtainable in large numbers, are reasonably homogeneous, and can be prepared in relatively pure form by washing them free of serum proteins. As is discussed in Chap. 13, electrophoretic mobility of erythrocytes is dependent upon the net electrostatic charge due principally to charged components of the membrane. The same is true of lymphocytes, granulocytes, and platelets. Since there is no nuclear material in erythrocytes, the analysis of proteins other than hemoglobin would seem to suggest that these nonhemoglobin erythrocyte proteins (NHEP) are uniformly distributed throughout the cell. Proteins unique to the erythrocyte membrane have been shown to exist, and although there may be some overlap between them and NHEP it is best for the moment to regard them as separate.

Nonhemoglobin Erythrocyte Proteins

The total soluble protein composition other than hemoglobin of the red cell (nonhemoglobin erythrocyte proteins, NHEP) has been studied by a number of investigators and lends itself to analysis by electrophoresis and immunoelectrophoresis [1, 4–6, 12–24, 26–36, 38–42]. NHEP is a light-yellow solution obtained either from thoroughly washed human erythrocytes by C.M. Sephadex column chromatography of stroma-free hemolysates (Fig. 11-1 and Fig. 11-2, fractions 5–9) [12] or by special electrophoretic methods using continuous particle electrophoretic equipment. NHEP contains a number of enzymes, and many of the enzymes and isoenzymes of glucose metabolism are present [12]. NHEP contains lactate dehydrogenase (LDH), malic dehydrogenase (MDH), glucose-6-phosphate dehydrogenase (G-6-PD), glutamic-oxalacetic transaminase (GOT), carbonic anhydrase, proteolytic enzymes, amylase, catalase, glutamic-pyruvic transaminase (GPT), and galactose-1-phosphate uridyl transferase (Table 11-1). All but the last four were found to be present as isoenzymes.

WASHING THE ERYTHROCYTES
(WASHED 6 TIMES WITH 12 VOL. SALINE)

LYSIS OF ERYTHROCYTES
(INSONATION - 128 WATTS FOR 10 SEC.)

REMOVE STROMA BY HIGH SPEED CENTRIFUGATION
(17,000g's, 25°C, 10 MIN.)

FRACTIONATE BY COLUMN CHROMATOGRAPHY
(CM-SEPHADEX GRADIENT ELUTION,
0.025–0.4 PHOSPHATE BUFFER, PH 7.0)

CONCENTRATION OF NHEP FRACTIONS
(ULTRAFILTRATION, 50 TIMES)

ELECTROPHORESIS OF NHEP
(ACRYLAMIDE & AGAR)

FIGURE 11-1

Outline of method for preparation of NHEP

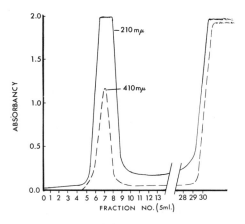

FIGURE 11-2

Absorbancy at 210 mμ and at 410 mμ of fractions from C.M. Sephadex column of hemolysates of human erythrocytes. Baseline separation of NHEP (fractions 5–9) from hemoglobin (fraction 28 and beyond) is evident. NHEP is light yellow and absorbs light at 410 mμ indicating heme pigments, but no peroxidase activity is evident.

Acid phosphatase, peroxidase, and cholinesterase are present in whole erythrocytes but not in NHEP. The last enzyme and probably acid phosphatase are stromal enzymes. The peak from fraction 28 and beyond (Fig. 11-2) from the C.M. Sephadex column contains large

TABLE 11–1

SUMMARY OF QUALITATIVE RESULTS OF TESTS FOR ENZYMES
ON FRACTIONS OF ERYTHROCYTES

Enzymes	Whole Erythrocytes	Hb	NHEP
Dehydrogenase			
Glucose-6-phosphate	+	—	+ 1
Lactate	+	—	+ 1
Malic	+	—	+ 1
Transaminase			
Glutamic oxalacetic	+	—	+
Glutamic pyruvic	+	—	+
Hydrolase			
Proteolytic	+	—	+ 1, 2
Amylase	+	—	+
Cholinesterase	+	—	—
Carbonic anhydrase	+	—	+ 1
Phosphatase			
Acid	+	—	—
Alkaline	+	—	—
Oxidase			
Peroxidase	+	+	—
Transferase			
Galactose-1-phosphate uridyl	+	—	+
Catalase	+	—	+

1, present as isoenzymes.
2, cysteine activation.

amounts of hemoglobin. Acrylamide-gel electrophoresis [37] of NHEP
discloses 10 to 15 protein components [15]. Two or three of them
have carbonic anhydrase activity [15]. Albumin is the only serum
protein detected in NHEP [15]. A water-soluble form of ABH blood
group substances also is present in NHEP [13]. No lipids or steroids
were detected by thin-layer or gas chromatography. Variability in the
isoenzyme pattern of carbonic anhydrase is found in NHEP electrophoretograms, and seven acrylamide electrophoretic NHEP patterns
emerge from a study of 169 hospitalized patients.

The procedure for preparation of NHEP is outlined in Figure 11-1.
Chromatographic separation, as shown in Figure 11-2, demonstrates
complete separation of NHEP (peak at fractions 5–9) from the other
substances of the red cell, particularly hemoglobin (peak at fractions
28 and beyond) [12]. Figure 11-3 shows acrylamide electrophoreto-

grams performed by disc electrophoretic technic on NHEP compared to protein stain of carbonic anhydrase isolated from human erythrocytes by ethanol-chloroform extraction (1), esterase stain of NHEP (2), protein stain of NHEP (3), and normal human serum (4) [14]. The bands in NHEP marked B and C have carbonic anhydrase activity and correspond to B and C carbonic anhydrase defined by Rickli et al. [38] in ethanol-chloroform extracts of erythrocytes. The band marked pre-B in Figure 11-3 is believed to be carbonic anhydrase.

Results of immunodiffusion and immunoelectrophoretic studies indicate that serum albumin is the only serum protein normally found in NHEP (Fig. 11-4); however, γG is occasionally found in hemolysates of thoroughly washed whole erythrocytes. The immunoglobulin is presumably picked up from the serum since treatment of cells with tannic acid or bromelin in the presence of serum proteins intensifies uptake by erythrocytes of albumin (see Chap. 12), γG-immunoglobin, and α_2-globulin [10, 11, 15]. Albumin is present in some but not all NHEP's. Variability in the concentration of albumin appears to be related to the age of the erythrocytes and disease of the patient (see Figs. 12-7 and 12-8 in chapter on analytical technics in electrophoresis and immunoelectrophoresis). Erythrocytes must be washed at least four times before all traces of serum protein can be displaced. Under controlled studies they are washed six times with 12 volumes of saline

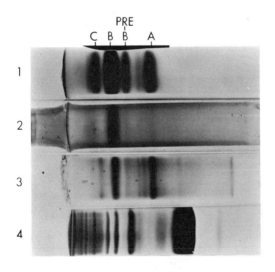

FIGURE 11-3

Four disc electrophoretograms from study of NHEP. Anode is on the right. (1) Chloroform-ethanol extract of human erythrocytes showing major carbonic anhydrase zones, A, B, C, and pre-B (Table 11-4). (2) Esterase stain of NHEP (α-naphthol acetate method). (3) Protein stain (Amido Black) of NHEP. (4) Protein stain (Amido Black) of normal human serum.

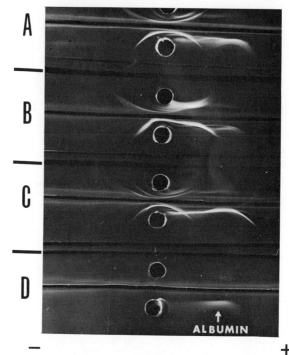

ALBUMIN

−   +

FIGURE 11-4

Four immunoelectrophoretograms against pooled NHEP. In each set the lower antigen well contains the concentrated NHEP and the upper well is a 1:4 saline dilution of the NHEP. In sets A, B, and C antisera used are from three separate rabbits each immunized with different whole washed erythrocytes. In set D the antiserum is goat anti–human serum. Note variability, and that in each instance albumin is depicted. Anode is on the right.

solution to insure removal of serum proteins. Antiserum used in studies from this laboratory is from rabbits immunized with human erythrocytes and human NHEP. In Figures 11-3 and 11-4 the complex composition of the minor protein components of human erythrocytes is evident. In numerous studies the positions of the minor components of hemolysates have been reported (Fig. 11-5). At this writing the nature of some of these components has been established (Fig. 11-5), and their position relative to some serum proteins and hemoglobins is known (Fig. 11-6). Variation of the NHEP pattern in 169 hospitalized individuals resolves itself into seven patterns, shown in Figure 11-7. Type II is the most frequent (Table 11-2). It is of interest that of seven cord bloods tested six were type V and one was type VII. Thus, the only significant change noted in NHEP referable to disease is the quantity of albumin [7]. Transferrin may on occasion be present although its unequivocal detection has not been accomplished. γG has been suspected in occasional preparations of NHEP.

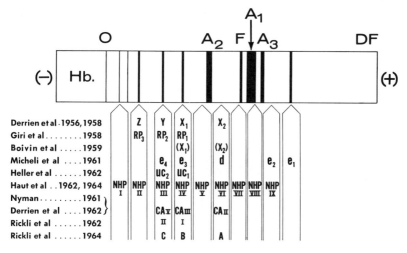

FIGURE 11-5

Drawing of starch-gel electrophoretogram of hemolysate showing position of hemoglobins and nonhemoglobin minor components that have been noted and studied by many authors. The author, year, and their designation of the minor component are shown.

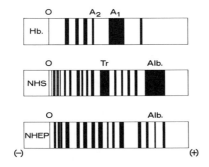

FIGURE 11-6

Comparison of protein bands in disc electrophoretograms of hemoglobin (Hb), normal human serum (NHS), and NHEP. Positions of hemoglobins A_1 and A_2, albumin (Alb.), transferrin (Tr), and origin (O) are depicted.

Zymographic studies of NHEP [8, 12, 15] disclose three to four LDH isoenzyme bands, two to three MDH bands, two G-6-PD bands, four esterase bands, one catalase band, several proteolytic zones, and, as already outlined, several zones of carbonic anhydrase activity. There appears to be considerable variability in the number of LDH bands, MDH bands, and G-6-PD bands.

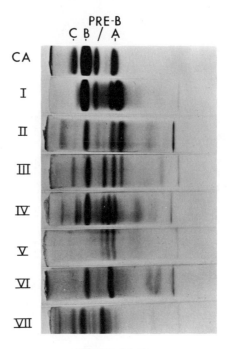

FIGURE 11-7

Seven NHEP patterns found in 169 individuals. The frequency of each is recorded in Table 11-2.

TABLE 11–2

ELECTROPHORETIC CLASSIFICATION OF NHEP

Group	Number	Frequency (%)
I	24	14.2
II	73	43.2
III	11	6.5
IV	13	7.7
V	11	6.5
VI	12	7.1
VII	25	14.8
Total	169	100.0

Inhibition studies [43] for blood group substances show that ABH is uniformly present in NHEP (Table 11-3), while the factors for Kell, Cellano, M, N, s, Do^a, Yt^a, D, C, E, c, e, and LW are absent and presumably are located in some portion of the stroma [13]. The Lewis

TABLE 11–3

INHIBITING CAPACITY OF NHEP

Blood Group of NHEP	Inhibition Titer for		
	A	B	H
A	32	0	2
B	0	2	4
AB	8	2	0
O	0	0	64

system, P_1, and P_2, are probably present. Jk^a and S may be present, although these results are equivocal. Fractionation of NHEP agar-gel electrophoretograms [9] disclose that the material inhibiting anti-A, anti-B, or anti-H (*Ulex europaeus*) has electrophoretic mobility comparable to that of β_2-globulin [13]. By acrylamide electrophoresis this fraction appears to travel just cathodic to carbonic anhydrase C isoenzyme. The amount of blood group substance present in the electrophoretogram is quite minute and is overshadowed by other material, such as carbonic anhydrase. It is not likely that the carbonic anhydrase has blood group specificity. The water-soluble ABH substances of NHEP may be under the same genetic secretor control as salivary blood group substances. Preliminary studies show that saliva and NHEP blood group substances are located in the same area of an agar-gel electrophoretogram. Some points remain to be settled, namely, the change in NHEP profile secondary to pathologic disorders and the relationship of blood group substances to the protein bands or enzyme bands of acrylamide electrophoretograms, as well as the significance of the secretor status of the patient.

With the initial work of Derrien et al. [17, 18], with starch-gel electrophoresis, the minor protein composition of the red cell, exclusive of hemoglobin, came under direct observation. They described four components designated as X_2, X_1, Y, and Z. About the same time Giri et al. [24] in 1958 observed these same minor proteins in agar-gel electrophoresis and designated three: RP_1, RP_2, and RP_3. All traveled cathodic to A_2 hemoglobin. In 1959 Boivin et al. [4], with an antiserum prepared against human erythrocyte proteins, demonstrated immunoprecipitin bands corresponding to X_2 and X_1. By immunoelectrophoresis carried out with antiserum raised in rabbits from hemolysis of carefully washed human erythrocytes, Rose et al. [40] (1960) designated six and sometimes eight distinct constituents of the hemolysate, each with different electrophoretic mobility; rabbit antihemolysate serum in some instances contained antihemoglobin antibodies. They were able to relate some of their immunoprecipitin bands to positions corresponding to X_1 and X_2 as well as hemoglobin A_1.

In 1961 Micheli and Grabar [34] designated five different components, some of which had esterase activity in erythrocyte hemolysates. The study was a combined immunoelectrophoretic and electrophoretic study. Immunoprecipitin bands with esterase activity and corresponding to esterase fractions were designated e_1 and e_2 for bands traveling ahead of A_3 hemoglobin. A zone designated as b had no esterase activity, traveled between A_1 hemoglobin and A_2 hemoglobin, and corresponded to the previously designated X_2 component. Bands e_3 and e_4, traveling behind A_2 hemoglobin, were believed to have the same electrophoretic mobility as did X_1 and Y respectively (Fig. 11-5). In 1962 Cerruti [16], with rabbit antiserum against the soluble proteins from human erythrocyte stroma, found six immunoprecipitin lines. By immunoelectrophoresis he designated these according to the Greek alphabet: α, β, γ, δ, ε_1, ε_2. He was not able to show any substantial difference between individuals although he thought there might be some difference in cases of thalassemia major in which an extra precipitin band was seen. Also in 1962 Heller et al. [28] designated two zones by agar-gel electrophoresis and immunoelectrophoresis in hemolysate agar-gel electrophoretograms as unknown components (UC), UC_1 and UC_2. Haut et al., in 1962 [27] and in 1964 [26], approached the problem of the number of minor protein components exclusive of hemoglobin of erythrocytes by removing the hemoglobin with cellulose, Sephadex, or ion exchange resin. The fluid obtained was designated nonhemoglobin protein (NHP). By starch-gel electrophoresis they originally described four major NHP zones—I, II, III, and IV—and observed that zone III contained catalase activity (in acrylamide catalase is just anodic to origin [7]) and that zone IV contained carbonic anhydrase. In the 1964 study, they extended their findings to nine and sometimes ten NHP fractions, some of which overlapped with the usual position of A_1 and A_2 hemoglobins. Zymograms revealed one to four zones of G-6-PD activity, five zones of LDH, and multiple esterase zones for the substrate α-naphthyl acetate. Our own studies, mentioned above, were reported in 1964. We separated minor erythrocyte proteins from hemoglobin by CM-Sephadex column chromatography and used disc electrophoresis for fractionation.

As early as 1961 Nyman [36] demonstrated the presence of three carbonic anhydrase (CA) zones—CA II, CA III, and CA V—in the solutions from erythrocytes by the method previously described by Derrien et al. [17, 18] and Laurent et al. [31], and in 1962 Derrien et al. [19] proved the identity of proteins X_1, X_2, and Y, isolated by chromatography with the three CA's described by Nyman [36]. Earlier (1960) Lindskog [32] had showed that there were multiple enzymes of CA and that these were individual variations. In 1963 Fine et al. [21], with isolated fractions X_1, X_2, and Y, reported on their electrophoretic mobility in starch and agar and their antigenic structure to specific antisera. The three proteins had essentially the same weight, they found, and were small enough not to be influenced by the sieving effect of starch gel. Antisera against X_1 reacted equally well with X_2 and very weakly with Y.

A relatively thorough immunochemical study of the nonhemoglobin proteins of human erythrocytes by Howe et al. [29] in 1963, using concentrated rabbit anti–whole human hemolysate, disclosed at least 12 antigenic constituents besides hemoglobin. Five were recognized as enzymes. In addition, four isoenzymes with LDH activity were encountered. Purified virus receptor substance, glycoprotein complex obtained by butanol treatment of stromal residue, and glycoprotein in aqueous extracts of stroma exhibited basic immunologic identity unrelated to blood groups. Besides M or N activity corresponding to the donor MN group, traces of blood group A activity were demonstrated in virus receptor substance and in butanol and aqueous fractions prepared from cells of group A_1. The blood group activity in NHEP may be glycolipid developed from the stroma during preparation of NHEP; however, the possibility of its being glycoprotein cannot be discounted, as pointed out in our studies with isolation of active principle in electrophoretograms.

In 1964 Rickli et al. [38] studied CA by electrophoretic analysis in starch gel from ethyl alcohol–chloroform extracts and designated two zones, B and C (previously called II and I respectively), traveling behind A_2 hemoglobin. The differences in nomenclature of CA were reconciled by agreement among the three laboratories doing most of the work, with acceptance of the designation of A, B, and C for the isoenzymes of CA (Table 11-4).

Bovine erythrocyte CA was reported by Gerov et al. [23] to vary with age, pregnancy, and bronchopneumonia. A variant of CA was noted in baboon erythrocytes by Barnicot et al. [3]. Bottini et al. [5, 6] disclosed 35 components in nonhemoglobin protein and human erythrocytes by two-dimensional starch-gel electrophoresis. They also noted some differences in NHEP electrophoretic patterns in thalassemic erythrocytes [5]. Fessas et al. [20] believe that uncombined α

TABLE 11–4

NOMENCLATURE AND PROPERTIES OF CARBONIC ANHYDRASE ISOENZYMES

Nyman et al.	V	III	II
Laurent et al. and Derrien et al.	Y	X_1	X_2
Rickli et al.	C	B	A
Enzyme activity	High	Low	Low
Protein staining	++	++++	+
Esterase activity	+	++++	+

chain in hemolysates travels in one of the zones of minor protein components, behind A_2 hemoglobin.

Erythrocyte Stroma Proteins

Erythrocyte stroma has been analyzed for proteins by starch-gel electrophoresis at acid pH after solubilization of delipidized membranes in urea and mercaptoethanol [1]. The reader must by this time be aware from the preceding historical review of the work done in the field of erythrocyte proteins that many authors used soluble stromal material from ghosts, and no clear distinction was made between nonstromal and stromal protein. The methods for preparation of NHEP may liberate some stromal proteins, but unquestionably others are too tightly bound and require more vigorous extraction methods [1, 33].

Azen et al. [1] recently reported on their findings with starch-gel electrophoresis of erythrocyte stroma. Erythrocyte membranes free of hemoglobin were extracted with ether repeatedly and then dried. The protein of the membrane was solubilized with aqueous urea and mercaptoethanol. Patterns were identical among 24 individuals. The investigators did note that one anodic band was increased if plasma proteins were involved. Fifteen or more protein bands were found, and the urea/aluminum lactate starch gel system proved superior for the task of revealing stromal proteins.

A number of investigators have studied the erythrocyte stromal proteins [2, 30, 35]. Stromatin, a protein component, was isolated and studied by Jorpes [30] and Ballentine [2]. Elinin, an ether extract from erythrocyte stroma, contains proteins, lipids, and carbohydrates [35]. ABO and Rh activity are present in elinin. Stromin, a similar glycolipid studied by Handa, also had blood group activity [25]. Marchise et al. [33] reported finding two proteins in erythrocyte stroma by disc electrophoresis, a major band and a minor band. An antiserum to the protein appears to react only with erythrocytes; it does not react with plasma protein or leukocytes. Their method of preparing ghosts is similar to others. The ghost proteins are solubilized by dialysis against adenosine triphosphate and 2-mercaptoethanol. The authors suggest the use of the term *spectrum* (Latin for ghost) for their stromal protein, but in view of the need to relate origin of protein to cell structure we have chosen to refer to proteins of this type as erythrocyte stroma proteins (ESP). This leaves room for future study of carbohydrates and lipids and complexes between lipid and protein in erythrocyte stroma. For example, the evidence suggests that lipoproteins are important membrane proteins. Bromelin, although forming a complex with cholesterol, also combines with delipidized lipoproteins isolated from erythrocytes [14].

Lymphocyte, Granulocyte, and Platelet Proteins

At this writing few studies are available on proteins of blood cells other than erythrocytes. As methods of preparing relatively pure populations of blood cells improve, a very active program of the proteins of lymphocytes, granulocytes, and platelets will be forthcoming. LDH isoenzyme studies of these cells are already well advanced (see Chap. 3).

References

1. Azen, E. A., Orr, S., and Smithies, O. Starch-gel electrophoresis of erythrocyte stroma. *J. Lab. Clin. Med.* 65:440, 1965.
2. Ballentine, R. Stromatin. *J. Cell. Comp. Physiol.* 23:21, 1940.
3. Barnicot, N. A., Jolly, C., Huehns, E. R., and Moor-Jankowski, J. A carbonic anhydrase variant in the baboon. *Nature* (London) 202:198, 1964.
4. Boivin, P., Hartmann, L., and Fauvert, F. Identification des protéines hydrosolubles des lysats d'hématiques humaines par l'électrophorèse à travers gel d'amidon et l'immunoélectrophorèse. *Rev. Franc. Étud. Clin. Biol.* 4:799, 1959.
5. Bottini, E., and Huehns, E. R. Peculiarities of the electrophoretic picture of non-hemoglobin protein fractions from thalassemic erythrocytes. *Boll. Soc. Ital. Biol. Sper.* 38:207, 1962.
6. Bottini, E., and Huehns, E. R. Electrophoretic analysis of the non-haemoglobin proteins of human red cells. *Clin. Chim. Acta* 8:127, 1963.
7. Cawley, L. P. Unreported studies, 1964.
8. Cawley, L. P. *Workshop Manual on Electrophoresis and Immunoelectrophoresis* (Rev. ed.). Commission on Continuing Education, Council on Clinical Chemistry. Chicago: American Society of Clinical Pathologists, 1966.
9. Cawley, L. P., and Eberhardt, L. Simplified gel electrophoresis: I. Rapid technic applicable to the clinical laboratory. *Amer. J. Clin. Path.* 38:539, 1962.
10. Cawley, L. P., and Eberhardt, L. Immunoelectrophoretic study of plasma proteins adsorbed by erythrocytes after treatment with tannic acid or bromelin. *Transfusion* 3:422, 1963.
11. Cawley, L. P., Eberhardt, L., and Goodwin, W. L. Agar-gel electrophoretic and analytic agar-gel electrophoretic study of bromelin. *Transfusion* 4:441, 1964.
12. Cawley, L. P., Eberhardt, L., Goodwin, W. L., Schneider, D., and Harrouch, J. Electrophoretic and zymographic study of non-hemoglobin erythrocyte proteins (NHEP). *Transfusion* 4:315, 1964.
13. Cawley, L. P., Eberhardt, L., Schneider, D., and Goodwin, W. L. Blood group substances in non-hemoglobin erythrocyte proteins (NHEP). *Transfusion* 5:375, 1965.

14. Cawley, L. P., and Goodwin, W. L.   Binding of bromelin by erythrocyte lipoprotein. *Fed. Proc.* 26:789, 1967.
15. Cawley, L. P., Goodwin, W. L., Schneider, D., and Eberhardt, L. Electrophoretic and immunologic studies of non-hemoglobin erythrocyte proteins (NHEP). *Fed. Proc.* 24:615, 1965.
16. Cerruti, M. P., Borrone, C., and Poggi, L.   Indagine immunoelettroforetica sulla costituzione antigenica degli stromi eritrocitari. *Pathologica* 54:398, 1962.
17. Derrien, Y., Laurent, G., and Borgomano, J.   Détection de la protéine d'accompagnement X, dans l'hémoglobine du nouveau-né. *C.R. Soc. Biol.* (Paris) 153:792, 1959.
18. Derrien, Y., Laurent, G., and Borgomano, J.   Sur une protéine accompagnant l'hémoglobine de l'homme adulte et sa concentration dans la fraction alcalino-résistante isolée de cette dernière. *C.R. Acad. Sci.* (Paris) 242:1538, 1956.
19. Derrien, Y., Laurent, G., Charrel, M., and Borgomano, M.   *Haemoglobin Colloquium,* Vol. 1. Stuttgart: Thieme, 1962.
20. Fessas, P., and Loukopoulos, D.   Alpha-chain of human hemoglobin occurrence *in vivo. Science* 143:590, 1964.
21. Fine, J. M., Boffa, G. A., Charrel, M., Laurent G., and Derrien, Y. Immunological and starch-gel electrophoresis studies of the carbonic anhydrases X_1, X_2 and Y isolated from human erythrocyte haemolysates. *Nature* (London) 200:371, 1963.
22. Fiorelli, G., Mannucci, P. M., and Lomanto, B.   Preliminary investigations of non-hemoglobin erythrocyte proteins. *Atti Accad. Med. Lombard.* 18:405, 1963.
23. Gerov, K., and Georgieva, R.   Carbonic anhydrase activity of calf blood according to age and health. *Chem. Abstr.* 63:6150, 1965.
24. Giri, K. V., Sathyanarayana, R., and Natarajan, S.   Agar electrophoretic patterns of red blood cell proteins. *Nature* (London) 182:184, 1958.
25. Handa, S.   Blood group active glycolipids from human erythrocytes. *Jap. J. Exp. Med.* 33:347, 1963.
26. Haut, A., Cartwright, G. E., and Wintrobe, M. M.   Electrophoresis of non-hemoglobin proteins from concentrated hemoglobin-free hemolysates. *J. Lab. Clin. Med.* 63:279, 1964.
27. Haut, A., Tudhope, G. R., Cartwright, G. E., and Wintrobe, M. M. The non-hemoglobin erythrocytic proteins studied by electrophoresis on starch gel. *J. Clin. Invest.* 41:579, 1962.
28. Heller, P., Yakulis, V. J., and Josephson, A. M.   Immunologic studies of human hemoglobins. *J. Lab. Clin. Med.* 59:401, 1962.
29. Howe, C., Avrameas, S., de Vaux St. Cyr, C., Grabar, P., and Lee, L. T.   Antigenic components of human erythrocytes. *J. Immun.* 91: 683, 1963.
30. Jorpes, E.   The protein component of the erythrocyte membrane or stroma. *Biochem. J.* 26:1488, 1932.
31. Laurent, G., Murriq, C., Nahon, D., Charrel, M., and Derrien, Y. Isolement des protéines "lentes" Y_1 X_1 et X_2 accompagnant l'hémoglobine humaine dans ses préparations. *C.R. Soc. Biol.* (Paris) 156: 1456, 1962.

32. Lindskog, S. Purification and properties of bovine erythrocyte carbonic anhydrase. *Biochim. Biophys. Acta* 39:218, 1960.

33. Marchise, V. T., and Steers, E., Jr. Selective solubilization of a protein component of the red cell membrane. *Science* 159:203, 1967.

34. Micheli, A., and Grabar, P. Étude immunochimique des hémolysates de globules rouges humains: IV. Les estérases carboxyliques et une protéase. *Ann. Inst. Pasteur* (Paris) 100:569, 1961.

35. Moshowitz, M., and Calvin, M. On the components and structure of the human red cell membrane. *Ex. Cell Res.* 3:33, 1952.

36. Nyman, P. O. Purification of carbonic anhydrase from human erythrocytes. *Biochim. Biophys. Acta* 52:1, 1961.

37. Ornstein, L., and Davis, B. J. *Disc Electrophoresis*, Parts I and II. Rochester, N.Y.: Distillation Products Industries, 1961.

38. Rickli, E. E., Fhazanfar, S. A. A., Gibbons, B. H., and Edsal, J. T. Carbonic anhydrase from human erythrocytes, preparation and properties of two enzymes. *J. Biol. Chem.* 239:1065, 1964.

39. Rieder, D. F., and Weatherall, D. J. A variation in the electrophoretic pattern of human erythrocyte carbonic anhydrase. *Nature* (London) 203:1364, 1964.

40. Rose, N., Peetoom, F., Ruddy, S., Micheli, A., and Grabar, P. Étude immunochimique des hémolysates des globules rouges humaines: I. Principaux constituants antigéniques. *Ann. Inst. Pasteur* (Paris) 98:70, 1960.

41. Shaw, C., Syner, F. N., and Tashian, R. E. New genetically determined molecular form of erythrocyte esterase in man. *Science* 138:31, 1962.

42. Tappan, D. V., Jacey, M. J., and Boyden, H. M. Carbonic anhydrase isoenzymes of neonatal and adult human and some animal erythrocytes. *Ann. N.Y. Acad. Sci.* 121:589, 1964.

43. Wiener, A. S. *Blood Groups and Transfusion*. Springfield, Ill.: Thomas, 1948. Pp. 272–294.

12

ANALYTICAL TECHNICS IN ELECTROPHORESIS AND IMMUNOELECTROPHORESIS

As each clinical laboratory becomes more involved in analyzing serum proteins and body fluid proteins by electrophoresis and immunoelectrophoresis, it may be well to go beyond the usual technics to search out new ones and actually plan modifications of one's own design. In this chapter it seemed logical to include a few of the technics which we have found valuable but which are not necessarily in use from day to day. It is also the purpose of this chapter to delve a little into what lies in the future and touch upon some of the important problems of biology that can be solved by an electrophoretic and an immunoelectrophoretic approach.

Disc Immunoelectrophoresis

Some years ago we became interested in purifying human chorionic gonadotropin (HCG), and early in our effort we attempted to define the position of the biologically active component in disc electrophoretograms using antiserum made against a somewhat impure HCG [16, 17]. Figure 12-1 shows two approaches. In the first, a so-called continuous immunoelectrophoresis, the disc electrophoretogram is sliced longitudinally, making two halves, one for protein stain and the other for immunoelectrophoresis. The latter is placed on a glass plate or plastic strip and covered by agar for the diffusion phase. Trenches dug on either side are used for antiserum. The second, the segmental system, is based upon segmenting the disc electrophoretogram along the lines of individual bands or at some standard specified segment. Eluates of some segments may be tested for biologic activity. In Figure 12-1 HCG was found by a test with male frogs to be in segment 3. Corresponding segments may be set to form a circle on glass or plastic and covered with agar or agarose. A central well serves as the source of antiserum.

Figures 12-2 and 12-3 represent results of continuous and segmental applications of disc immunoelectrophoresis to the study of

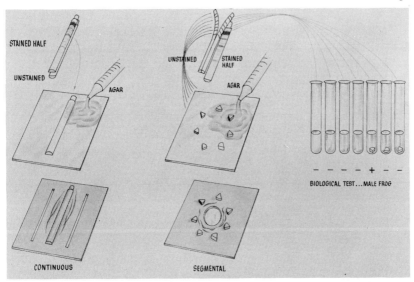

FIGURE 12-1

Drawing showing two types of disc immunoelectrophoresis, continuous and segmental. On the right, the fragments are shown in test tubes. The results of biologic testing for HCG in the frog are indicated beneath as + or —. Fraction 3 shows biologic activity in the frog.

HCG [16]. The continuous approach (Fig. 12-2) was set up against a number of known commercially available HCG preparations. Each was then tested against goat anti–normal human serum and rabbit anti-HCG. All HCG preparations were free of serum proteins, but each disclosed two bands by anti-HCG. Because of the restriction in the pore size of acrylamide, diffusion from the gel is very poor, as may be seen in the upper pattern, where normal human serum was tested against anti–normal human serum. Only one large band is present; it is albumin, showing tailing from the point of application. The fact that no other proteins are seen indicates trapping of larger molecules within the gel. The marked tailing of albumin is a reflection of the trapping of proteins by the gel. This property of the gel can be an asset in evaluating molecular size of a particular protein.

The segmental study is somewhat more informative (Figure 12-3). The small segments, labeled 1 through 11, placed radially about the central well serve several useful functions in this type of study. From Figure 12-1 the biologic activity is seen in the third tube from the right. This corresponds to segment 3 of Figure 12-3, where a large broad band is seen commencing between positions 2 and 3 and extending beyond position 4 and toward 5. A second large band commencing at 3 and extending to 9 is also seen, and a minor component at positions 10 and 11. It is clear that there are at least three separate

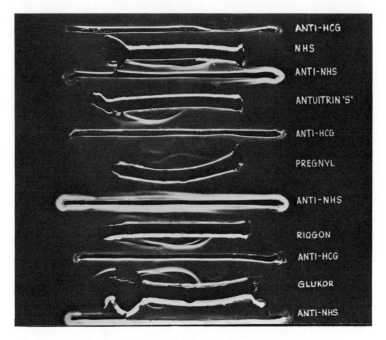

FIGURE 12-2

Series of tests against a number of commercial preparations of HCG. Anti–normal human serum (NHS) and anti-HCG are employed against each preparation. NHS also is tested with the same antiserum (*top*). No activity is noted against NHS by anti-HCG. See text for discussion. Anode is on the right.

components, one of which corresponds to the biologically active component and is the major immunoprecipitin band midway between the antigen segment, 3, and the antiserum well. To be certain that HCG is incorporated within the band, it should be tested for biologic activity in the frog. The technics involved are allied to the Echo technic introduced by Nace [25]. The band is first dissected free from the electrophoretogram and the protein eluted into low- or high-pH fluid. At these two pH extremes the antigen-antibody complex disassociates and will not re-form unless the pH is adjusted to neutral. On injection of this material into a frog, a desired response should develop, although some inactivation of HCG by anti-HCG recombining in the lymph sac may take place.

The Echo technic [25] was designed to isolate, from a mixture of experimentally produced antitissue antibodies, a single antitissue antibody and pinpoint its corresponding tissue antigen by fluorescent microscopy. The individual bands, along with a little agar, are removed, and each is placed on the surface of a frozen section of the

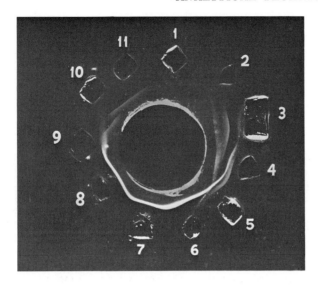

FIGURE 12-3

Results of segmental immunoelectrophoresis from HCG preparation of disc electrophoretogram as depicted on the right of Figure 12-1

tissue which was used to make the antibodies. Lowering or raising the pH causes the antibody-antigen complex to disassociate. The antiserum freely diffuses into the frozen section where, after a suitable period of time, the pH is adjusted to neutrality. Afterward the fluorescent-labeled antiserum combines with specific tissue antigen and can be localized by fluorescent microscopy.

The same principles of immunoelectrophoresis outlined above are applicable to gels of all types including agarose and agar gel. In the case of agar gel or agarose the transverse segments of agar or agarose may be shaped into small rolls and then either set in agar plates, extracted to release protein for testing, or placed in a disc electrophoretogram for electrophoresis [5, 6, 8]. Such a procedure was used in the study of the effects of bromelin on albumin. Two components were easily separated by this technic and studied separately with specific antiserum against albumin (see Chap. 9).

Segment Analysis for Nonimmunologic Components

Removal of segments of an electrophoretogram followed by extraction of lipids and chromatographic studies by gas-liquid and thin-layer methods disclosed that cholesterol is the dominant lipid regardless of where in the electrophoretogram the fraction is taken; however, it is most abundant in the β-lipoprotein zone (see Chap. 4) [15]. Figure 12-4 schematically outlines a method of obtaining fractions

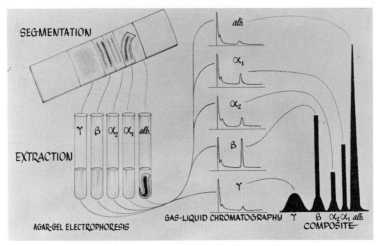

FIGURE 12-4

Outline of steps in analyzing segments corresponding to protein zones
for cholesterol by gas-liquid chromatography and relating the information
to scan of electrophoretogram [15]

from an electrophoretogram and determining cholesterol on each by
gas-liquid chromatography.

Analysis of Patterns of Electrophoretograms

In the clinical laboratory analysis of the electrophoretogram by
gross inspection is often sufficient. In some pathologic situations such
inspection usually is enough; however, a densitometric scan is often
helpful. This can take many forms. For example, 35 mm. photographs
of a disc electrophoretogram can be mounted in the microscanning
attachment of the Analytrol to obtain a useful scan.

Figure 12-5 shows a scan and corresponding disc electrophoretogram from a study of nonhemoglobin erythrocyte protein [6]. Although the scan does not add greatly to the overall pattern for the
purpose of interpretation, it does permit a method of quantitation.
Besides densitometric scanning and fluorometric methods of scanning,
segmental analysis by cutting stained zone electrophoretograms into
segments corresponding to the six protein zones and determination
of the amount of dye present in each zone by use of Autoanalyzer or
Robot Chemist is most applicable for large numbers of electrophoretograms.

A flying spot scanner has been proposed as a method of obtaining
rapid readout for electrophoretograms [27]. Figure 12-6 is a photograph of such an instrument made in the author's laboratory. It consists of cathode-ray tubes so arranged as to scan the transparent agar
or agarose electrophoretograms after staining. The interruption of the

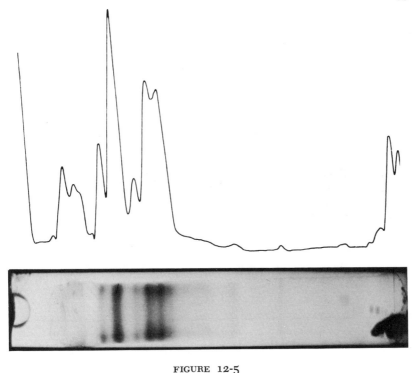

FIGURE 12-5

Analytrol scan of 35-mm. color slide. Disc electrophoretogram of non-hemoglobin erythrocyte proteins (NHEP). See text and Acrylamide Electrophoresis (in Chap. 15) for details. Anode is on the right. (From L. P. Cawley [2].)

scanning beam is then picked up by a photomultiplier tube, which drives a second cathode-ray tube. The latter generates the curves seen in Figure 12-7. The upper portion is the integrated curve, the height of each peak representing the area under each fraction of the scan directly beneath. The actual electrophoretogram is at the base of the photograph. Analytrol scan of this electrophoretogram is not unlike that seen in the flying spot scan. Albumin, α_1-, α_2-, β-, and γ-globulin are identified. A small peak is seen at the beginning of γ-globulin which represents a γM monoclonal peak. The scan shows six protein zones, including the monoclonal peak, which correspond to the six integrated curves. The flying spot scanner was equipped with a number of operational amplifiers, permitting discrimination of the beginning and end of each peak for the purpose of integration. Although the instrument is operational and is extremely fast, the readout is a photograph; however, visual inspection of patterns is also available. Modifications of such a device probably could result in a rapid readout system for the clinical laboratory.

FIGURE 12-6

Flying spot scanner developed in the author's laboratory. Scan as shown in Figure 12-5 is visualized through the circular opening on the upper right side. The black box on the shelf to the right houses the photomultiplier tube. Strips are fed into the slot at the base of the photomultiplier housing. The photomultiplier tube then drives a second cathode-ray tube, which generates a scan.

Multiple Electrophoretic and Immunoelectrophoretic Studies

Sometimes studies are run side by side for comparative reasons —studies of fractions from a liquid chromatographic column, for instance. An attempt to define the amount of serum proteins absorbed to bromelin-treated erythrocytes was carried out on a large sheet of plastic in such a way that direct comparison between various parts of the study could be easily visualized (Fig. 12-8) [9]. For example, cells were washed with saline solution in the first set of tests, then treated with tannic acid or bromelin either in saline or in serum; finally, heat eluates, supernatant, or cells were tested against rabbit anti–human erythrocyte serum (Fig. 12-8) [4, 13]. SS represents saline supernatant, SeSS represents serum saline supernatant, and SC represents saline-treated cells. SE stands for saline eluate. The cells were hemolyzed either by freezing with distilled water or by sonifica-

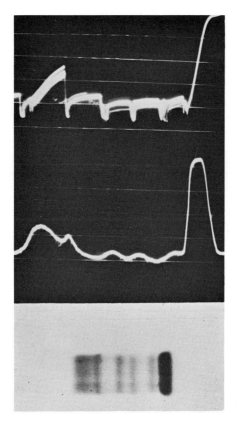

FIGURE 12-7

Scan from cathode-ray tube of flying spot scanner shown in Figure 12-6. The upper curve is the integration curve and the bottom curve is the scan. The stained electrophoretogram is shown at the bottom of the photograph. Anode is on the right.

tion. At the bottom of Figure 12-8 the S stands for serum. Against serum the rabbit anti–human erythrocyte serum detects albumin and several α_2-globulins, several β-globulins, and a single γ-globulin, presumably γG. No γA or γM is present [11, 13]. Bromelin supernatant (BS) or saline supernatant (SS) results in release of A_1 hemoglobin, which is depicted as a single arc by the antiserum. In the case of the saline cells (SC) no γG is present. In contrast, bromelin cells (BC) do contain γG and other components not present in the serum control. The amount of albumin is also increased.

The variability of reaction of several individuals with the same antiserum is shown in Figure 12-9. Here the reaction of red cell hemolysates against rabbit anti–human erythrocyte serum is compared to

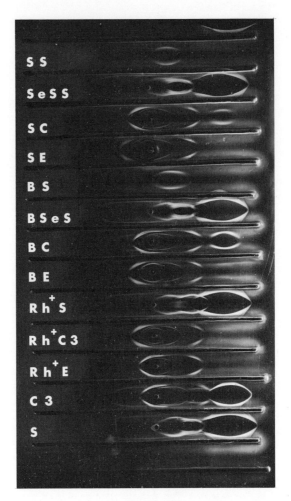

FIGURE 12-8

Sheet of immunoelectrophoretograms developed with rabbit anti–human erythrocyte serum. Anode is on the right. See text for details.

their reaction against cells of five individuals and normal human serum. The very dense band is against hemoglobin [13]. The band to the right is albumin. The red cells were washed six times with 12 volumes of saline solution each time. The hemolysate of the washed cells was tested against rabbit anti–human erythrocyte serum and, as shown (Fig. 12-9), in some instances albumin could still be identified as a component in the hemolysate. In rare instances γ-globulin could be identified. The red cells of individuals or samples tested are, from top to bottom, albumin (not labeled), autoimmune disease

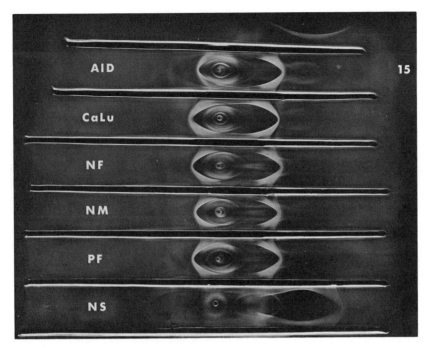

FIGURE 12-9

Sheet of immunoelectrophoretograms in a comparative study of one anti–human erythrocyte serum against hemolysate prepared from thoroughly washed human erythrocytes from five individuals. Antiserum was also compared against normal serum (NS). The antiserum was obtained from rabbit 15. From top to bottom patients had the following distinction: autoimmune disease (AID), female; carcinoma of the lung (CaLu), male; normal female (NF); normal male (NM); and pregnant female (PF). Note that antiserum detects presence of albumin, α_1- and α_2-globulins, and γ-globulin in normal serum. There is increasing concentration of γ-globulin and albumin erythrocytes in individuals with certain diseases such as AID and CaLu.

(AID), carcinoma of the lung (CaLu), normal female (NF), normal male (NM), pregnant female (PF), and normal serum (NS). Close observation of these patterns shows several subtle differences. Also, each of several antisera showed different and distinct patterns. The antigenic composition of whole human erythrocytes is very complex, and more studies using standardized procedures are needed to clarify the nature of the protein composition of erythrocytes. Some findings can be stated clearly: Albumin does appear to be intimately associated with the erythrocyte in some instances, usually in cells from patients with lymphoproliferative or autoimmune disease [13]. Further, albumin is found in nonhemoglobin erythrocyte protein (NHEP) (see

Chap. 11), which may mean that it is not exclusively attached to the surface of the cell but may be located within the cell as well. However, tannic acid and bromelin treatment modify the red cell causing uptake from serum of albumin and γG (see Fig. 12-8). Other plasma proteins may also be involved.

Addition of Substances to Protein or Gel

To increase resolution or for specific studies some component may be added to the agar or attached to the proteins and the effect on the electrophoretic mobility of protein examined. Starch added to agar or agarose slightly reduces the pore sizes, with slight improvement in resolution; the starch also can serve as substrate for amylase (see Chap. 8). Addition of formaldehyde to the protein or the agar markedly influences electrophoretic mobility of γ-globulin but has little effect upon the electrophoretic mobility of albumin. This finding implies that there are more peptide groups in the γ-globulin than in albumin.

Another approach is to set up a system as one would a closed system of immunoelectrophoresis (see Chap. 15, under Standard Method for Immunoelectrophoresis) but instead of using an antigen-antibody system, substitute other reagents. This approach is demonstrated in Figure 12-10: Polybrene, polyglycine, and polylysine are placed in

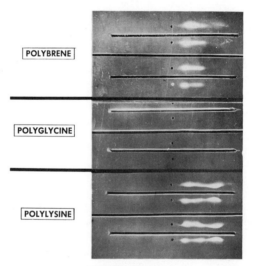

FIGURE 12-10

Polybrene added to the trenches and permitted to diffuse out toward separated normal human serum by electrophoresis shows precipitation in the zones occupied by α_1-, α_2-, and β-globulin, whereas polylysine in the bottom portion of the photograph depicts albumin as well. Polyglycine is without activity. In no instance did Polybrene or polyglycine depict γ-globulin.

the antiserum trenches and permitted to diffuse inward toward serum protein which has been separated by electrophoresis. Polybrene and polylysine cause precipitation of the proteins from albumin back to the point of application. However, γ-globulins are not precipitated. Likewise a reverse of the situation is possible wherein the other reactant is placed in the antigen well and separated by electrophoresis and reacted against normal human serum. Separation of bromelin followed by diffusion of normal human serum in from the side results usually in formation of two precipitin zones, one at the point of application and one on the cathodic side matching the reactive zone of bromelin [12]. If normal human serum is subjected to electrophoresis followed by diffusion against bromelin from the antisera trenches, there is a zone of action depicted in the β-globulin area of the serum protein. Lipid-staining discloses positive reaction indicating that the serologic active principle of bromelin combines with serum β-lipoprotein [12]. By electron microscopy bromelin has been shown to combine with erythrocyte lipoprotein [9, 10].

These observations are of some significance biologically since it has been shown that the extract of *Phaseolus vulgaris* also combined with β-lipoprotein [24]. Recent demonstration of the reactions between lectins and only a certain portion of monoclonal peaks points up an example in which the plant protein reacts with a carbohydrate moiety of serum protein (glycoprotein) rather than a serum lipoprotein [22]. Earlier studies utilizing the approach outlined above demonstrated that certain plant extracts reacted very strongly with macroglobulins [20].

The idea of studying specific markers or proteins by gel diffusion and immunoelectrophoretic principles using materials not normally reacted with serum proteins is very appealing. α_2-Globulin neutralizes the hemagglutinin activity of *Maclura pomifera* and, since this lectin is neutralized by N-acetylgalactosamine and galactase, α_2-globulin must, therefore, contain these sugars as glycoproteins [14, 21]. The extract from the bean *Phaseolus vulgaris* has a strong hemagglutinin which is likewise neutralized by N-acetylgalactosamine. Some years ago, in trying to compare the antigenicity of seeds in reference to the pulp of certain foods, an attempt was made to determine whether or not the seeds were better antigens for skin testing to food allergies than pulp by using antisera prepared in rabbits against seeds and pulp [12]. When checking the antisera and the preimmunized rabbit sera against the seed or pulp we were surprised to find that the original serum reacted with extracts of seed and pulp. The immunized sera did contain bands in excess of those present before immunization. Saline extracts made of a variety of seeds and tested for proteolytic activity, agglutinating activity, and formation of precipitin bands in gel diffusion plates against human serum yielded a host of very significant bands (Table 12-1). In some instances bands between plant

TABLE 12-1

STUDY OF SERUM PROTEIN REACTIONS WITH PLANT SEED EXTRACTS

Seed Extract	Agg. Coated Cells	Gel Diffusion Bands with Human Serum	Proteolytic Activity
Chocolate	−	5	0
Rice	−	1	0
Grape	−	7	0
Almond	−	1	0
Peach	−	1	0
Soy bean	−	1	0
Bromelin	+	1	+
Bean (Great Northern)	+[a]	0	0
Apple	+[a]	2	0
Maclura pomifera	+[a]	0	0
Ulex europaeus	+ (H)	1 NHS, 2 LES	0

[a] Agglutinates coated (incomplete anti-D) and normal cells.

NOTE: H = hemolysis; NHS = normal human serum; LES = lupus erythematosus serum.

proteins in addition to bands with plasma proteins were detected (Fig. 12-11). Extract of chocolate and grape reacted with serum from human, horse, rabbit, lamb, chicken, pig, and cow [12]. No reaction

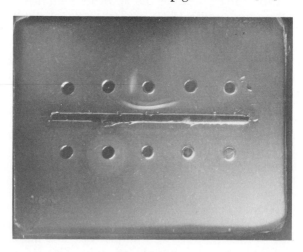

FIGURE 12-11

An immunodiffusion plate comparing plant seed extracts diffused from the wells against normal human serum. Ten plant proteins were tested. Only the third one from the right at the top caused activity. This is from grape seed and gave rise to a series of bands as depicted. It also reacted with the adjacent plant protein (squash). See Table 12-1 and text for details.

was detected against lactalbumin, egg white, or bovine albumin. Spurious bands which sometimes form in gel diffusion plates may be a reaction between two proteins such as hemoglobin and albumin or, in the examples cited above, between plant products (not necessarily protein) and human serum protein. The bands are not immunologic reactions but reactions causing bands with the same shape and form as immunoprecipitin bands.

Histochemical Reactions on Immunoelectrophoretograms

Application of histochemical reactions to nonimmunoprecipitin bands, described above, or bands of immunologic origin has resulted in some interesting findings. An investigation using an antiserum against glucose oxidase prepared in our laboratory will suffice to demonstrate the technic and the findings. Anti–glucose oxidase was studied from the standpoint of whether or not it caused neutralization of glucose oxidase and whether or not it would react with galactose [7]. We also were interested in a study of substrate specificity of the glucose oxidase using various substrate overlays or various substrates incorporated into the agar or agarose. By immunoelectrophoresis the rabbit anti–glucose oxidase described three components against glucose oxidase (Fig. 12-12). Band 1 corresponds to the electrophoretic mobility of the zone with enzyme activity [7] and contains glucose oxidase activity (Fig. 12-13). This was determined by first removing all traces of nonreacted glucose oxidase by soaking the immunoelectrophoretogram for 72 hours in a saline wash and then applying substrate (glucose) to the gel surface along with peroxidase and o-dianisidine. Failure of an antiserum to neutralize its enzyme antigen is not unusual and has prevented the use of enzyme antibody reaction to describe the sites of specific activity on the enzyme molecule. On the other hand, the persistence of the activity of the enzyme after combination with the antibody can serve as a marker to determine titers of the antibody.

A recent study used lactate dehydrogenase (LDH) activity for determination of the strength of the anti-LDH$_1$ and anti-LDH$_5$. It

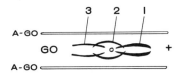

FIGURE 12-12

Drawing of immunoelectrophoretogram developed with rabbit anti–glucose oxidase (A-GO). Three bands are shown: 1, 2, and 3. Enzyme activity was disclosed to rest with position occupied by immunoprecipitin band 1.

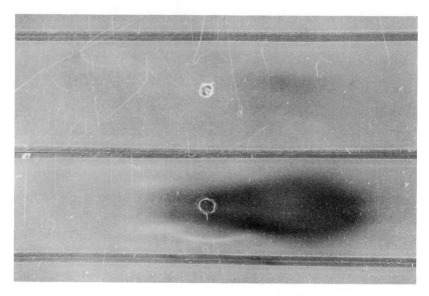

FIGURE 12-13

Photograph of actual gel immunoelectrophoretogram of glucose oxidase (*bottom*), and galactose oxidase (*top*). Both antigens were reacted against rabbit anti–glucose oxidase. Overlay staining technic utilizing glucose as substrate discloses that the band corresponding to position 1 of Figure 12-11 contains glucose oxidase activity; bands 2 and 3 are negative. Note slight trace of enzyme activity with galactose. The antiserum did not describe a detachable immune precipitin band against galactose oxidase in the gel. The enzyme stain, however, did bring out that a small amount of protein traveling in essentially the same position as glucose oxidase enzyme is present in the preparation of galactose oxidase and reacts with glucose. It apparently was trapped in the gel by the glucose oxidase antiserum, since it was not removed following a suitable period of washing in saline solution. Anode is on the right.

was carried out by making serial dilutions of the antisera, reacting it with LDH_1 and LDH_5, and after incubation submitting the mixture to electrophoresis and staining the completed electrophoretogram for the enzyme [28]. The enzymes coupled to proteins for the purpose of depicting an end point may bear further exploration [3].

Solid-Phase Labeling

Fluorescein isothiocyanate (FITC) chemically coupled to antibodies is a well-established, chemically sound, and sensitive method of labeling protein [26, 30]. The principle of chemically coupling a detector substance to protein seemed well suited to serum proteins fixed in a dry gel matrix of an electrophoretogram. Agarose or agar-

gel electrophoretograms on single strips or multiple electrophoretograms on sheets of thin plastic (Mylar, Cronar, etc.), according to the
method developed in our laboratory, are fixed and dried [2, 3]. The
protein in the dry fixed electrophoretograms is coupled to FITC. The
final product shows strong fluorescence of all protein fractions and
compares favorably with that of standard methods of staining electrophoretograms with Thiazine Red R or Amido Black B stains. Following electrophoresis the strip is fixed in alcohol, dried, and coated
with the coupling agent (in this case fluorescein in carbonate buffer
pH 9 for 10 minutes). The excess fluorescein is removed by washing.
The wet strip is scanned in a recording fluorometer (a Turner III
coupled to a Photovolt recorder) [3].

The results of 20 serum electrophoretograms examined by usual
Thiazine Red R staining method and by fluorescent labeling are sum-
marized below. A normal serum is shown in Figure 12-14. The thiazine stain is on the right; the fluorescein is on the left. Figure 12-15
shows the scans of serum patterns from a patient with agammaglobulinemia. On the right is the Thiazine Red R stain and on the left the

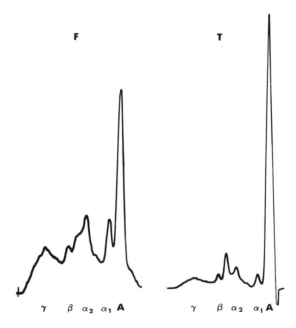

FIGURE 12-14

Two scans, Thiazine Red R (T) on the right and fluorescein (F) on the
left. Patterns are from a normal serum and show in general the features of
the fluorescein stain compared to Thiazine Red R. Albumin is less prevalent,
β_1 appears less prevalent, and γ-globulin appears more prevalent in the
fluorescein.

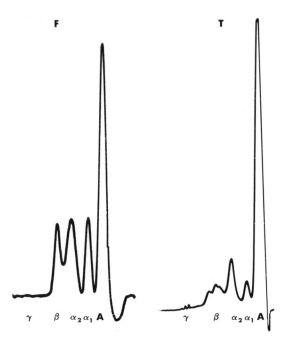

FIGURE 12-15
Serum patterns from a patient with agammaglobulinemia. Note the increased quantity of α_1 in the fluorescein stain preparation and the increase of α_2 fraction; γ-globulin in both instances is negative. Anode is on the right.

fluorescein stain. Both show high concentrations of albumin and a complete absence of γ-globulin. The differences in the scans shown here are quite general. In comparison to Thiazine Red R, fluorescein appears to have proportionally less binding to albumin and a greater affinity for the globulin fractions, particularly α_1-, α_2-, and γ-globulins. β-Globulin does not react with fluorescein stain to the same extent as the other globulins.

Reviewing the work of Strickland et al. [32], who weighed the various fractions from agar-gel electrophoretograms, one would be led to conclude that in a normal population approximately 63–64 percent of the serum protein is made up of albumin and approximately 14 percent consists of γ-globulin. The Thiazine Red R stain matches this particular figure relatively closely, whereas the fluorescein stain shows a disproportionate increase of the globulins. The reason will be discussed later.

A very unusual electrophoretogram from a patient (also discussed in Chap. 4) with long-standing rheumatoid arthritis who was admitted with rectal abscesses (Figs. 4-3, 4-4, 4-5, and 4-6) is seen in Figure 12-16. A monoclonal peak in the post-γ-globulin area was

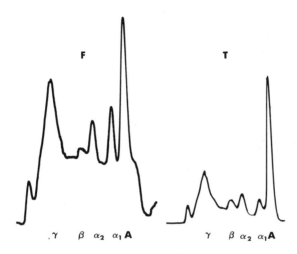

F T

γ β α₂ α₁ A γ β α₂ α₁A

FIGURE 12-16

Electrophoretogram from a patient with rheumatoid arthritis. Note the small post-γ monoclonal peak on the left of both scans, the decreased albumin in both scans, but the greater absolute increase detected by the fluorescein of the γ-globulin fractions and of the α_1 fraction as compared to the thiazine stain. Anode is on the right.

delineated on the first electrophoretogram processed. Our interests were: Was this a γ-globulin fraction? Was it related to the patient's disease? Would it take the fluorescein label? The scans for fluorescence and Thiazine Red R of the patient's electrophoretogram are shown in Figure 12-16. Notice that the patient did have an increased concentration of γ-globulin by the Thiazine Red R stain and a decrease in the albumin concentration. Also note the position of the postmonoclonal peak. Compare this now with the fluorescein stain. There is rather marked accentuation of the globulin fraction and decreased concentration of albumin. The α_1 and α_2 are also elevated by the fluorescein method, and the β-globulin fraction seems to be decreased. As can be seen, the small monoclonal peak was delineated by the fluorescein method.

Table 12-2 gives the percent of total protein in each fraction as determined by Thiazine Red R, fluorescein, and gravimetric analysis. Since Thiazine Red R couples to proteins by adsorption, one might expect the stain concentration to be roughly proportional to the amount of protein present. Such a relationship does exist, as can be seen from the data. The coupling of FITC with proteins, however, involves actual bond formation of the thiocyanate with specific groups in the protein molecule. The fluorescein concentration would then be pro-

TABLE 12-2

PERCENT DISTRIBUTION OF PROTEINS

Fraction	Thiazine	Weight	FITC
Alb.	57.5	64.6	33.5
α_1	4.5	6.1	10.4
α_2	11.0	11.3	18.1
β	11.5	5.0	16.2
γ	15.5	13.2	21.8

portional to the number of nonsterically-hindered active sites in the molecule, which varies in the different protein fractions. This active site is believed to be that of the ε-amine groups of lysine, but other functional groups present can also undergo thiocyanate reactions. FITC is far more efficient in labeling the globulin fractions than albumin. This disproportional tagging is the primary advantage to this method since the fractions normally present in the smallest concentrations are the fractions most efficiently tagged.

Other detector systems have been tried with modest success. They include labeling of electrophoretograms with ferritin and peroxidase. The concept of solid-phase coupling to augment detection and quantification of proteins in electrophoretograms appears to be an area worthy of further exploration [3]. This subject is discussed further in the Appendix.

Molecular Sieve Electrophoresis

It is not the purpose of the present text to describe in detail the use of this form of electrophoresis since it is primarily a research tool. However, its methods are becoming useful in restricted areas in the laboratory, particularly in the laboratory that desires to investigate in more detail protein components from serum or cells (see Chap. 11). Smithies' classic method of starch-gel electrophoresis is excellent [31], and its counterpart in acrylamide electrophoresis is the technic developed by Raymond et al. [29]. A microstarch electrophoretic method developed by Marsh, Jolliff, and Payne [23] is shown in Figure 12-17. The patterns are of polyclonal and monoclonal gammopathies and show a number of components not seen in zone electrophoretic patterns (see Figs. 2-8 and 2-9 in chapter on monoclonal, polyclonal, and dysclonal gammopathies). The disc electrophoresis technic developed by Ornstein [27] with slight modification in our laboratory [2] is discussed in Chap. 15 under Acrylamide Electrophoresis.

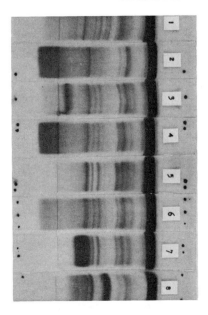

FIGURE 12-17

Seven serum proteins separated by a microstarch-gel technic developed by Marsh, Jolliff, and Payne [23]. (1) Normal human serum, (2) cirrhosis of liver, (3) macroglobulinemia, (4) cryoglobulinemia, (5) agammaglobulinemia, (6) lupus erythematosus, (7) γG myeloma, (8) γA myeloma. Anode is on the right. (Photograph courtesy of Carl Jolliff, Lincoln, Nebraska.)

References

1. Borberg, H., Woodruff, J., Hirschhorn, R., Gesner, B., Mieschee, P., and Silber R. Phytohemagglutinin: Inhibition of the agglutinating activity by N-acetyl-D-galactosamine. *Science* 154:1019, 1966.
2. Cawley, L. P. *Workshop Manual on Electrophoresis and Immunoelectrophoresis* (Rev. ed.). Commission on Continuing Education, Council on Clinical Chemistry. Chicago: American Society of Clinical Pathologists, 1966.
3. Cawley, L. P., Dibbern, P., and Eberhardt, L. Fluorescent labeling of electrophoretograms by fluorescein isothiocyanate. *Clin. Chem.* 13: 701, 1967.
4. Cawley, L. P., and Eberhardt, L. Immunoelectrophoretic study of plasma proteins absorbed by erythrocytes after treatment with tannic acid or bromelin. *Transfusion* 3:422, 1963.
5. Cawley, L. P., Eberhardt, L., and Goodwin, W. L. Agar-gel electrophoretic and analytic agar-gel electrophoretic study of bromelin. *Transfusion* 4:441, 1964.
6. Cawley, L. P., Eberhardt, L., Goodwin, W. L., Schneider, D., and Harrouch, J. Electrophoretic and zymographic study of non-hemo-

globin erythrocyte proteins (NHEP). *Transfusion* 4:315, 1964.

7. Cawley, L. P., Eberhardt, L., and Musser, B. O. Zymographic and gas-lipid chromatographic study of substrate specificity of glucose oxidase. *Amer. J. Clin. Path.* 47:366, 1967.

8. Cawley, L. P., Eberhardt, L., Schneider, D., and Goodwin, W. Blood group substances in non-hemoglobin erythrocyte proteins (NHEP). *Transfusion* 5:375, 1965.

9. Cawley, L. P., and Goodwin, W. L. Immunochemical study of reaction between erythrocytes and bromelin. *Fed. Proc.* 25:236, 1966.

10. Cawley, L. P., and Goodwin, W. L. Binding of bromelin by erythrocyte lipoprotein. *Fed. Proc.* 26:789, 1967.

11. Cawley, L. P., Goodwin, W. L., Schneider, D., and Eberhardt, L. Electrophoretic and immunologic studies of non-hemoglobin proteins (NHEP). *Fed. Proc.* 24:615, 1965.

12. Cawley, L. P., and Hale, R. Unpublished studies, 1961.

13. Cawley, L. P., Houser, C. P., and Riner, A. Simplified gel immuno-electrophoresis and some applications in the study of hemolysates of erythrocytes. *Transfusion* 3:148, 1963.

14. Cawley, L. P., Jones, J. M., and Teresa, G. W. The distribution and characterization of the erythrocyte receptors in man and animals for lectins from *Maclura pomifera*. *Transfusion* 7:343, 1967.

15. Cawley, L. P., Musser, B. O., and Eberhardt, L. Gas-liquid chromatographic analysis of cholesterol in fractions of serum protein separated by means of agar-gel electrophoresis. *Amer. J. Clin. Path.* 40:429, 1963.

16. Cawley, L. P., Sanders, J. A., and Eberhardt, L. Disc electrophoretic and disc immunoelectrophoretic analysis of human chorionic gonadotropin (HCG). *Amer. J. Clin. Path.* 40:429, 1963.

17. Cawley, L. P., Tretbar, H. A., Houser, C. P., and Riner, A. Immunologic Comparison Between Menopausal Gonadotropins and Human Chorionic Gonadotropin. Delivered at the Joint Annual Meeting of the American Society of Clinical Pathologists and College of American Pathologists, October 4–5, 1961.

18. Cawley, L. P., Wiley, J. L., Schneider, D., and Harrouch, J. Immunoelectrophoretic and immunochemical characterization of antibodies against bromelin. *Transfusion* 4:316, 1964.

19. Hale, R., Cawley, L. P., Holman, J. G., and Minard, B. Radio-immunoelectrophoretic studies of antigen binding capacity of atopic sera. *Ann. Allerg.* 25:88, 1967.

20. Harris, H., and Robson, E. B. Precipitin reactions between extracts of seeds of *Canavalia ensiformis* (Jack Bean) and normal and pathological serum proteins. *Vox Sang.* 8:348, 1963.

21. Jones, J. M., Cawley, L. P., and Teresa, G. W. The lectins of *Maclura pomifera*, zymographic studies, distribution in the developing plant and production in tissue cultures of epicotyls. *J. Immun.* 98:364, 1967.

22. Leon, M. A. Concanavalin A reaction with human normal immunoglobulin G and myeloma immunoglobulin G. *Science* 158:1325, 1967.

23. Marsh, C., Jolliff, C. R., and Payne, L. C. A rapid micromethod for starch-gel electrophoresis. *Amer. J. Clin. Path.* 41:217, 1964.

24. Mellman, W. J., and Rawnsley, H. M. Blastogenesis in peripheral blood lymphocytes in response to phytohemagglutinin and antigens. *Fed. Proc.* 25:1720, 1966.
25. Nace, G. W. Tissue localization of multiple forms of enzymes. *Ann. N.Y. Acad. Sci.* 103:980, 1963.
26. Nairn, R. C. *Fluorescent Protein Tracing.* Baltimore: Williams & Wilkins, 1964.
27. Ornstein, L. *Disc Electrophoresis,* Part I. Rochester, N.Y.: Distillation Products Industries, 1961.
28. Rajewsky, K., Rottlander, E., Peltre, G., and Muler, B. The immune response to a hybrid protein molecule. *J. Exp. Med.* 126:581, 1967.
29. Raymond, S., and Nakowichi, M. Electrophoresis in synthetic gels: I. Relation of gel structure to resolution. *Anal. Biochem.* 1:23, 1962.
30. Rees, V. H., Filds J. E., and Laurence, D. J. R. Dye-binding capacity of human plasma determined fluorometrically and its relation to determination of plasma albumin. *J. Clin. Path.* 7:336, 1954.
31. Smithies, O. An improved procedure for starch-gel electrophoresis, further variations in the serum of normal individuals. *Biochem. J.* 71: 585, 1959.
32. Strickland, R. D., Mack, P. A., Guriele, F. T., Podleski, T. R., Salome, O., and Childs W. A. Determining serum proteins gravimetrically after agar electrophoresis. *Anal. Chem.* 31:1410, 1959.

13

ELECTROPHORETIC SEPARATION
OF BLOOD CELLS BY CONTINUOUS
PARTICLE ELECTROPHORESIS

From the standpoint of the physical chemist, the most logical way to obtain zeta potential measurements of particulate matter is by electrophoresis. Where the particle is of molecular size, moving boundary electrophoresis is quite adequate, but for larger particles gravity becomes a factor and moving boundary electrophoresis is not applicable.

The microscopic method of particle electrophoresis is applicable for study of large particles [1]. In this method the individual cell, organism, or particle is observed through a microscope, and its rate of migration is measured in a known electrical field. Biologic cells of all types have been studied thus. With a suitable apparatus the method is rapid, and it has a wide range of applications. In the microscopic method the cylindrical or rectangular chamber is made of glass, and the rate of migration of particles in the chamber is found by timing their migration over a standard distance.

One of the most widespread applications of particle electrophoresis has been to the study of blood cells. Erythrocytes, lymphocytes, and granulocytes each have individual electrophoretic mobility permitting separation [6]. For example, erythrocyte electrophoretic mobility is 12.7×10^{-5} cm.^{2} per volt second. Lymphocytes have electrophoretic mobility of 10.1×10^{-5} cm.^{2} per volt second, and granulocytes have a mobility of 8.7×10^{-5} cm.^{2} per volt second. Comparing this to the electrophoretic mobility of albumin, 6.6×10^{-5} cm.^{2} per volt second, we see that the erythrocyte has nearly twice the electrophoretic mobility of albumin. In disease there is considerable modification of the electrophoretic mobility, not only of the erythrocytes but also of the other components [6]. Studies under way in many laboratories, utilizing the microscopical system, have disclosed that malignancy of the lymphoid tissue associated with lymphatic leukemia results in increased electrostatic charge of the surface of lymphocytes, thus giving rise to greater electrophoretic mobility [6]. Mobility then approaches that of the erythrocyte, and considerable overlapping of migration patterns occurs. The same is true of the granulocytes in chronic granulocytic or acute granulocytic leukemia [6]. In lymphosarcoma and a number of other conditions there is evidence of increased electrophoretic mobility [6]. Malignant cells of one type or another have all been shown to have more rapid electrophoretic mobility [1]. Following treatment with neuraminidase malignant cells show an essentially normal electrophoretic distribution pattern [1]. This has been attributed to the fact that the enzyme splits off sialic acid from the surface of the cell, thus reducing the cell's negative charge [9].

Microscopic particle electrophoresis applied to cells has received wide attention and continues to contribute to our knowledge concerning surface charge on biologic cells. It has long been the wish of many investigators to have available a system that would furnish a large number of cells of a single electrophoretic mobility for purposes of more advanced analytical studies. Bulk supply of cells has not been easy to come by, and the microscopical system is not applicable.

Continuous particle electrophoresis (CPE) has recently become available in an instrument manufactured by Beckman Instruments (Fig. 13-1) [10]. Briefly, particles in liquid suspension usually have a surface charge which may originate as ionization of surface groups, absorption of ions from the medium, or ion exchange processes. The density of the surface charge, as well as the ionic and dielectric properties of the medium, determines the mobility of the particles in an electrical field. It is known that the surface charge of different particles may vary widely. They can be modified by several techniques which may improve separation.

Although CPE is relatively new, a large body of evidence exists on the availability of applicable theory to explore separation of particles with different zeta potentials [5]. The CPE system works as follows: A suspension of cells is introduced as a fine, steady stream into a vertical flowing curtain of electrolyte (Fig. 13-1). A horizontal direct current electrical field is applied to the curtain. Each particle has two components of motion: a vertical component, which is the same as the curtain velocity, and a horizontal component, which is proportional to the electrophoretic mobility or zeta potential of the particle. If one imagines the curtain as being a thin waterfall which carries the particles by gravity as a thin stream down into the fraction-collecting area (Fig. 13-1E), it is not difficult to picture what happens when a voltage is applied at right angles to the path of the particles. The particles separate into fine streams if there is variability of surface charge among the particles. In the case of blood cells a multiple of bands appear in the window (Fig. 13-1D) and can be recorded by photography (Figs. 13-2, 13-5, 13-6, and 13-7) or by an optical scanner attached to a strip chart recorder.

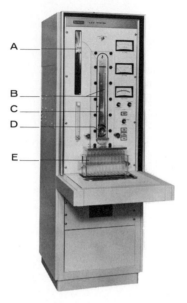

FIGURE 13-1

Photograph of continuous particle electrophoresis (CPE) system. Sample is introduced through top of electrophoretic cell via small plastic tube from reservoir at A. Electrodes (B) are on either side of the electrophoretic cell (C). The window (D) at the base of the electrophoretic cell is designed for visualization of particle streams, which are illuminated by an indirect lighting system mounted behind the window. Fractions are collected into 48 tubes (E). Various dials shown on the instrument govern the voltage, current, and volts per centimeter. The front plate of the electrophoretic cell is chilled by circulating fluid and permits close control of temperature during operation. (Courtesy Beckman Instruments, Inc., Fullerton, California.)

Erythrocyte Electrophoretic Patterns

At this stage in the development of continuous cell electrophoresis, it is desirable to pick specific diseases and determine the electrophoretic patterns of different blood cells. Our own effort at this time has been predominantly related to study of erythrocytes in various diseases. Figures 13-2, 13-5, 13-6, and 13-7 show photographs of erythrocyte electrophoretograms from a number of pathologic conditions. Effect of Keflin and incomplete anti-D is shown in Figure 13-2. In contrast to the findings of Sachtleben [8], incomplete anti-D did change electrophoretic mobility. As reported by Ruhenstroth-Bauer [6], lymphocytes, granulocytes, and erythrocytes each have individual electrophoretic mobilities. However, as shown in Figures 13-3 and 13-4, considerable overlapping of zones is evident, and there are at least two streams for each cell type. In a case of granulocytosis a marked increase of major streams of granulocytes is seen (Fig. 13-4).

FIGURE 13-2

Experimental study of erythrocytes modified by Keflin and incomplete anti-D. Anode is on the right. Erythrocyte electrophoretogram of normal cells (A). Same cells treated with Keflin (B) and incomplete anti-D (C). In the normal (A), as stated in the text, two cell streams are usually found. The major band has a mobility of 30 mm. on the scale. Keflin greatly increases the electrophoretic mobility of both bands. Many streams, all with fast electrophoretic mobility, are shown with cells treated with incomplete anti-D.

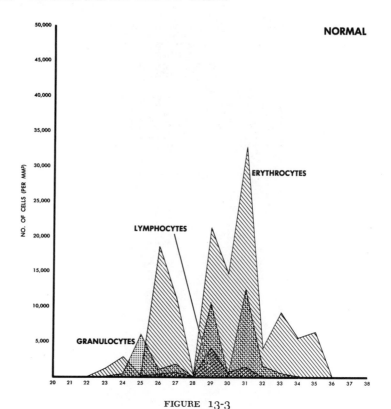

FIGURE 13-3

Quantitation of cell fractions collected from normal blood. Note wide spread of cell types. Anode is on the right.

Figure 13-5A is from a patient with sickle cell disease, and Figure 13-5B shows the same cells following treatment with bromelin. Figure 13-5C is from a patient with a small γM monoclonal peak and leukocytosis. Normally two streams, a major and minor component, are seen in each erythrocyte electrophoretogram. The major stream usually travels more cathodically than the minor stream, and approaches the 30-mm. mark or is just to the anodic side of the 30-mm. mark. Bromelin, crude protein from the stem of pineapple containing proteolytic enzymes, reacts with erythrocytes. It binds to erythrocyte lipoproteins and lipids, presumably cholesterol [2, 3]. Following bromelin treatment normal cells usually show decrease in mobility, and more streams develop [4]. Bromelin treatment does not always result in lower mobility, as is shown in a sickle cell disease pattern (Fig. 13-5A), which has two streams, the major band being the slower; both streams are just to the cathodic side of the 30-mm. mark. Following bromelin treatment the cells increase in mobility in contrast to normal. A patient with a small γM monoclonal peak and leukocytosis is shown in

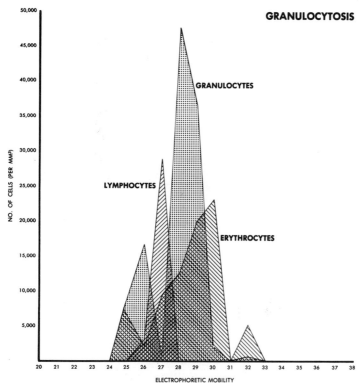

GRANULOCYTOSIS

FIGURE 13-4

Cell fractions from blood with granulocytosis. Note marked increase of granulocyte peak. Considerable overlapping is evident. Anode is on the right.

Figure 13-5C and discloses four bands, two major and two minor, all of which are more rapid in their electrophoretic mobility than normal erythrocytes.

Figure 13-6A shows the erythrocytes from a patient with auto-immune hemolytic disease of the warm antibody type. Cells have high negative charge and migrate far to the right in comparison to normal cells. Four components are seen—three minor and one major. Figure 13-6B shows the erythrocytes from a patient with myelofibrosis of relatively long standing. Many components are seen, all of them having higher negative charge density than normal cells. Figure 13-6C is an erythrocyte electrophoretogram from a healthy individual after bromelin treatment of erythrocytes. A whole host of erythrocyte components is found, all with a charge density less than normal.

Figure 13-7A shows the erythrocyte electrophoretic pattern from a pregnant patient. Notice the large number of components, all with electrophoretic mobility slightly greater than that of normal cells.

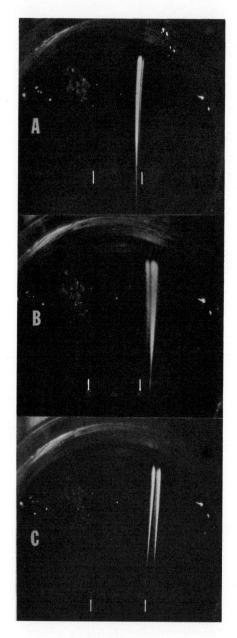

FIGURE 13-5

Erythrocyte electrophoretograms of three erythrocyte preparations. Anode is on the right. A. Patient with sickle cell disease. B. Same cells following treatment with bromelin. C. Patient with γM monoclonal peak and leukocytosis.

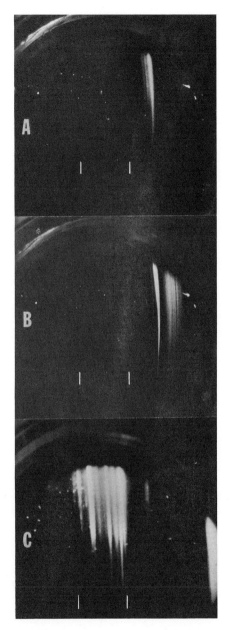

FIGURE 13-6

Erythrocyte electrophoretograms from three patients with various diseases.
Anode is on the right. A. Autoimmune hemolytic disease of the warm antibody type. B. Myelofibrosis of long standing. C. Normal patient whose
cells show a great number of erythrocyte populations after treatment with
bromelin, all of which have electrophoretic mobility less than normal.

FIGURE 13-7

Erythrocyte electrophoretograms of various conditions. Anode is on the right. A. Pregnancy. B. Chronic lymphatic leukemia. C. Chronic myelogenous leukemia.

The pattern returns to normal following delivery. Figures 13-7B and C, from chronic lymphatic and chronic myelogenous leukemia, respectively, show marked changes in erythrocyte mobility. Isolation and collection of a stream of cells in the fraction collector for purposes of carrying out analysis to determine differences in enzymes and hemoglobin, has not been altogether satisfactory. Erythrocyte, lymphocyte, and granulocyte zones show considerable overlapping, as is exemplified in Figures 13-3 and 13-4. As the fractionating system improves, abnormalities of various cell populations in various diseases can be analyzed.

Lymphocyte and Granulocyte Electrophoretic Separation

Lymphocytes have electrophoretic mobility between that of erythrocytes and that of granulocytes. Figures 13-3 and 13-4 are hemocytoelectrophoretograms of leukocyte-rich blood. It is too early to predict the full impact of this tool in the study of blood cells, but electrophoretic charge density is greatly disturbed on erythrocytes, lymphocytes, and granulocytes in various diseases. Two populations of small lymphocytes in thoracic duct lymph were recently shown by Ruhenstroth-Bauer and Lucke-Huhle by cell electrophoresis [7].

Other Uses for Cell Electrophoresis

One other area seems of some interest to us, involving separation of tagged antibody from untagged antibody and the tagging material. This has been particularly worthwhile for electron microscopy where ferritin-tagged antibodies are difficult to prepare in pure state. The electrophoretic distribution of these components, as checked by the fractions with immune antisera, shows that by and large the label, plus the antibody, renders the molecule separable from the other two components. The separation is based on a difference in charge density of the components of the system. The instrument, working as a continuous device, gives rise to fractions which can be collected in great bulk, and therefore a large volume of labeled antisera may be separated by this technique.

References

1. Ambrose, E. J. *Cell Electrophoresis*. Boston: Little, Brown, 1965.
2. Cawley, L. P., and Goodwin, W. L.   Binding of bromelin by erythrocyte lipoprotein. *Fed. Proc.* 26:789, 1967.
3. Cawley, L. P., and Goodwin, W. L.   Electron microscopic investigation of the binding of bromelin by erythrocytes. *Vox Sang.* 13:393, 1967.

4. Cawley, L. P., and Goodwin, W. L. Electrophoretic behavior of bromelin-treated erythrocytes. *Fed. Proc.* 27:724, 1968.
5. Pollack, W., Hager, H. J., Reckel, R., Toren, D. H., and Singher, H. O. A study of the forces involved in the second stage of hemagglutination. *Transfusion* 5:158, 1965.
6. Ruhenstroth-Bauer, G. The Normal and Pathological Haemocytopherogram of Man. In Ambrose, E. J. (Ed.), *Cell Electrophoresis.* Boston: Little, Brown, 1965. Pp. 66–72.
7. Ruhenstroth-Bauer, G., and Lucke-Huhle, C. Two populations of small lymphocytes. *J. Cell Biol.* 37:196, 1968.
8. Sachtleben, P. The Influence of Antibodies on the Electrophoretic Mobility of Red Blood Cells. In Ambrose, E. J. (Ed.), *Cell Electrophoresis.* Boston: Little, Brown, 1965. Pp. 100–114.
9. Seaman, G. O. F., and Cook, G. M. W. Modification of the Electrophoretic Behavior of the Erythrocyte by Chemical and Enzymatic Methods. In Ambrose, E. J. (Ed.), *Cell Electrophoresis.* Boston: Little, Brown, 1965. Pp. 48–65.
10. Strickler, A. Continuous particle electrophoresis: A new analytical and preparative capability. *Separation Sci.* 2:335, 1967.

14

CREATINE PHOSPHOKINASE
ISOENZYMES

A s has been outlined in other chapters dealing with isoenzymes, the mere possibility that each organ has a unique enzyme or isoenzyme profile which, when demonstrated in the serum, calls forth supposition that the tissue of origin has been injured is appealing to the clinical pathologist because of the diagnostic potential involved. In the case of lactate dehydrogenase (LDH) isoenzyme, much of this type of reasoning has now become fact, and the LDH isoenzyme determinations are routine in many laboratories. LDH is found in almost all tissues of the body, and there is considerable overlapping in the profile from one organ to another, particularly in reference to the erythrocyte and the myocardium, both having a predominance of LDH_1. Thus, hemolysis, sometimes unavoidable, can contribute to a false interpretation in distinguishing myocardial damage from nonmyocardial damage.

This is the main reason many laboratory investigators began to look for organ-specific enzymes. With the introduction of a serum test for creatine phosphokinase (CPK, EC 2.7.3.2) by Dreyfus et al. [6], a specific test related to muscle tissue became available. CPK is located in cardiac muscle, skeletal muscle, and brain, and is not found in significant concentration in erythrocytes [1, 3, 6, 8, 10, 16, 17]. It is, therefore, possible to relate increased amounts of serum CPK to damage of either brain or muscle tissue. There is no difficulty clinically in distinguishing brain damage and myocardial damage. Muscular exercise is associated with transient increase of CPK in the serum. The enzyme is known to be liberated into the peripheral blood shortly after myocardial infarction, reaching a peak presumably within 18 to 30 hours [1, 3, 16, 17]. The rise, however, is transient [3, 16, 17]. The enzyme is not stable either at room temperature or at refrigerator temperature or at $-20°$ C. Thus, the procedure should be done quickly, preferably within three hours after blood collection, although storage with sulfhydryl compounds such as glutathione and cysteine preserves the enzyme activity [8].

Evidence of isoenzymes of CPK has been published [2, 4, 5, 7, 9, 12, 14, 15, 19]. The electrophoretic procedure for detection is described in Chap. 15 and is based on a biochemical technic developed by Oliver [11] coupled to a tetrazolium salt (Nitro-blue) for visualization, or, as an alternate method, the fluorescence of nicotinamide-adenine dinucleotide reduced (NADH) [11].

CPK Isoenzymes

Van der Veen et al. [15] published evidence dealing with isoenzymes of CPK in serum in patients with myocardial infarction. They described three fractions, with fast (I), medium (II), and slow (III) mobility. Their designation will be used in this text. The CPK zymographic determinations in Figure 14-1 represent the human tissues

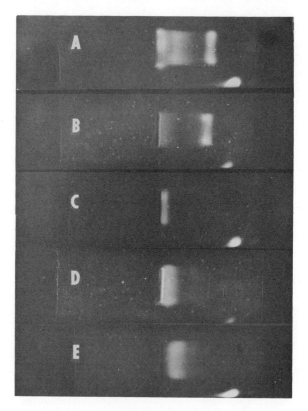

FIGURE 14-1

CPK zymograms of human tissue extracts of mixture of brain, skeletal muscle, liver, and heart (A), of brain (B), of skeletal muscle (C), of liver (D), and of heart (E). Anode is on the right. Mobility of albumin is displayed by small round marker on the lower edge of each electrophoretogram.

studied in our laboratory and are similar to those studied in other laboratories. Brain, heart, liver, and muscle do show more or less organ-specific isoenzymes. The zymogram for the heart (E) extract shows a major CPK band (fraction III) which has the mobility of a β_1-globulin. Liver CPK (D) is slightly slower than that of the heart. Muscle CPK (C) is slower still and travels just in front of the point of application, whereas brain CPK (B) has very fast mobility comparable to that of an α_1-globulin (Fig. 14-1). The combination of all of these tissues is shown in A of Figure 14-1 and discloses at least three well-defined zones of activity. In normal serum a CPK zymogram even using 15 µl. of serum is usually negative because the amount of the enzyme is very small. Following a myocardial infarction, however, a sufficient amount of enzyme is present.

In 1964 Sjovall [14] showed, by agar-gel electrophoresis, heterogeneity of the CPK enzyme of serum and of organs. By starch-gel electrophoresis four bands for CPK were found for each of the major tissues examined, such as brain and muscle, whereas by agar-gel electrophoresis only one major band for each of these was seen. Sjovall noticed the difference between the electrophoretic mobility of CPK of muscle and brain, and Deul et al. [5] confirmed these findings. In 1966 Van der Veen et al. [15] reported that with an agar-gel electrophoretic system similar to that of Wieme [18] they could detect three CPK isoenzymes. As noted before, they labeled these fractions I, II, and III, with electrophoretic mobility of albumin, α_2-globulin, and γ-globulin, respectively. As is shown in Figure 14-1, heart contains two components, muscle one, and brain two—a fast and a slow (fractions I and III). The method appears specific, since without the substrate (creatine phosphate) the bands do not appear except in the position where fraction II appears (α_2-globulin), and under these conditions the band is not discrete but is a diffuse band, probably because of a myokinase activity. In one patient with myocardial infarction and high levels of CPK the zymogram taken early in the disease was shown by Van der Veen et al. [15] to be different from that taken at 24 hours. In the first analysis fractions II and III were evident whereas in 24 hours only fraction III was present. In Figure 14-1 it should be noted that the heart zymogram shows a predominance of fraction III. Van der Veen et al. also studied two patients with progressive muscular dystrophy. In one patient only fraction III, in the other fractions II and III, were demonstrable. This can be explained only if the heart muscle was damaged in the second patient. The effect of hemorrhagic shock in dogs was investigated too, and band III was found to be elevated.

It is too early at this time to determine whether the CPK test will be useful in the clinical laboratory. Only time and clinical trial will give us the answer.

References

1. Blough, M. E., Crow, E. W., and Cawley, L. P. Myocardial infarction; heart proteins, including enzymes in serum following myocardial infarction. *J. Kansas Med. Soc.* 67:545, 1966.
2. Burger, A., Richterich, R., and Aebi, H. Die Heterogenität der Kreatine Kinase. *Biochem. Z.* 339:305, 1964.
3. Cook, V. P. Serum creatine phosphokinase in myocardial damage. *Amer. J. Med. Techn.* 33:275, 1967.
4. Dawson, D. M., Eppenberger, H. M., and Kaplan, N. O. Creatine kinase: Evidence for dimeric structure. *Biochem. Biophys. Res. Commun.* 21:346, 1965.
5. Deul, D. H., and Van Breemen, J. F. L. Electrophoresis of creatine phosphokinase from various organs. *Clin. Chim. Acta* 10:276, 1964.
6. Dreyfus, J. C., Schapira, G., Resnais, J., and Scebat, L. La créatine-kinase sérique dans le diagnostic de l'infarctus myocardique. *Rev. Franc. Étud. Clin. Biol.* 5:386, 1960.
7. Eppenberger, H. M., Eppenberger, M., Richterich, R., and Aebi, H. Ontogeny of creatine kinase isozyme. *Develop. Biol.* 10:1, 1964.
8. Hughes, B. P. A method for the estimation of serum creatine kinase and its use in comparing creatine kinase and aldolase activity in normal and pathological sera. *Clin. Chim. Acta* 7:597, 1962.
9. Kar, N. C., and Pearson, C. M. Creatine phosphokinase isoenzymes in muscle in human myopathies. *Amer. J. Clin. Path.* 43:207, 1965.
10. Nichol, C. J. Serum creatine phosphokinase measurement in muscular dystrophy studies. *Clin. Chim. Acta* 11:404, 1965.
11. Oliver, I. T. A spectrophotometric method for the determination of creatine phosphokinase and myokinase. *Biochem. J.* 61:116, 1955.
12. Rosalki, S. B. Creatine phosphokinase isoenzymes. *Nature* (London) 207:414, 1965.
13. Sherwin, A. L., Siber, G. R., and Elhilali, M. M. Fluorescence technique to demonstrate creatine phosphokinase isozymes. *Clin. Chim. Acta* 17:245, 1967.
14. Sjovall, K. Creatine-phospho-transferase isozymes. *Nature* (London) 202:701, 1964.
15. Van der Veen, K. J., and Willebrands, A. F. Isoenzymes of creatine phosphokinase in tissue extracts and in normal and pathological sera. *Clin. Chim. Acta* 13:312, 1966.
16. Vincent, W. R., and Rapaport, E. Serum creatine phosphokinase in the diagnosis of acute myocardial infarction. *Amer. J. Cardiol.* 15:17, 1965.
17. Warburton, F. G., Bernstein, A., and Wright, A. C. Serum creatine phosphokinase estimations in myocardial infarction. *Brit. Heart J.* 27:740, 1965.
18. Wieme, R. J. *Agar Gel Electrophoresis.* Amsterdam: Elsevier, 1965.
19. Wood, T. Adenosine triphosphate—creatine phosphotransferase from ox brain: 1. Purification and isolation. *Biochem. J.* 87:453, 1963; and 2. Properties and function. *Biochem. J.* 89:210, 1963.

15

METHODS

Standard Method of Preparation of Gel-Coated Plastic Strips for Electrophoresis

PRINCIPLE

A mixture of electrolyte and agarose or agar is boiled and while still hot pipetted evenly on strips or sheets of thin plastic film (Cronar, Mylar, etc.) (Fig. 15-1A) and allowed to harden. Surface tension holds the fluid agarose on the film and no runoff occurs [15].

Alternate methods of forming gel-coated plastic sheets or strips are shown in Figure 15-1 and include thin layer by pull-through method (B), spraying (C), thin layer with spreader (D), and mold (E). References dealing with other methods for preparation of agar or agarose suitable for electrophoresis include Wieme [93], Smith [81], Grabar et al. [33], Nerenburg [63], and Crowle [21].

EQUIPMENT AND REAGENTS

1. Photographic film P40B, 35 mm. (E. I. DuPont de Nemours & Co. or HCS Corp.).*
2. Hot plate or Bunsen burner.
3. Beakers and pipettes.
4. Agarose (Bausch and Lomb).*
5. Buffer.
   a. Barbital buffer, pH 8.6, ionic strength 0.05 (Fisher Scientific Co. or Beckman Instruments, Inc.).* Dissolve one package of barbital buffer, pH 8.6, ionic strength 0.075, in 1500 ml. distilled water.
   b. Barbital buffer, pH 8.6, ionic strength 0.025. Dilute one part barbital buffer, pH 8.6, ionic strength 0.05, with one part distilled water.

PROCEDURE

1. Cut photographic film base in 15-cm. lengths. Mark with liquid tip pen—Blaisdell—on the convex side according to procedures for point of origin and anodic migration point.

*See Appendix.

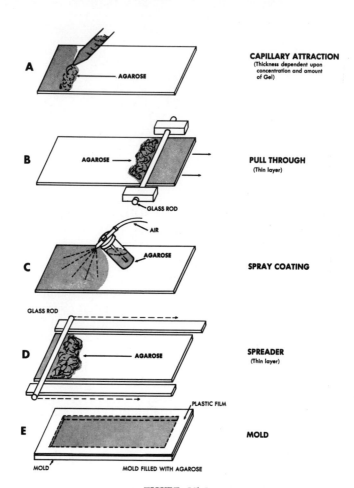

FIGURE 15-1

Methods of coating plastic sheets, strips, or glass plates with agarose, agar, or any other substance which gels after contact with surface of plastic or glass. A. Capillary attraction, with thinness of gel layer depending upon the amount and concentration of agar or agarose added. B. Pull through, which results in a very thin layer. Large plates or thin strips may be coated this way and it works very well. Six or seven feet of Cronar film can be coated in a small amount of time. Starch gel can also be applied this way. C. Spray coating, which results in a very thin and relatively uniform layer. The presence of small air bubbles gives a "frosted glass" pattern to the finished product. This pattern, however, clears during electrophoresis, and the resultant patterns are satisfactory. D. Spreader technic, a system taken from thin-layer chromatography and somewhat analogous to method B. It too is very satisfactory. E. Molding technics of one type or another. The mold shown is simply a mold with a standard depth representing the thickness to be used. The mold is precoated with a material that prevents the agarose from becoming attached. When the mold is removed, the gel will remain fixed to the plastic film, and the completed strip is ready for electrophoresis.

2. Coat the strips concave side up.
3. Prepare gel by weighing the desired amount of agarose, adding other additives if necessary, and measuring desired buffer specified according to procedure.
4. Bring the solution to boiling point as quickly as possible (covering container prevents evaporation).
5. Coat each strip with 5 ml. of hot buffered agarose solution spreading the solution evenly over the entire surface.
6. Allow to harden. This takes 5–10 minutes.

RESULTS

Strips are ready for use or may be stored in moist, tightly sealed containers in the refrigerator for long periods before use. Antibacterial agents added to the agarose reduces bacterial growth.

Trench Method for Coating Gel Strips

PRINCIPLE

The practice of placing the sample (usually 1.5µl. of serum) in gel with a razor blade is applicable for fluids with protein concentration comparable to that of serum. However, in a number of body fluids the satisfactory outcome of electrophoresis can be assured only if an adequate quantity of reactants is present. In the case of urine, spinal fluid, and many serum enzymes a higher volume of sample than the usual 1.5µl. is required to ensure that the end point is strong enough for detection and quantitation. A trench or slot in the gel creates an opening sufficient for 10µl. to 20µl. of sample. Attempts to place more than 3µl. in the gel with the razor blade method (see Methods of Sample Application, below, Fig. 15-2G) cause uneven spread of sample in the slit and a flowing of the sample out of the slit. With the stripper method (Methods of Sample Application, Fig. 15-2B) 6µl. is the upper limit of sample size which can be applied to the gel surface and still obtain a satisfactory line of inoculation [15]. Methods to prepare suitable-sized slots in agar or agarose gels are discussed in several standard references [33, 63, 81, 93]. A capillary trench method is described below [15]. An alternate method using small lengths of metal rods or wire is also described.

Short capillary tubes are placed in the agarose while it is still fluid and removed after gel formation. The trenches formed by the capillary tubes to receive the material for electrophoresis have sharp, even edges; ragged edges of trenches cause distortion of protein pattern. There are many methods for formation of trench for sample, as shown in Figure 15-3 in Methods of Sample Application.

EQUIPMENT AND REAGENTS

1. Standard gel electrophoretic equipment (see Standard Gel Electrophoretic Method, below).

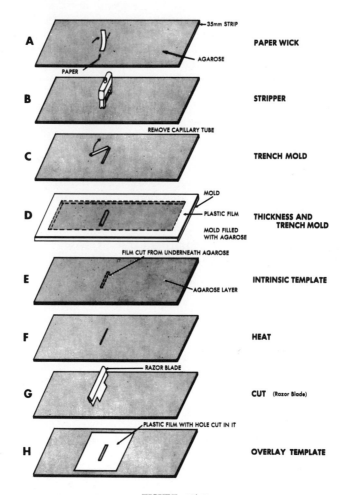

FIGURE 15-2

Eight methods for application of sample for electrophoresis. A, B, and C. Self-explanatory; used in many standard procedures described in this text. D. Molding technic that can be adapted to suit any laboratory need. E. Unique in that the sample slot is precut into the plastic and the agarose is then overlaid. The strip is inverted so that the agar surface faces the buffer surface. The sample is applied to the agarose through the precut sample slot in the plastic. This method has been tried and is reasonably satisfactory. It is related to method H, overlay template. F. Method using a hot device such as a glass slide or wire which causes instantaneous melting of the gel. Liquid remaining can be removed with a capillary pipette or filter paper. G. Described in the text and diagrammatically demonstrated in Figure 15-4.

2. Standard gel electrophoretic reagents. See page 236.
3. Capillary tubes, outside diameter 1.3–1.4 mm. cut in 1.5-cm. lengths.
4. Standard agarose-buffer mixture. Buffer of the pH and ionic strength desired for a particular type of electrophoretic procedure may be used.

PROCEDURE

1. Prepare film strip as in standard method.
2. Coat the strip with 5 ml. of hot buffered agarose solution. The percentage of agarose and the kind of buffer depend upon the desired firmness, the pH, and the ionic strength of the finished product for the particular electrophoretic pattern.
3. While the agarose is still fluid, place a 1.5-cm. length of capillary tubing horizontally on the strip at the midpoint line.
4. Allow the agarose to harden five minutes or more.
5. Immediately before using remove the tube with fingertips or tweezers by flexing the side of the strip.

RESULTS

An even trench is formed that will hold up to 20μl. of material to be analyzed by electrophoresis.

ALTERNATE PROCEDURE*

1. Prepare film strip as in standard method.
2. Coat the strip with 5 ml. of hot buffered agarose solution.
3. While the agarose is still fluid, place a 1-cm. length of smooth-surfaced wire the size of #1 paper clips horizontally on the strip at the midpoint line.
4. Allow the agarose to harden five minutes or more.
5. Immediately before using remove the wire lengths by holding a magnet directly above but not touching the agar surface. The magnet will remove the wire lengths, leaving an even trench that will hold 5μl. of material to be analyzed by electrophoresis.

Methods of Sample Application

Many methods have been described for sample inoculation for electrophoresis in several standard references [15, 21, 33, 63, 81, 93]. In Figure 15-2, eight variations of sample application are diagrammed for electrophoresis, and in Figure 15-3, seven variations of antigen and antiserum application are diagrammed for immunoelectrophoresis. Each of these will be discussed briefly.

*Suggestion of James B. McCormick, M.D., Hinsdale, Ill.

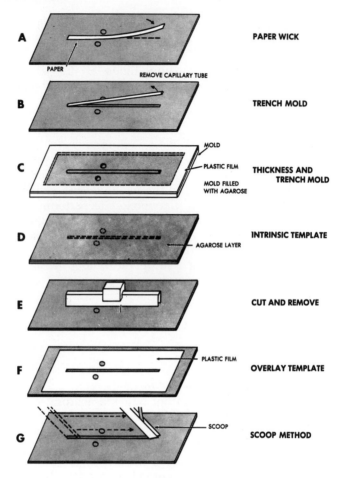

A — PAPER WICK
PAPER — REMOVE CAPILLARY TUBE

B — TRENCH MOLD

C — MOLD — PLASTIC FILM — THICKNESS AND TRENCH MOLD — MOLD FILLED WITH AGAROSE

D — INTRINSIC TEMPLATE — AGAROSE LAYER

E — CUT AND REMOVE

F — PLASTIC FILM — OVERLAY TEMPLATE

G — SCOOP — SCOOP METHOD

FIGURE 15-3

Methods for creating antiserum trenches. These are comparable in many instances to those shown in Figure 15-2 for sample inoculation. A, B, and C. Self-explanatory. D. Intrinsic template in which the groove for the trench is cut into the plastic and the agar or agarose is overlaid. Application is made from the reverse side into the agarose through the preformed trench. E. Standard method used in many laboratories. F. Template overlay, which is reasonably satisfactory. G. Standard method used in the laboratory and described in the text.

Paper wicks are used for inoculation of electrophoretic strips (Fig. 15-2A) by saturating the filter paper with the sample. The wick is then placed at the midpoint across the precoated agarose strip. When a paper wick is used for immunoelectrophoresis (Fig. 15-3A), the paper is saturated with the antiserum and placed along the mid-longitudinal axis.

The "stripper" method is a widely used technic for sample application, particularly in cellulose acetate and paper electrophoresis. The stripper (Fig. 15-2B) consists of two small, closely spaced, parallel wires joined together at each end. The sample is placed on the wires by means of a micropipette. The sample is then gently touched to the agar surface and released. The stripper has a capacity of up to 6µl. of sample.

The trench mold has been described in the preceding section. The trench mold (Fig. 15-2C) leaves a well-defined trench in the gel strip which has a capacity of up to 20µl. of sample. This method is used for serum isoenzyme detection and for samples with low protein or reactant concentration. The immunoelectrophoretic trench method is described in Figure 15-3B. The trench is made by placing a capillary tube (outside diameter, 1.3 mm.; length, 3.5 inches) along the mid-longitudinal axis for the antiserum trench.

The thickness-trench mold (Fig. 15-2D) is a modification of the third method. The mold is designed to cast an agar slab of any desired thickness with a preformed trench. The trench can vary in size according to substance used to create it (fine wire to capillary tube). The hot agarose solution is poured into the mold, and a strip or sheet of thin plastic film is immediately rolled down on it. The agarose is allowed to harden. The mold-strip is turned upside down and the mold removed leaving an agarose-coated strip with a preformed trench. The trench mold for immunoelectrophoresis (Fig. 15-3C) is formed in the same basic way as the electrophoretic strip except that the trench is along the mid-longitudinal axis and glass beads are used on either side of the trench to make an indentation for the antigen wells.

An intrinsic template (Fig. 15-2E) can be utilized for electrophoretic sampling. A small rectangular trench is cut in the plastic film with a very sharp knife before coating with the gel solution. The rectangular slit is covered with a piece of masking tape or it may be fitted in a thickness-trench mold with the trench form protruding through the cut rectangular trench. There must be a tight seal or the hot gel solution will leak underneath the strip. After the strip is coated with gel solution, either it is removed from the mold or the masking tape is removed and the strip is placed in the electrophoretic cell with the gel-coated surface down. The strip is inoculated from the plastic side. An immunoelectrophoretic intrinsic template (Fig. 15-3D) is formed in essentially the same way except that both antigen wells and antibody trenches are cut into the plastic. The principles of these methods may serve as the background for automation of sample introduction.

Two rather similar methods for electrophoretic inoculation are the heat (Fig. 15-2F) and the trench method (Fig. 15-2C). The precoated agar strip is placed on a firm level surface and a hot object such as a microscope cover slip (1 cm. wide) is placed straight down into the hardened agarose strip and removed after a few seconds [93]. This

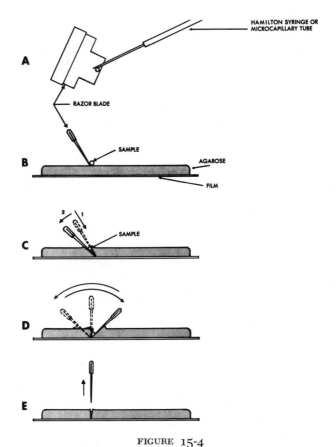

FIGURE 15-4
Razor blade technic for applying sample—method G of Figure 15-2. Each
laboratory trying this technic will soon find shortcuts. In some instances it
has been found best to spread the sample along the edge of the razor blade
although a single application is all that is necessary.

leaves a well-defined trench just like one made by a capillary tube,
only smaller. The same method can be used for antibody trenches in
immunoelectrophoresis with increase in the dimensions of the glass
to the size needed for the trenches. In another technic for immunoelectrophoretic trenches, shown in Figure 15-3E, the parallel cutter
blades cut the trenches without heat. The gel plugs are removed by
aspiration.

The razor blade method of inoculation (Fig. 15-2G) is basic for
inoculation of samples with a normal to high quantity of protein, such
as serum proteins and hemoglobins. It uses a small amount of inoculum (1.5µl.–3µl.) per strip. The resolution of serum protein fractions
applied through the gel instead of to the gel surface is excellent.

A single-edged razor blade is cut to a 1-cm. width and 1.5μl. of sample is placed on the center and spread slightly along the cutting edge of the blade (Fig. 15-4A). The gel-coated strip is placed on a flat level surface with long axis pointed away from the person doing the procedure. The razor blade, with sample facing the person, is pressed down through the surface of the gel to the plastic strip at an angle greater than 90 degrees (Fig. 15-4B, C). In a forward-backward motion the sample fills the cut trench by capillary attraction (Fig. 15-4D). The blade is then lifted up and out (Fig. 15-4E), and the strip is placed on the cell. Objects with a thickness greater than the razor blade can be used for creating the trench, but the wider the sample inoculum, the poorer the separation of proteins.

The last method illustrated for electrophoretic application is the overlay template (Fig. 15-2H). The template, made of plastic, light steel, or aluminum, is a small piece about 4 cm. by 1.5 cm. with a pre-cut slit [92]. The template is laid on the gel surface and the sample inoculated in the slit. After electrophoresis the template can be removed. Figure 15-3F shows how the application of antigen and antibody can be done in a similar manner using a larger overlay cut with antigen wells and antibody trenches [21].

The agar scoop method is used to prepare trenches for immuno-electrophoretic procedures used in this text (Fig. 15-3G). The gel-coated strip or sheet is placed on the Immuno-Gel board, where positions of the antigen wells and the antibody trenches are clearly shown. The trench is scooped out by means of a sharp stainless-steel scoop. Trenches so prepared are smooth and uniform. The piece of agar removed by the scooping motion slides up inside the scoop and is easily removed by gently tapping the scoop. The antigen wells are cut or actually sucked out by an 18-gauge needle (ground off flush and sharpened) attached to a water aspirator pump.

Each method has some advantages and disadvantages. The two types of sample application of electrophoretograms described in routine methods are the razor blade and the trench method.

Standard Gel Electrophoretic Method

PRINCIPLE

Zone electrophoresis is the separation of migrating particles in an electrolyte solution by an electrical current. The particles separate into fractions because of their electrostatic charge in an inert stabilizing medium. Agarose and agar have been very successfully employed as stabilizing media for electrophoresis and in immunoelectrophoresis [15, 21, 30, 33, 63].

Agarose electrophoresis is a simple, rapid, inexpensive, versatile system for analysis of proteins, hemoglobins, lipoproteins, and isoenzymes [15]. The gel may be mounted on glass (microscope slides, for example) [63, 92], glass plates [33, 81], or transparent plastic [15]. The dried agarose plastic strips are durable and show little deterioration upon storage. The agarose plastic strip method is a flexible system and can be enlarged or made smaller for use in any electrophoretic system [15].

Quantitation of the gel strip is easily accomplished with densitometry on account of the clarity of the finished electrophoretogram [15, 33, 63, 92]. Alternately, the stained electrophoretic fractions can be cut apart, the color eluted into an alkaline solution, and its concentration determined by colorimetry. The latter method can be automated by use of Robot Chemist or Autoanalyzer.

EQUIPMENT AND REAGENTS
1. Standard gel electrophoretic equipment (HCS Corp.)* or equivalent.
   a. Electrophoretic cell and power supply.
   b. Photographic film P40B, 35 mm.
   c. Staining pans.
   d. Drying oven—forced hot air 60° C.
   e. Hamilton syringe—1μl. to 10μl. or 1μl. to 50μl. capacity.
   f. Hot plate or Bunsen burner.
   g. Beakers and pipettes.
   h. Analytrol with B-2 cam or equivalent densitometer.
2. Standard gel electrophoretic reagents.
   a. Barbital buffer, pH 8.6, ionic strength 0.05.
   b. Fixation solution, 10 percent (V/V) acetic acid in absolute methanol.
   c. Protein stain, 1 gm. Thiazine Red R (Allied Chemical Corp.)* dissolved in 100 ml. of 10 percent (V/V) aqueous acetic acid.
   d. Destaining solution, 2 percent (V/V) aqueous acetic acid.
3. Standard agarose-buffer mixture. Add 250 mg. agarose to 50 ml. of barbital buffer and dissolve by bringing to a boil with gentle stirring.

PROCEDURE
1. Prepare agarose-gel strips using either the standard method of preparation of gel-coated strips or the trench method for coating gel strips.
2. Fill each buffer compartment of the electrophoretic cell with equal volumes of barbital buffer. It is important that the buffer level be the same on anodic and cathodic sides.
3. Inoculate strips by method designated and place strips in the

*See Appendix.

electrophoretic cell with the anodic mark on the anodic side of the cell.

4. Perform electrophoresis at 150 volts until the leading edge of the stained albumin reaches the 3.2-cm. anodic migration point (usually 30–35 minutes). Two technics are used for detecting anodic migration point. The first one is to add 2 drops of 2 per cent (W/V) aqueous bromphenol blue to 50 ml. of the agarose-buffer solution. The albumin is stained as it migrates during electrophoresis, provided protein concentration is sufficient. The second method is to place a small dot of saturated bromphenol blue–albumin solution at the marginal edge of the origin point. During electrophoresis two stained areas become visible. The leading edge is bromphenol blue and the second spot is the stained albumin. This method is used when protein concentration is low or when dyed gel would interfere in end results, particularly isoenzyme or peroxidase reactions.

5. Remove the strips from the electrophoretic cell and stain them according to the specific procedure for protein, isoenzyme, or special stain.

6. Dry, scan, and quantitate if desired.

7. Agarose electrophoretic strips can be scanned in most densitometers with some or no modifications. The Spinco Analytrol in our experience is a satisfactory densitometer. The B-5 cam used for paper electrophoresis is replaced with a B-2 cam. The slit width is adjusted to 0.5 mm. by a length of 1 cm. using the midportion of the slit. This gives a scan of the electrophoretic pattern except the extreme lateral margins. The 500-mμ filters are used in both front and rear positions for thiazine-stained electrophoretograms. The dry stained electrophoretogram is inserted into the strip guide with the agarose side toward the front photocell, mainly for ease of feeding. In our laboratory strips are always scanned from the anodic to the cathodic side. Thus albumin always is on the right.

8. After the electrophoretogram is scanned, draw a perpendicular line from the intersection of the peak curves through the integration marks. Count the integration marks below each fraction and total them. Divide the number of marks per fraction by the total number of integration marks to give the percentage of that specific fraction.

$$\frac{\text{No. marks/fraction}}{\text{Total no. marks}} \times 100 = \% \text{ of fraction}$$

9. An Olivetti Programma 101 desk computer considerably reduces the calculation time in preparing an electrophoretic report. Six or seven entries into the computer consisting of the number of

integration marks for each protein fraction (albumin and α_1-, α_2-, β_1-, β_2-, and γ-globulins) and the total serum proteins in grams per 100 ml. yields the percentage of each fraction and, directly beneath each fraction, the value in grams per 100 ml.

10. The agarose strips can be stapled to the finished report for filing if desired. The electrophoretograms can be preserved indefinitely.

NOTES

1. Uniform reproducible results are obtained if standardization of procedure is maintained. The amount of inoculum, migration time, fixation and staining periods, and uniformity of agarose strip should be consistent to yield reproducible results.

2. The fixative solution performs a threefold task. First, and most important, it fixes the protein in the strip. Second, it dehydrates the agarose-buffer layer for quicker drying. Third, it removes the buffer salts so that there is no buffer crystallization to interfere with the quantitation and interpretation of end results. In special stains such as lactate dehydrogenase (LDH) isoenzymes the strips are stained before fixation or soaking so the enzyme will not be destroyed. The fixative or soaking solutions remove excess buffer salts. These strips will require a longer drying time because they are not dehydrated, as are those processed in the acid-methanol solution.

3. Regardless of which method of inoculation is used, it is imperative that the pipette or syringe and the razor blade, etc., be rinsed or cleaned thoroughly between samples. The simplest practice with the razor blade method is to have a Petri dish of distilled water for rinsing.

4. A multiple inoculating device consisting of 6–12 razor blades mounted to a steel frame has been developed for applying many samples at one time to a large sheet of agarose-coated film.

Serum Protein Electrophoresis

PRINCIPLE

Zone electrophoresis in agarose is simple, inexpensive, rapid, and a reproducible method for separating, identifying, and quantitating the six main fractions of serum proteins (Fig. 15-5) [15, 17, 33, 81, 93].

EQUIPMENT AND REAGENTS

1. Standard gel electrophoretic equipment (see Standard Gel Electrophoretic Method).

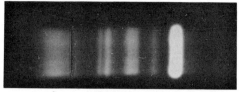

FIGURE 15-5

Agarose-gel electrophoretogram with normal serum showing the six components usually found in serum: albumin, α_1-globulin, α_2-globulin, β_1-globulin, β_2-globulin, and γ-globulin. Notice the small band just anodic to β_1. As noted elsewhere β_2-globulin corresponds to complement and lipoproteins. Anode is on the right.

2. Razor blade as described under Methods of Sample Application.
3. Standard gel electrophoretic reagents (see Standard Gel Electrophoretic Method).
4. Standard agarose-buffer mixture (see Standard Gel Electrophoretic Method).

PROCEDURE

1. Prepare standard agarose strips with 5 ml. of gel stained with bromphenol blue or, for short strips (4 inches), use 3.5 ml. of gel-stain mixture. The amount of gel is based on 1 ml. of gel per 10 sq. cm. of surface area. Mark the point of origin (midpoint) and the anodic migration point.
2. Fill electrophoretic cell with barbital buffer.
3. Inoculate strips with 1.5 μl. of serum by razor blade method (see description of inoculation under Methods of Sample Application, Fig. 15-4).
4. Place strips in cell.
5. Migrate at 150 volts until leading edge of stained albumin reaches the 3.2-cm. anodic migration mark.
6. Fix agarose strips for 20 minutes, then dry.
7. Stain for 30 minutes.
8. Destain in three changes of destaining solution, and dry.
9. Scan strips in Analytrol at 500 mμ and quantitate as for paper.

NORMAL

See Table 15-1.

SOURCES OF ERROR

1. Excess protein will cause a distortion in the albumin percentage because of limitations of the densitometer. This can be corrected

TABLE 15-1

MEAN AND RANGE FOUND IN 25 HEALTHY INDIVIDUALS

	Percent				Gm./100 ml.		
	2 S.D. (M) (±)	Mean	Range	2 S.D. (P) (±)	Mean	Range	2 S.D. (P) ±
Total protein		—	—		6.7	(5.8–7.9)	1.12
Albumin	5.75	57.5	51.5–66.0	7.0	3.85	—	
Globulin							
α_1	0.45	4.5	2.0–6.0	2.0	0.72	—	
α_2	1.1	11.0	6.5–16.5	4.0	0.11	—	
β_1	0.65	6.5	4.5–9.0	3.0	0.43	—	
β_2	0.50	5.0	3.0–11.0	4.0	0.36	—	
γ	1.55	15.5	10.0–21.0	7.0	1.04	—	

2 S.D. (P) = standard deviation of population.
2 S.D. (M) = standard deviation of method. Based on 10% of calculated population mean. Method of establishing reasonable limits to precision of procedure based on observation that relative standard deviation computed on basis of physiologic mean is usually 5% or less for those procedures that are acceptable.

by resetting the baseline with one or more neutral-density filters on the front side of the lamp until the recorder pen remains on the chart throughout the scan.

2. For uniform results strips need to be fixed, stained, and destained under standard, controlled, and uniform conditions.

3. Fibrinogen peak, if not recognized, could be mislabeled as a monoclonal peak and included in the calculations of γ-globulin.

Concentration of Cerebrospinal Fluid (CSF) and Urine for Electrophoresis

PRINCIPLE

Concentration of proteins is required in solutions which have a low protein content such as CSF and urine. Concentration by ultrafiltration is satisfactory and reasonably rapid [30, 40]. The membrane permits passage of fluid crystalloids but restrict proteins. Atmospheric pressure is directed to the fluid by applying vacuum on side arm

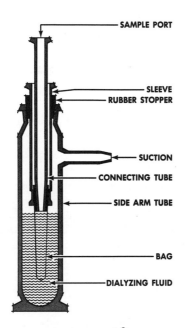

FIGURE 15-6

Diagram of the ultrafiltration unit of Schleichier and Schuell Company, Keene, New Hampshire. Suction arm must be connected to a suitable vacuum.

according to the diagram of the dialysis system of Schleichier and Schuell (Fig. 15-6).

1. Dialysis system—S & S Collodion Bags and Glass Suction Apparatus (Carl Schleichier & Schuell Co., Keene, N.H.).
2. Saline solution, 0.9 percent.

PROCEDURE

1. Pull the collodion bag over the plastic sleeve of the inner glass tube for about 1 cm. of the bag's length and fasten it tightly to the inner tube by the aid of the outer tube.
2. Insert the assembled parts vertically into the suction vessel filled with normal saline solution. The saline level inside the suction vessel must not be above the lower edges of the inner and outer glass tubes. The unit can be attached to a laboratory stand.
3. Fill the collodion bag with the biological fluid (10 ml. of centrifuged urine or 5 ml. of spinal fluid) by means of a pipette via the inner glass tube. Apply a vacuum to the suction vessel by means of a vacuum pump.
4. Concentrate the specimen to 0.1 ml. or less and remove with a pipette.
5. The collodion bag may be reused if it is rinsed with saline and stored in the suction vessel submerged in saline.

Cerebrospinal Fluid (CSF) Protein Electrophoresis

PRINCIPLE

Separation of the CSF proteins by agarose or agar-gel electrophoresis for the qualitative and quantitative evaluation of the proteins is clinically useful [15, 79, 93]. Concentration of the fluid is required, and the method is described under Concentration of Cerebrospinal Fluid (CSF) and Urine for Electrophoresis, above.

EQUIPMENT AND REAGENTS

1. Standard gel electrophoretic equipment (see Standard Gel Electrophoretic Method).
2. Standard gel electrophoretic reagents (see Standard Gel Electrophoretic Method).
3. Standard agarose-buffer mixture (see Standard Gel Electrophoretic Method).
4. A marker dye, if desired, to determine migration. Albumin

saturated with bromphenol blue. (Dot midpoint at the edge of the strip if the protein concentration is so low that the bromphenol blue in the agarose does not mark the albumin position.)

PROCEDURE

1. Coat the strips as per trench method with 5 ml. of warm buffered agarose (standard mixture). Strips should be marked 1.5 cm. cathodic to the midpoint and 3.2 cm. anodic to the midpoint.
2. Fill each buffer compartment with barbital buffer.
3. Place strips in the cell with the anodic mark on the anodic side of the electrophoretic cell.
4. Fill the trenches with 15μl. of the concentrated spinal fluid.
5. Migrate at 150 volts until the stained albumin reaches the 3.2-cm. anodic migration mark. (The albumin fraction will stain as it migrates. If the marker dye is used, the second or slow migrating dot is the albumin.)
6. Place the strips in the fixative solution for 20 minutes and then dry.
7. Stain the strips for 30 minutes, then destain in three different changes of destain solution. Dry the strips.
8. Scan with the Analytrol using 500-mμ interference filters and quantitate the spinal fluid protein fractions.

RESULTS

1. Normal CSF protein electrophoretogram.

Total protein	0–50 mg./100 ml.
Prealbumin, prominent	$3.3\% \pm 1.1\%$
Albumin, predominant	$55.6\% \pm 2.6\%$
α_1-globulin	$5.6\% \pm 2.3\%$
α_2-globulin	$8.4\% \pm 0.4\%$
β-globulin—slow transferrin	$16.5\% \pm 3.8\%$
γ-globulin	$10.5\% \pm 3.0\%$

2. Extra CSF protein or serum protein pattern—reflects the serum abnormalities, such as γG paraprotein (multiple myeloma).
3. Capillary permeability pattern—increase in the normal serum proteins in the CSF pattern with an increase in the total protein. The β_1-globulin is elevated primarily.
4. Degenerative pattern—marked by an increase in the β_2-globulin which is greater than the β_1-globulin fraction.
5. Immunoglobulin pattern—increased γ-globulin without corresponding serum changes. The immunoglobulin specificity by immunoelectrophoresis does not show paraprotein changes. γG is increased in multiple sclerosis while γM is increased in neurosyphilis and trypanosomiasis.

Urine Protein Electrophoresis

PRINCIPLE

Separation of urinary proteins by agarose electrophoresis for qualitative and quantitative evaluation of the proteins, as in the case of spinal fluid, requires concentration since the protein content is less than can be detected in an electrophoretogram by the usual staining methods. By concentrating urine 100 times and increasing the quantity of sample to 15μl., sufficient protein is detected in the electrophoretogram to assure good preparation for interpretation. Information derived from urine protein electrophoresis is clinically very useful [9, 15, 79, 93].

EQUIPMENT AND REAGENTS

1. Standard gel electrophoretic equipment (see Standard Gel Electrophoretic Method).
2. Standard gel electrophoretic reagents (see Standard Gel Electrophoretic Method).
3. Standard agarose-buffer mixture (see Standard Gel Electrophoretic Method).

PROCEDURE

1. With 5 ml of warm buffered agarose, coat strips as for trench method. Mark at the midpoint (origin) and 3.2-cm. anodic migration point.
2. Fill each buffer compartment with barbital buffer.
3. Place strips in the cell with the anodic mark on the anodic side of the electrophoretic cell.
4. Fill the inoculation trenches of each strip with 15μl. of the concentrated urine.
5. Migrate at 150 volts until the stained albumin reaches the 3.2-cm. anodic mark. (The stained albumin fraction can be readily identified during migration. If the marker dye is used, the second or slow migrating dot is the albumin.)
6. Place the strips in the fixative solution for 20 minutes.
7. Dry the strips.
8. Stain the strips for 30 minutes, destain in three separate changes of destain solution, and dry.
9. Scan in Analytrol and quantitate the urinary protein fractions as for serum proteins.

RESULTS

1. Normal urinary electrophoretogram—predominantly albumin with trace amounts of $\alpha_1 G$, $\alpha_2 G$, and βG, and indistinct γG. Normal total urine protein—0–10 mg. per 100 ml.
2. Glomerular permeability pattern—predominantly albumin with

increased amounts of βG and $\alpha_1 G$. Total urine protein is increased.

3. Prerenal pattern—marked by an increase of low-molecular-weight proteins (Bence Jones, hemoglobin, myoglobin, cationic protein [CP]), which may give increased α_1-, α_2-, β-, or γ-globulin depending upon what particular protein is increased in the serum. Total protein is increased.

4. Tubular pattern—marked increases in the α_2-globulin and β_2-globulin with some increase in the slow γ-globulin. The albumin is decreased. Total protein is increased.

5. Chyluria—protein pattern similar to a serum electrophoretogram with the urine having a milky appearance. Increased total protein.

6. Postural proteinuria—in an upright position, pattern comparable to that of glomerular permeability. In a recumbent position the urine pattern is normal.

7. Exercise proteinuria—following exercise, urine pattern same as that of an accentuated normal pattern.

Bence Jones Protein

PRINCIPLE

Bence Jones protein precipitates between 40° and 60° C., dissolves at 100° C. in an acid pH, and reprecipitates when urine is cooled to 60° C. Bence Jones proteins are the light chains of the immunoglobulin molecule, which can be typed with specific antisera by immunodiffusion as $\varkappa$ type (60 percent), λ type (30 percent), or unidentified (10 percent) [68, 77, 82].

Bence Jones protein can be qualitatively detected by a heat extraction test [62]. The uroproteins are precipitated with trichloracetic acid, then are dried on filter paper, and the Bence Jones is extracted from the other uroproteins into a hot acetate buffer solution. When the acetate buffer is cooled and acidified with sulfosalicylic acid, the Bence Jones protein reprecipitates. Other uroglobulins and albumin remain fixed to the filter paper during the extraction procedure.

REAGENTS

1. Screening procedure for the presence of protein.
   a. Sulfosalicylic acid, 5 percent (W/V)—5 gm. sulfosalicylic acid dissolved in 100 ml. distilled water.
2. Differentiation procedure for Bence Jones specificity.
   a. Trichloracetic acid, 50 percent (W/V)—50 gm. trichloracetic acid dissolved in 100 ml. distilled water.
   b. Ammonium hydroxide, 27 percent (W/V) (U.S.P.).

c. Sodium chloride–acetate buffer, pH 4.3.
NaCl, 10 percent (W/V)   100 ml.
Sodium acetate, 2M   14 ml.
   Anhydrous salt, 164.1 gm. per liter
   Trihydrated salt, 272.1 gm. per liter
Acetic acid, 2M   36 ml.
   Glacial acetic acid, 114.6 ml. per liter
Distilled water to   1000 ml.
d. Sulfosalicylic acid, 20 percent (W/V)—20 gm. sulfosalicylic acid dissolved in 100 ml. distilled water.

PROCEDURE

1. Screening procedure for detectable protein.
   a. Mix equal parts of a centrifuged urine and 5 percent sulfosalicylic acid.
   b. If precipitate or cloudiness occurs, uroproteins are present, and extraction is required to determine whether Bence Jones protein is also present. If there is no cloudiness, the test is negative for uroproteins including Bence Jones protein.
2. Extraction procedure for differentiation of Bence Jones protein from other uroproteins.
   a. Depending on the amount of precipitate formed in the screening procedure, add 50 percent trichloracetic acid to another aliquot of the urine as follows:
      More than 2+ precipitate (easily seen as turbidity):
        5 ml. urine and 0.5 ml. acid.
      Less than 2+ precipitate (cloudy or slightly turbid):
        10 ml. urine and 1.0 ml. trichloracetic acid.
   b. After the uroproteins have precipitated, centrifuge at 2500 rpm for 10 minutes.
   c. Decant the supernatant, drain, and wipe the inside of the tube dry.
   d. Resuspend the precipitate in 3 drops of 27 percent ammonium hydroxide by stirring with a glass rod.
   e. Drain the resultant suspension on a strip of filter paper (½″ by 3″) and dry at room temperature.
   f. In a test tube heat 1.5 ml. of NaCl-acetate buffer in a boiling water bath. Roll up the dry filter paper, drop it in the hot buffer, and leave for 10 minutes.
   g. Cool the test tube by immersion in cold water. Remove the paper with an applicator stick.
   h. Add 3 drops of 20 percent sulfosalicylic acid to the cooled NaCl-acetate buffer.
   i. Observe the solution for cloudiness or precipitate.

RESULTS

Bence Jones negative: The buffer will be clear or unchanged after the addition of the sulfosalicylic acid.

Bence Jones positive: The buffer will become cloudy or have a precipitate after the addition of the sulfosalicylic acid.

NOTES

1. A urine should be examined for Bence Jones protein by the heat test and by electrophoresis. If the electrophoretogram and the Bence Jones heat test are both positive (monoclonal peak in the urine and a positive Bence Jones test), an immunoelectrophoresis should be done using specific light-chain antiserum to determine the type of light chain; approximately 10 percent of Bence Jones protein will not be typable.

2. For completeness of study urines should have both an electrophoretogram and a Bence Jones determination. Occasionally a monoclonal peak in the urine will not be associated with a positive test for Bence Jones proteins. For this reason it is our practice to include γG, γA, γM, and immunoelectrophoretic precipitin (IEP) antisera in the immunoelectrophoretogram along with x and λ antisera as denoted in the pattern for immunoelectro-plate described below under Urine Protein Immunoelectrophoresis.

Standard Method for Immunoelectrophoresis

PRINCIPLE

Immunoelectrophoresis consists of two phases: (1) gel electrophoresis and (2) immunodiffusion (see Chap. 1) [15, 33]. The antigen (usually serum protein in the clinical laboratory) is first separated by electrophoresis, and antiserum is placed in trenches parallel to the axis of electrophoresis for the immunodiffusion phase. Immunoprecipitin bands form in the gel where the antigen and specific antigen meet. The bands are useful in the identification of specific antigens.

EQUIPMENT AND REAGENTS

1. Standard immunoelectrophoresis equipment.
    a. Standard gel electrophoretic equipment.
    b. Immuno-Gel board or equivalent method for making immunoelectro-plates.
    c. Incubation chamber, room temperature or equivalent (HCS Corp.).*
    d. Mylar D or Cronar, 0.004-0.005 inch thick, 5″ by 7″ sheets,

*See Appendix.

or equivalent (E. I. DuPont de Nemours & Co. or HCS
Corp.).*
    e. Solar Immuno-Glo (HCS Corp.)* camera with 4-inch lens
        or equivalent system for recording immunoprecipitin bands,
        such as a Polaroid MP-3 camera.
    f. Immuno-Glo (HCS Corp. or Hyland Laboratories)* or
        equivalent view box.
    g. Wintrobe pipette.
    h. Plastic film support (HCS Corp.)*—5″ by 7″ Lucite plastic
        ⅛-inch thick sheets.
    i. Water aspiration pump.
2. Standard gel immunoelectrophoretic reagents.
    a. Barbital buffer, pH 8.6, ionic strength 0.05.
    b. Normal saline solution, 0.9 percent.
    c. Protein stain—1 gm. Thiazine Red R (Allied Chemical
        Corp.)* in 100 ml. of 10 percent aqueous acetic acid solution.
    d. Destaining solution—2 percent aqueous acetic acid solution.
    e. Tannic acid solution, 1 percent (W/V) (Mallinckrodt Chem-
        ical Works).*
3. Standard agarose-buffer mixture for immunoelectrophoresis.
    Dissolve by gently boiling 350 mg. of agarose (Bausch &
    Lomb)* and 2 drops of 2 percent bromphenol blue (Allied
    Chemical Corp.)* in 25 ml. barbital buffer and 25 ml. distilled
    water.

PROCEDURE

1. Fill electrophoretic cell buffer chambers with barbital buffer.
2. Preparation of immunoelectro-plate.
    a. Place Mylar or Cronar sheet on the Immuno-Gel board.
        Label the sheet with the name and/or number, and also
        label the antibody trenches. Mark marginal midpoint and
        an anodic 3.2-cm. migration limit point.
    b. Coat the sheet with 30 ml. of the warm buffered agarose
        and permit to harden.
    c. Using the guidelines and the slide, cut the trenches on the
        5-inch axis according to the desired pattern with the gel
        scoop (Fig. 15-7). See Figure 15-8 for open and closed
        antigen pattern.
    d. Line up the slide so that the punched wells of the slide
        correspond with the diagram (Figs. 15-7 and 15-8). Using
        a Wintrobe pipette, connected to water aspirator suction,
        cut antigen wells and remove plug of gel in one operation
        (Fig. 15-9).
    e. Place the immunoelectro-plate in the electrophoretic cell
        filled with barbital buffer as shown in Figure 15-10.

*See Appendix.

FIGURE 15-7

Immuno-Gel board used to control position of groove in agarose by scoop method for immunoelectrophoresis

3. Electrophoretic phase.
   a. Inoculate the antigen wells with 2μl. of specimen using a Hamilton syringe (Fig. 15-11).
   b. Migrate at 150 volts until the stained albumin has migrated 3.2 cm. from point of origin.
   c. Remove the immunoelectro-plate from the electrophoretic cell and place it on the plastic film support.
4. Immunodiffusion phase.
   a. Fill the corresponding antibody trenches with 100μl. of the antiserum.
   b. Place the immunoelectro-plate in the incubation chamber for 16 hours (Fig. 15-12).
5. Photography.
   a. Put the immunoelectro-plates in 5″ by 7″ film hangers and place them in tannic acid for 10 minutes to intensify the precipitin bands.
   b. Immediately place in water.
   c. Photograph the immunoelectrophoretogram twice with a Solar Immuno-Glo camera with 4-inch lens or a Polaroid MP-3 camera (4-inch lens f22, one second) and dark-field illumination box (Immuno-Glo). See chapter on monoclonal, polyclonal, and dysclonal gammopathies (Figs. 2-5 and 2-7).

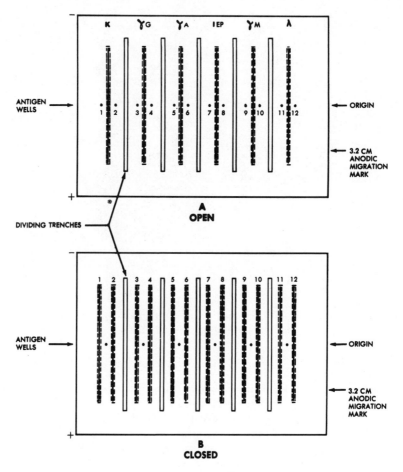

FIGURE 15-8

Immunoelectro-plates. Designs are of two types, open and closed. Dividing trenches are cut with the scoop method at specified intervals to separate each testing unit. In the open plate method the usual practice is to include trenches for light chain ($\varkappa$ and λ) and γA, γG, γM, and polyvalent (IEP). In the closed plate method one antigen well is surrounded by two antiserum trenches. Therefore, 12 antisera may be used against six antigens.

    d. The Polaroid prints are sufficient for completion of the test, and the actual immunoelectrophoretograms may be discarded or, if desired, processed for staining.

6. Soaking and staining.

    a. Soak the immunoelectro-plates in distilled water for 24 hours (to remove tannic acid), then in normal saline solution for 48 hours. The soaking phase removes excess protein and buffer salts.

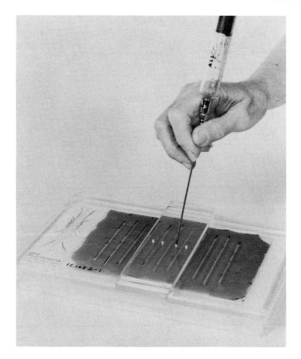

FIGURE 15-9
The slide on the Immuno-Gel board guide being used to position aspirating
pipette for creation of antigen wells

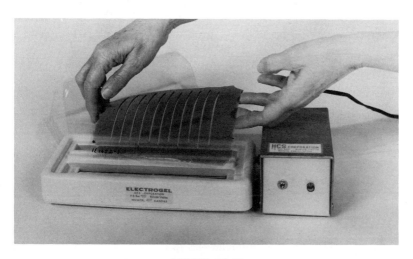

FIGURE 15-10
Completed immunoelectro-plate of the open type being inserted into an
electrophoretic cell

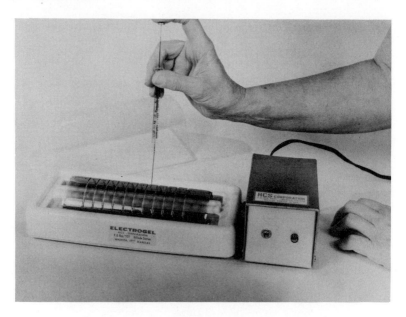

FIGURE 15-11
Immunoelectro-plate being inoculated with antigen by Hamilton syringe

FIGURE 15-12
Incubation chamber for immunodiffusion phase of immunoelectrophoresis.
In order to prevent loss of moisture from the immunoelectro-plate, a small
beaker of water is usually placed in the incubation chamber.

b. Dry the immunoelectro-plates in a drying oven at 60° C.
c. After the plates are completely dry, they may be processed exactly as protein electrophoretograms are by staining with Thiazine Red R for 30 minutes, destaining, and drying.

REPORT

The Polaroid print is used for interpretation, and a duplicate may be attached to the report and labeled according to antisera and antigen.

Serum Protein Immunoelectrophoresis

PRINCIPLE

Immunoelectrophoresis of serum proteins is useful in identifying specific antigens [15, 21, 33, 81]. Abnormalities are shown by the absence of a specific antigen, by a decrease of a specific antigen, and/or by an increase of a specific antigen. An increase of a specific antigen is depicted usually as a broadening of the precipitin band. Paraproteins can be demonstrated by displacement, bowing, or broadening of the immunoprecipitin line. (See Chap. 1.)

EQUIPMENT AND REAGENTS

1. See Standard Method for Immunoelectrophoresis, above.

PROCEDURE

1. Prepare open antigen immunoelectro-plate as per general immunoelectrophoresis instructions (Fig. 15-8). One sheet is designed for a single serum, which is tested for γG, γA, and γM. It is also reacted with IEP (antisera to whole human serum). In each system the serum is used both undiluted and diluted 1:10.
2. Inoculate antigen wells 3, 5, 7, 9 from left to right with 2μl. of serum according to open antigen pattern (Fig. 15-8). Inoculate 4, 6, 8, 10 antigen wells with 2μl. of a 1:10 dilution of serum in saline solution. (A normal control serum may be used in place of the dilution.)
3. Continue electrophoresis until albumin reaches 3.2-cm. anodic migration mark.
4. Remove sheet from electrophoretic cell and fill the antiserum trenches with 100μl. of specified antisera according to Figure 15-8 (open antigen pattern).
5. Following a suitable period of incubation for the immunodiffusion phase (usually overnight), the immunoelectrophoretograms are intensified and photographed.

REPORT

The Polaroid print is used for interpretation and reporting, and a duplicate print is attached to the report.

NOTES

1. The dilution of the serum serves as a rough quantitative measurement of the amount of antigen present.
2. Prozone effect is sometimes encountered in high concentration of antigen, and immunoprecipitin bands may be dissolved. The 1:10 dilution of the serum will overcome this effect and thus prevent having to repeat the test on a subsequent day with a more dilute sample.
3. The addition of 2-mercaptoethanol to serum is occasionally needed in high concentrations of γM. The γM will become tightly bound in the gel and will neither migrate nor form a definite precipitin band. One drop of 2-mercaptoethanol to 0.3 ml. of serum before inoculation of the immunoelectrophoretogram will break the disulfide bands of the macromolecule and permit it to migrate and diffuse freely.

Cerebrospinal Fluid (CSF) Protein Immunoelectrophoresis

PRINCIPLE

Immunoelectrophoresis for CSF protein is used as a tool for identifying the presence of abnormal or increased quantities of protein [15, 26, 51, 79, 93]. In Figure 15-13 is a CSF immunoelectrophoretogram on spinal fluid from a patient with multiple scelerosis developed with rabbit anti-CSF made against pooled spinal fluid from 10 cases of multiple sclerosis.

EQUIPMENT AND REAGENTS

See Standard Method for Immunoelectrophoresis, above.

PROCEDURE

1. Prepare immunoelectro-plate as per general immunoelectrophoresis instructions. The closed antigen pattern (Fig. 15-8) is used. Routinely used antisera are γG, γA, γM, β_1C/β_1A, transferrin, IEP, hemopexin, ceruloplasmin, fibrinogen, α_2-macroglobulin, β-lipoprotein, anti-CSF.
2. Inoculate the antigen wells with 2µl. of concentrated CSF.
3. Continue electrophoresis until albumin reaches 3.2-cm. anodic migration mark.
4. Fill antiserum trenches with 100µl. of specific antisera.

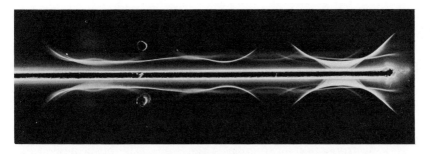

FIGURE 15-13

Pattern of a spinal fluid immunoelectrophoretogram showing the delicate nature of the bands formed. The anode is on the right. Notice the pre-albumin. In the γ-globulin zone notice the discrete number of components, one of which is γG. See text for more detail. This is a spinal fluid from a patient with multiple sclerosis. The immunoelectrophoretogram was developed with spinal fluid antiserum raised in a rabbit. The spinal fluid used for production of the antiserum was pooled from several cases of multiple sclerosis.

5. Diffuse, intensify, and photograph the immunoelectrophoreto-gram.

REPORT

Total protein, color, and cells of spinal fluid are helpful parameters in interpretation of electrophoresis and immunoelectrophoresis.

The patterns of CSF electrophoretograms can be correlated with the immuno patterns as follows:

Extra CSF protein pattern—identification of increased serum antigen in spinal fluid, such as identification of a serum monoclonal peak (γG) also present as a γG gammopathy in the spinal fluid.

Capillary permeability pattern—characterized by an increase of the fast transferrin.

Degenerative pattern—characterized by an increase of the slow transferrin.

Increased immunoglobulin pattern—characterized by an increase, but not a dysprotein, of one of the immunoglobulins. γG is increased in multiple sclerosis and γM increased in neurosyphilis.

The presence of hemopexin, α2-macroglobulin, indicates a non-specific abnormal spinal fluid. β-Lipoprotein has been reported in cases of brain or spinal cord tumors.

Urine Protein Immunoelectrophoresis

PRINCIPLE

Immunoelectrophoresis is used to identify the specific proteins found in a urinary electrophoretogram [9, 15, 77, 93]. It most often is utilized to type monoclonal peaks found in urine. Monoclonal peaks in the urine can be composed of the intact immunoglobulin, light chain (Bence Jones, ϰ or λ type), or heavy chain.

EQUIPMENT AND REAGENTS

See Standard Method for Immunoelectrophoresis, above.

PROCEDURE

1. Prepare immunoelectro-plate according to general immuno-electrophoresis instructions (Fig. 15-8). The open antigen pattern (Fig. 15-8) is used with a set for each of the following antisera: ϰ, γG, γA, IEP, γM, λ.
2. Inoculate the antigen wells from left to right: 1, 3, 5, 7, 9, 11 —2μl. of 100-fold concentrated urine; 2, 4, 6, 8, 10, 12—2μl. of urine, unconcentrated.
3. Migrate albumin to 3.2-cm. anodic mark.
4. Fill the antiserum trenches with 100μl. of the specified anti-serum.
5. Incubate immunoelectrophoretogram in moist chamber for 16 hours (overnight) at room temperature.
6. Intensify and photograph the immunoelectrophoretogram as outlined under general immunoelectrophoretic instructions.

RESULTS

Interpret the Polaroid photograph with reference to electrophoretic pattern.

Haptoglobin Electrophoresis and Haptoglobin Genotyping

PRINCIPLE

Haptoglobins are a group of plasma proteins which have the property of combining with free hemoglobin. Polymorphism of the haptoglobins can be demonstrated by molecular sieve electrophoresis (Fig. 15-14). Biologically the haptoglobins serve to bind free hemoglobin released from ruptured red cells. Following hemolysis, as in hemolytic anemia, the haptoglobins are decreased. To detect the level of haptoglobin in a patient's serum a known amount of hemoglobin is added to the serum in excess of the haptoglobin-binding capacity. The mixture is subjected to zone electrophoresis, and the hemoglobin is detected by staining for peroxidase activity [12, 59, 63, 93]. The percentage of each fraction is calculated from the area under each peak obtained from an Analytrol scan. The hemoglobin fraction migrates in the post-β zone while the haptoglobin-hemoglobin complex migrates

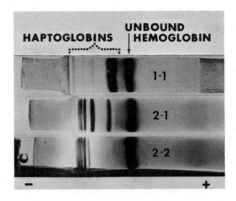

FIGURE 15-14

Disc electrophoretograms of haptoglobin type. Upper—type 1-1; middle —type 2-1; lower—type 2-2.

between the β and the α_2 zone. If the above mixture is subjected to molecular sieve electrophoresis, haptoglobin genotypes 2-2, 1-1, and 2-1 can be demonstrated (Fig. 15-14).

EQUIPMENT AND REAGENTS

1. Standard gel electrophoretic equipment (see Standard Gel Electrophoretic Method).
2. Barbital buffer, pH 8.6, ionic strength 0.05.
3. Glycine buffered gel—1.0 gm. glycine and 250 mg. agarose (Bausch & Lomb)* dissolved in 50 ml. barbital buffer, pH 8.6, ionic strength 0.025, by bringing to a boil.
4. Hemolysate—prepared from normal erythrocytes, as for hemoglobin electrophoresis, except that concentration is adjusted to approximately 3 gm. per 100 ml. After filtering hemolysate measure hemoglobin concentration of solution.
5. Peroxidase stain.
   a. Saturate 50 ml. of 10 percent aqueous acetic acid with benzidine.
   b. Just before use add 0.1 ml. of 30 percent hydrogen peroxide.
   c. To check for reactivity of the stain, touch an applicator stick with whole blood and pour a small amount of prepared stain onto it. If the blood-stained applicator stick turns blue to black quickly, the stain is well saturated with benzidine and is reacting correctly. If not, prepare more stain from fresh materials.

*See Appendix.

PROCEDURE

1. Collect 5 ml. of unhemolyzed venous blood from candidate and also from a normal control. Allow to clot and centrifuge. Remove the serum.
2. Using standard method, coat strips marked at midpoint with 5 ml. of the warm glycine buffered gel. Place in an electrophoretic cell containing barbital buffer, pH 8.6, ionic strength 0.05.
3. Measure 0.9 ml. of candidate's serum into one tube and 0.9 ml. of control serum into another tube. Add 0.1 ml. of the hemolysate to each tube, and incubate for 30 minutes at 37° C.
4. Inoculate strips with 1.5μl. of the prepared serum-hemolysate mixture by the razor blade method. Electrophoresis should continue for 60 minutes at 150 volts.
5. Remove the strips from the cell and place immediately in enough peroxidase stain to cover strips for 5–10 minutes.
6. Remove the strips from the stain and place in several different washes of distilled water for a total of 90 minutes.
7. Dry and scan in Analytrol with 550 mμ filters. Quantitate percentages of hemoglobin fractions and hemoglobin-haptoglobin fraction from the integration data. Calculate milligrams per 100 ml. of hemoglobin in each fraction based on known hemoglobin concentration of the original hemolysate.

> EXAMPLE: 4.0 gm. known hemoglobin concentration of starting hemolysate
> 60% hemoglobin fraction
> 40% hemoglobin-haptoglobin fraction
> 1:10 dilution of original hemolysate = hemoglobin concentration of 400 mg./100 ml.
> $400 \times .60 = 240$ mg./100 ml.
> $400 \times .40 = 160$ mg./100 ml.

REPORT

1. _____ mg. per 100 ml. of serum—hemoglobin fraction.
2. _____ mg. per 100 ml. of serum—hemoglobin-haptoglobin complex.

NORMAL

Average normal mean is 300 mg. per 100 ml. in our laboratory with range of mean ± 100 mg. per 100 ml. Each laboratory should establish its own normal range.

HAPTOGLOBIN GENOTYPES

Haptoglobin typing has been frequently employed in paternity testing, in anthropologic studies, and in genetic investigation. A brief description of a simple method seems in order. The hemoglobin-binding capacity as measured by electrophoresis does not take into consideration the difference of binding of the three major types of haptoglobin (2-2, 1-1, and 2-1). To evaluate the haptoglobin binding accurately, Ferris et al. [31] proposed a combined approach of determining the phenotype and hemoglobin binding in the same electrophoretogram using an acrylamide slab. For genetic studies we have used disc electrophoresis, as outlined under the section on methods of acrylamide electrophoresis. The hemolysate is prepared as for haptoglobin electrophoresis—10μl. of the hemoglobin mixture mixed with 100μl. of serum to be tested and incubated for 30 minutes at 37° C. Procedure for disc electrophoresis of serum up to fixation and staining is followed. Staining as for haptoglobin electrophoresis is followed. Results as shown in Figure 15-14 are to be expected.

Hemoglobin Electrophoresis

PRINCIPLE

Hemoglobin types can be separated because the difference in primary structure of α and β chains imparts enough difference in net charge from the natural form to make possible separation by electrophoresis [15, 59, 63, 81, 93]. Electrophoresis in a glycine-gel system shows that hemoglobin variants can be identified by their relative mobility compared to that of known controls (Fig. 15-15).

EQUIPMENT AND REAGENTS

1. Standard gel electrophoretic equipment.
2. Barbital buffer, pH 8.6, ionic strength 0.05.
3. Barbital buffer, pH 8.6, ionic strength 0.025.
4. Buffered gel—250 mg. agarose (Bausch and Lomb)* and 1.0 gm. of glycine added to 50 ml. of barbital buffer, pH 8.6, ionic strength 0.025, and brought to a boil.
5. Fixative solution—glacial acetic acid–absolute methanol solution, 10 percent (V/V).
6. Protein stain—1 gm. Thiazine Red R (Allied Chemical Corp.)* in 100 ml. of 10 percent (V/V) aqueous acetic acid.
7. Destaining solution—2 percent (V/V) aqueous acetic acid solution.
8. Sodium chloride, 0.85 percent (W/V).
9. Toluene.

PROCEDURE

1. Preparation of hemolysate.
   a. Collect 5 ml. of blood in ethylenediamine tetra-acetic acid

*See Appendix.

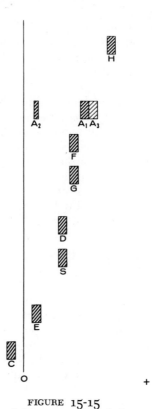

FIGURE 15-15

Relative mobility of hemoglobin types found by agarose electrophoresis. Not all types, but those most commonly encountered are shown.

(EDTA) and separate the cells from the plasma. Prepare an AA control and an F control in addition to the patient. An AS control is also desirable.

b. Wash the cells four times with normal saline solution.

c. Adjust the hemoglobin concentration to approximately 5 gm. per 100 ml. with distilled water. Electrophoretic separation appears to be better at this hemoglobin concentration.

d. Add 1 ml. toluene and shake vigorously for three to five minutes. Centrifuge at 3000 rpm for 15 minutes. Discard the toluene layer at the top. Filter the hemolysate. The hemolysate should be free of plasma and red blood cell (RBC) stroma since their presence will cause smudging in the electrophoretic pattern.

2. Place origin mark at 1.5 cm. cathodic from midpoint, coat film strips with 5 ml. of agarose-glycine buffered solution as for general method, and allow gel to harden.

3. Inoculate each strip with 1.5µl. of hemolysate using the razor blade at the 1.5-cm. cathodic mark. Place strips on the electrophoretic cell containing barbital buffer, pH 8.6, ionic strength 0.05.
4. Migrate at 150 volts for 90 minutes.
5. Fix, dry, and stain as for protein electrophoresis, except increase fixation time to overnight or until glycine crystals are removed.
6. When dry, scan in Analytrol with 500 mµ interference filter and calculate the percentage of the different hemoglobin fractions.

REPORT

Report the percentage of the F control, AA control, patient, and any other controls that may have been run. Sickle cell preparation and alkali-resistant hemoglobin tests are often necessary to clarify the nature of the hemoglobin type and should be routinely performed.

RESULTS

The more common hemoglobin variants migrate toward the anode in agarose gel at alkaline pH on account of differences in electric charge arising from slight differences in amino acid composition in the following order: $A_3 > A_1 > F > S + D > A_2 + E > C$. These relative electrophoretic patterns are shown in Figure 15-15.

Hexokinase Isoenzymes of Erythrocytes

PRINCIPLE

Hexokinase (adenosine triphosphate [ATP]: D-hexose-6-phosphotransferase, 2.7.1.1) [29] catalyzes conversion of glucose to glucose-6-phosphate. Glucose-6-phosphate dehydrogenase (G-6-PD) augments breakdown of glucose-6-phosphate. The enzyme hexokinase exists in erythrocytes in multiple forms which are demonstrated histochemically in agarose electrophoretograms by the formation of an insoluble blue-black formizan as the final hydrogen acceptor [38, 41]. The linked reactions are as follows:

Glucose + ATP $\xrightarrow{\text{HK}}$ glucose-6-phosphate + ADP

Glucose-6-phosphate + NADH + PMS $\xrightarrow{\text{G-6-PD}}$ 6-phosphogluconate + NAD + red. PMS

Red. PMS + Nitro-BT $\longrightarrow$ formizan $\downarrow$ + PMS

EQUIPMENT AND REAGENTS

1. Standard gel electrophoretic equipment.
2. Buffer for tank and strips.
   Barbital buffer, pH 8.6, ionic strength 0.05   1000.0   ml.

      Sodium ethylenediamine tetracetate

          (Na$_2$EDTA), 1.0 mM. per liter   0.372 gm.

      2-Mercaptoethanol, 5 mM. per liter   0.39 gm.

3. Phosphate buffer for hemolysate, pH 7.0, M/150.

      Phosphate buffer, pH 7.0, M/15   10.0   ml.

          KH$_2$PO$_4$, M/15   19 ml.

          Na$_2$HPO$_4$, M/15   31 ml.

      Distilled water   90.0   ml.

      2-Mercaptoethanol, 5 mM. per liter   0.039 gm.

      Na$_2$EDTA, 5 mM. per liter   0.186 gm.

      Glucose, 0.01M per liter   0.18 gm.

4. Saline solution for washing cells.

      Sodium chloride   9.0   gm.

      Sucrose, 0.01M   0.18 gm.

      Distilled water to   1000.0   ml.

5. Substrate.

      Phosphate buffer, pH 7.4, M/10   10.0   ml.

      Nicotine adenine dinucleotide phosphate

          (NADP)*   38.3   mg.

      MgCl$_2$ 15.2   mg.

      Adenosine triphosphate (ATP)*   32.3   mg.

      Glucose-6-phosphate dehydrogenase

          (G-6-PD)*   40.0   E.U.†

      Potassium cyanide (KCN)   1.3   mg.

      Phenazine methasulfate (PMS)* (1 mg./ml.) 0.4   ml.

      Nitro Blue Tetrazolium (NBT)*   4.0   mg.

      Glucose (1 mg./ml.)   0.1   ml.

6. Buffered gel. To 50 ml. of barbital buffer for strips add 250 mg. of agarose and bring to a boil.

PROCEDURE

    1. Collect blood in heparin or EDTA.

    2. Centrifuge; remove plasma and buffy coat.

    3. Wash cells three times in saline solution containing glucose and pack by centrifugation.

    4. Resuspend cells in an equal volume of phosphate buffer for hemolysate.

    5. Lyse by freezing and thawing.

    6. Centrifuge to remove stroma.

    7. Prepare agarose strips as for trench method and place in electrophoretic cell.

    8. Inoculate with 15μl. of the hemolysate.

    9. Apply dot of bromphenol blue–albumin as starting marker.

  10. Migrate at 150 volts until marker has reached the 3.2-cm. line.

*See Appendix (Sigma Chemical Co.).

†E.U. = enzyme unit.

11. Remove from cell and cover with substrate. Incubate for 90 minutes at 37° C.
12. Soak in distilled water 30 minutes to remove buffer salts and dry.

RESULTS

Enzyme activity is demonstrated by the appearance of blue-black bands.

Myoglobin Electrophoresis

PRINCIPLE

Myoglobin in the urine is frequently confused with hemoglobin because of the dark or reddish color imparted to the urine. Detection of myoglobin may be done spectroscopically, although hemoglobin may cause interference. Electrophoretically, hemoglobin and myoglobulin can be separated and localized histochemically by staining for peroxidase activity [27, 28, 93]. Myoglobin is less negatively charged than hemoglobin and travels less rapidly than hemoglobin in an electrophoretic field.

EQUIPMENT AND REAGENTS

1. Standard gel electrophoretic equipment (see Standard Gel Electrophoretic Method).
2. Standard gel electrophoretic reagents (see Standard Gel Electrophoretic Method).
3. Standard agarose-buffer mixture (see Standard Gel Electrophoretic Method).
4. Peroxidase stain. To 50 ml. of 10 percent acetic acid saturated with benzidine add 0.1 ml. of 30 percent hydrogen peroxide. Make up immediately before using and check reactivity with whole blood.
5. Myoglobin control (muscle extract purified by column chromatography).

PROCEDURE

1. Prepare strips as for trench method. Mark midpoint and anodic migration point 4.5 cm. from midpoint.
2. Place the strips in the electrophoretic cell containing barbital buffer, pH 8.6, ionic strength 0.05.
3. Inoculate the strips with 15μl. of the following:
   a. Control urine or serum.
   b. Patient's urine or serum.
   c. Patient's hemolysate.
   d. Myoglobin control (if available).

4. Apply a small dot of bromphenol blue–albumin at the margin of the midpoint.
5. Migrate at 150 volts until the stained albumin reaches the 4.5-cm. mark.
6. Immediately place the strips in the fresh peroxidase stain for 5–10 minutes.
7. Place the strips in several changes of water for a total of 90 minutes.
8. Dry, scan in Analytrol with 550 mμ filter, and quantitate.

RESULTS

The peroxidase-positive fraction cathodic to the A_1 hemoglobin band and anodic to A_2 hemoglobin is the myoglobin fraction. The patient's hemolysate is included so an unsuspected hemoglobinopathy does not lead to an erroneous conclusion of myoglobinuria. Myoglobin can be distinguished from hemoglobin with specific antiserum by gel diffusion.

Lipoprotein Electrophoresis

PRINCIPLE

Lipoproteins are carriers of triglycerides, cholesterol, fatty acids, and phospholipids and are detected in an electrophoretogram by lipid staining. These lipoproteins can be electrophoretically separated into four fractions [15, 32, 49, 50]. They are: chylomicrons at the origin, β-lipoproteins, pre-β-lipoproteins, and α-lipoproteins. An agarose system is used in which albumin and EDTA have been added to the buffering system. The electrophoretograms are stained for fat. Phenotypes of lipoproteins, according to Fredrickson and Lees, can be obtained by this method. Genetic polymorphism of β- and α-lipoproteins has been described (see Chap. 4).

EQUIPMENT AND REAGENTS

1. Standard gel electrophoretic equipment (see Standard Gel Electrophoretic Method).
2. Barbital buffer, pH 8.6, ionic strength 0.05.
3. Buffered gel. Dissolve 250 mg. agarose and 0.15 gm. EDTANa₄ in 50 ml. barbital buffer, pH 8.6, ionic strength 0.025. After the agarose has cooled to 55°–65° C., add 1.7 ml. of 30 percent bovine albumin (smaller amount may be used).
4. Fixative and destain solutions—55 percent (V/V) reagent alcohol (95 percent ethanol, 5 percent isopropyl).
5. Sudan Black B stain (Allied Chemical Corp.)*—10 gm. of Sudan Black B dissolved in 1 liter of warm 55 percent (V/V) reagent

*See Appendix.

alcohol. Filter immediately before using. NaOH may be added to improve some lots of stain.

PROCEDURE

1. Mark the midpoint and anodic migration point, which in this case is 3.2 cm. from origin. Using the trench method, coat each strip with 5 ml. of the warm agarose solution.
2. Place the strips in the cell filled with barbital buffer, pH 8.6, ionic strength 0.05.
3. Fill the trenches with 5μl. to 12μl. of serum. Apply a dot of bromphenol blue at the margin of each strip at the midpoint.
4. Perform electrophoresis at 150 volts until the leading edge of the bromphenol blue reaches the 3.2-cm. mark.
5. Remove the strips from the cell and place in the fixative solution for 15 to 30 minutes.
6. Dry the strips.
7. Stain the strips with the Sudan Black B for 30 to 60 minutes.
8. Destain in several changes of destain solution and dry.
9. Scan strips in Analytrol with 550 mμ filter and quantitate.

REPORT

PATTERN INTERPRETATION. Serum concentration of cholesterol, triglycerides, phospholipids, and fatty acid are helpful for pattern interpretation. In almost all instances only cholesterol and triglycerides are needed.

Normal—β- and α-lipoprotein with the β in excess of the α.

Abetalipoproteinemia—absence of β-lipoprotein; decrease of cholesterol, phospholipids, and plasma triglycerides.

Familial high-density lipoprotein deficiency (Tangier disease)— absence of α-lipoprotein; decrease of cholesterol and phospholipids with slight increase of triglycerides.

PHENOTYPES.

Phenotype I (familial fat-induced hyperlipemia)—increased chylomicrons, slight increase of serum cholesterol and triglycerides.

Phenotype II (familial hyperbetalipoproteinemia)—increased β-lipoprotein; increased serum cholesterol; normal to decreased triglycerides.

Phenotype III—binding of β-lipoprotein and pre-β-lipoprotein into a single broad band designated *floating beta;* increased serum cholesterol and triglycerides.

Phenotype IV (familial hyperprebetalipoproteinemia)—increased pre-β-lipoprotein; normal to decreased serum cholesterol and triglycerides.

Phenotype V (familial hyperchylomicronemia with hyperprebetalipoproteinemia)—increased chylomicrons, increased β-lipoprotein, and increased pre-β-lipoprotein; increased serum cholesterol and triglycerides.

Glycoprotein Electrophoresis

PRINCIPLE

Seromucoids are a heterogeneous class of carbohydrate-containing proteins which can be divided into two groups, the mucoproteins and the glycoproteins. The mucoproteins are defined as proteins with which an acid mucopolysaccharide (a polysaccharide containing hexosamine but no amino acid) is combined in a linkage that is relatively easy to split [36]. The glycoproteins are defined as proteins containing hexosamine linked to the amino acids of the proteins by a very stable bond. Mucoproteins contain more than 4 percent hexosamine whereas glycoproteins contain less than 4 percent hexosamine.

The electrophoretic fractions of serum are stained with periodic acid–Schiff reagent [8, 42]. In this stain the periodic acid oxidizes the closed aldehyde groups in the polysaccharide conjugate, permitting the fuchsin sulfite to stain the polysaccharide a violet color.

EQUIPMENT AND REAGENTS

1. Standard gel electrophoretic equipment (see Standard Gel Electrophoretic Method).
2. Barbital buffer, pH 8.6, ionic strength 0.05.
3. Barbital buffer, pH 8.6, ionic strength 0.025.
4. Agarose-buffer mixture. Dissolve 250 mg. agarose (Bausch & Lomb)* in 50 ml. of barbital buffer, pH 8.6, ionic strength 0.025, by bringing to a boil.
5. Fixative solution—95 percent reagent alcohol.
6. Periodic acid (G. Frederick Smith Chemical Co.)* (Fisher Scientific Co.).*† Dissolve and mix the following: 2 gm. periodic acid, 75 ml. distilled water, 175 ml. reagent alcohol, and 0.68 gm. sodium acetate. Make fresh each time it is used.
7. Schiff's reagent (Beckman Instruments, Inc.)* (Fisher Scientific Co.).*† Dissolve 6 gm. basic fuchsin in 1200 ml. of 90° C. distilled water; cool to 50° C. and filter. Add 30 ml. of 2N hydrochloric acid and 4 gm. potassium metabisulfite to the filtrate. Stopper and allow to stand in a cool, dark place overnight. Add 3 gm. powdered animal charcoal, then mix thoroughly and allow to stand for one minute. Filter. Add 40 ml. of 2N hydrochloric acid. Keep in a refrigerator. Discard the stain when it turns pink.

*See Appendix.
†Fisher Stain Pak for PAS. Commercial reagents which can be substituted for periodic acid, Schiff's reagent, and sulfite rinse.

8. Sulfite rinses (Fisher Scientific Co.).*† Dissolve 4 gm. potassium metabisulfite in 1000 ml. distilled water. Add and mix 10 ml. concentrated hydrochloric acid. Discard rinses after use.

PROCEDURE

1. Using trench method, coat film strips with 5 ml. of the warm buffered agarose. The strips are premarked with a midpoint and a 3.2-cm. anodic point.
2. Place the strips in the electrophoretic cell filled with barbital buffer, pH 8.6, ionic strength 0.05.
3. Apply 15μl. of serum to each strip. Dot with bromphenol blue–albumin solution.
4. Perform electrophoresis until the stained albumin portion of the marker reaches the 3.2-cm. anodic mark.
5. Fix the strips in the fixative solution for 30 minutes and dry in the oven.
6. Place the strips in the periodic acid solution for five minutes.
7. Wash the strips in running tap water for 10 minutes.
8. Place the strips in the Schiff reagent for 30 minutes.
9. Rinse the strips in the sulfite rinse for 10 minutes.
10. Wash the strips again in running water for 10 minutes.
11. Again immerse the strips in the sulfite rinse for 10 minutes.
12. Re-dry the strips.
13. Scan with Analytrol using 550 mμ filters and quantitate the strips.

REPORT

The percentages for each of the five fractions are reported.

NORMALS

Approximate normal percentages for the fractions are as follows:

Albumin	10.9%
$\alpha_1 G$	18.6%
$\alpha_2 G$	29.6%
βG	21.6%
γG	19.3%

SUGGESTION

The strips can be sprayed with a plastic spray after the final drying stage to prevent fading of the pink color. The strips then may be kept for long periods with little loss of intensity.

*See Appendix.
†Fisher Stain Pak for PAS. Commercial reagents which can be substituted for periodic acid, Schiff's reagent, and sulfite rinse.

Immunochemical Characterization of Serum β_1C (Complement 3–C'3)

PRINCIPLE

The very complex nature of the complement system and the difficult technic of analysis have prevented application of complement analysis in the diagnostic laboratory. It is now clear that complement 3 (C'3) is present in serum in relatively high concentration, that it is the same as β_1C globulin isolated by Muller-Eberhard et al. [60], and that it can be identified by immunoelectrophoresis and quantitated by immunodiffusion. By immunodiffusion, Klemperer et al. [44] have found that a measurement of serum β_1C correlated with the serum hemolytic titer of the complete complement system. In hereditary angioneurotic edema, during attacks, and in some apparently normal persons the complement titer is low while the β_1C content is normal. In the former disease there is an inborn defect of the complement which is reflected in decrease of C'4 and C'2. The decrease of serum complement titer in patients with acute glomerulonephritis is associated with decrease of serum β_1C. In lupus erythematosus (LE) both the serum complement and β_1C are decreased. A measure of β_1C by immunodiffusion thus has been established as a practical procedure to aid in the diagnosis of renal disease and of some autoimmune diseases such as LE.

The normal serum concentration of β_1C is 120–160 mg. per 100 ml. It should be noted that β_1C, after storage, undergoes structural changes associated with *loss of complement*. This change is noted electrophoretically as increase in mobility. The new mobility corresponds to a protein previously designated β_1A by immunoelectrophoretic analysis.

The procedure to be described is based on titration of serially diluted serum by double gel diffusion. The method is calibrated against fresh pooled normal serum (at least three sera).

Complement levels may also be quantitated by prepared quantitative plates obtained commercially which contain known standards. The use of single gel diffusion with the antibody in the gel, as described under Quantitation of Immunoglobulins, below, is also described in this section.

EQUIPMENT AND REAGENTS FOR DOUBLE GEL DIFFUSION METHOD

1. Goat anti–human β_1C/$_1$A (Hyland Laboratories).*
2. Cronar or Mylar sheet, 5″ by 7″.
3. Kodak film hangers, 5″ by 7″.
4. Incubation chamber.

*See Appendix.

5. Agarose.
6. Veronal buffer, pH 8.6, ionic strength 0.05. A phosphate buffer, pH 7.5, M/15 is perhaps better since precipitin bands form better near a pH of 7.0.

PROCEDURE FOR DOUBLE GEL DIFFUSION METHOD

1. Cover plastic sheet with 30 ml. agarose buffer solution, as for general immunoelectrophoresis.
2. With metal scoop, cut a trench 6 inches long in the middle on the long axis.
3. At 1-cm. intervals, along each side of the antiserum trench, and 0.5 cm. from the trench, cut 10 antigen wells by aspiration with a 18-gauge needle.
4. The finished product should have paired antigen wells approximately 1 cm. apart, separated by the central antiserum trench.
5. The right side is used for control and the left side for the unknown. The serum to be studied and the control serum are serially diluted with saline to final dilution of 1:512. Each of the 10 wells is then charged with 2μl. of the corresponding dilution.
6. Use anti-$\beta_1C/_1A$ serum in the trench (approximately 200μl.) and place the plate in an incubator for 16 hours (overnight).
7. Intensify the plate in 1 percent tannic acid for seven minutes and then soak it in water. Photograph the finished product, consisting of agarose-coated plastic sheet in a film hanger.

INTERPRETATION

The results are semiquantitative, but unknowns with end points three dilutions less than the control are considered to have $\beta_1C/_1A$ concentration less than 80 mg. per 100 ml. The end point is the last dilution showing an immunoprecipitin band. In our experience glomerulonephritis serum will usually have an end point at least three dilutions less than control. Experience in one's own laboratory, however, is necessary to arrive at a satisfactory system of evaluating the content of β_1C by immunodiffusion, as described above.

EQUIPMENT AND REAGENTS FOR SINGLE GEL DIFFUSION

1. Quantitative Immunoplate (Hyland Laboratories).* Each kit comes with three standards—high, mean, low—and six plates for the $\beta_1C/_1A$ antibody.
2. Hamilton syringe, 1μl. to 10μl.
3. Incubation chamber.
4. Solar Immuno-Glo camera and Immuno-Glo with quantitative grid (HCS Corp. or Hyland Laboratories).*

*See Appendix.

PROCEDURE FOR SINGLE GEL DIFFUSION

1. The plate is labeled according to its antigen specificity.
2. Each plate contains six antigen wells which are inoculated with 8μl. of sample as follows:
   a. $\beta_1C/_1A$ high standard.
   b. $\beta_1C/_1A$ mean standard.
   c. $\beta_1C/_1A$ low standard.
   d, e, f. Unknowns.
3. The Immunoplate is re-covered and placed in a moist incubator at room temperature for 16 hours (overnight).
4. The plates are intensified in a 1 percent tannic acid solution for five minutes.
5. The plates are gently washed with distilled water and placed on the quantitative grid (1-mm. squares) and a Polaroid print is taken (see Quantitation of Immunoglobulins, Fig. 15-20).
6. The millimeter marks can then be counted on the print and recorded.
7. It is convenient to have a clear plastic overlay or envelope to draw the curve on with a wax pencil so that the system may be reused. The diameters of the standards are plotted on the X axis against log concentration on the Y axis (see Quantitation of Immunoglobulins, Fig. 15-23). The curve is usually linear except at low concentration. Also the well size causes some error.
8. From the calibration curve the concentration of the unknowns can be interpolated.

RESULTS OF SINGLE GEL DIFFUSION

Normal serum complement level is 120–160 mg. per 100 ml.

Ceruloplasmin

PRINCIPLE

Ceruloplasmin is a protein with molecular weight of 151,000 which binds copper and has an agarose electrophoretic mobility equivalent to α_2-globulin. Ceruloplasmin acts as a polyphenoloxidase (monamine, oxygen oxidoreductase; 1.4.3.4) [29]. Ceruloplasmin can be localized immunoelectrophoretically with specific antiserum or by its reaction with o-dianisidine. This reagent is commonly used to detect hemoglobin and hemoglobin-haptoglobin complexes after zone electrophoresis. Phenylenediamine is also employed for the detection of ceruloplasmin in electrophoretograms and has been used extensively by Uriel [87, 89]. The oxidation of o-dianisidine by ceruloplasmin is presumably due to the oxidase properties of the protein. The modifications reported by Owen et al. [69] are used.

EQUIPMENT AND REAGENTS

1. Standard gel electrophoretic equipment (see Standard Gel Electrophoretic Method).
2. Barbital buffer, pH 8.6, ionic strength 0.05 (see Standard Gel Electrophoretic Method).
3. Standard agarose-buffer mixture (see Standard Gel Electrophoretic Method).
4. Solution A. Place 1 gm. of o-dianisidine (Sigma Chemical Co.)* and approximately 50 ml. of water in a flask. Add dilute hydrochloric acid (0.1N) until the o-dianisidine just dissolves. Make the volume up to 100 ml. with water (the solution will keep at room temperature two to three months).
5. Solution B.

Solution A		10 ml.
Acetate buffer, pH 5.6, 0.1M		10 ml.
Acetic acid, 0.2M	4.8 ml.	
Solution of sodium acetate, 0.2M	45.2 ml.	
Distilled water up to	100.0 ml.	
Ethanol		30 ml.
Distilled water up to		100 ml.

PROCEDURE

1. Coat strips marked at midpoint and 3.2-cm. anodic migration point using the trench method with 5 ml. of buffered gel solution.
2. Inoculate with 15μl. of serum.
3. Migrate at 150 volts until bromphenol blue–albumin reaches the 3.2-cm. anodic mark.
4. Remove from the cell and place in staining solution of o-dianisidine.
5. Incubate two hours at 37° C.
6. Soak in water to remove buffer salts and dry.

RESULTS

Ceruloplasmin is a single protein, stains brown, and migrates in the α_2-globulin area (Fig. 15-16). Ceruloplasmin also oxidizes benzidine.

Amylase Isoenzymes

PRINCIPLE

Alpha amylase (α-1,4-glucan 4-glucanohydrolase, 3.2.1.1) [29], which exists in body tissues, particularly salivary gland organ fluids, hydrolyzes the breakdown of polysaccharides, particularly starch, to

*See Appendix.

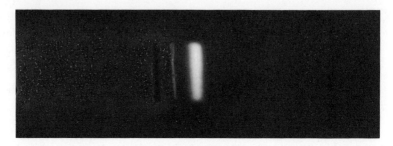

FIGURE 15-16

Demonstration of ceruloplasmin in an agar-gel electrophoretogram. Anode is on the right. Notice dense zone corresponding to the α_2-globulin position and a very light zone corresponding to the β-globulin zone. The latter represents lipoproteins.

maltose. Further enzyme hydrolysis is necessary to complete the breakdown to glucose.

Electrophoretically, amylase has been described in the literature as associated with all protein fractions [10, 53]. However, it is found only in the γ-globulin zone [1, 39, 65, 94]. The enzyme can be determined in the various protein zones by segmentation of the completed electrophoretogram into respective protein zones and analysis of each fraction for the enzyme. Serologic specificity of the enzyme has been reported [53]. Rapidity of analysis appears to be enhanced if the electrophoretogram is placed in direct contact with suitable substrate. Incorporation of the substrate in the gel prior to electrophoresis has reduced the time required to perform the analysis. The mixture of starch and agarose solidifies firmly and gives a "frosted glass" pattern. Following electrophoresis of the sample the strip is incubated for a suitable period of time.

Two methods are available for positive identification of the zone of activity of the enzyme. One is observance of clearing of the frosted pattern, which is quickly and easily spotted, and the other is staining to detect the starch with periodic acid–Schiff (PAS) stain or with iodine reaction. Both of these stains show false localization due to combining of the starch with the protein or combining of the iodine with the protein so that multiple zones of amylase activity appear to be present, whereas only one enzyme zone is present.

Two cathodic amylase bands have been found in urine; in serum there is only one. The faster one in urine corresponds with the serum amylase. This band has the same mobility as salivary amylase. The slower-moving band is from the pancreas.

EQUIPMENT AND REAGENTS

1. Standard electrophoretic equipment (see Standard Gel Electrophoretic Method).

2. Gram's iodine—diluted 1:10 in distilled water.
   1 gm. iodine
   2 gm. potassium iodide
   300 ml. distilled water
3. Starch-agarose gel.
   250 mg. agarose
   150 mg. cornstarch
   50 ml. barbital buffer, pH 8.6, ionic strength 0.05

PROCEDURE

1. Prepare strips by the trench method using starch-agarose gel marked at midpoint and 3.2-cm. anodic migration point.
2. Place in electrophoretic cell. Inoculate with 15μl. of material.
3. Migrate at 150 volts until bromphenol blue–albumin reaches the 3.2-cm. anodic mark.
4. Incubate at 37° C. for 30 minutes.
5. Remove from cell and fix in 1 percent acetic acid in absolute methyl alcohol (V/V) for 15 minutes. The strips can be examined for amylase activity without further delay.
6. If desired, the strips may be stained by placing in a dilute Gram's iodine.

RESULTS

Amylase activity is demonstrated by cleared areas where the starch has been removed by the enzyme. The starch background in the strip turns a deep-purple color with iodine stain. Most of the amylase activity corresponds to the first part of the γ-globulin area. Since the starch fails to stain with iodine in the areas occupied by serum protein, particularly albumin, these areas differ from the zone of true enzyme activity in still having a "frosted glass" appearance. The enzyme zone is perfectly transparent since the starch has been hydrolyzed.

IMMUNOELECTROPHORETIC APPLICATION

In this instance substrate should be in a separate sheet, applied directly, and left in contact for a suitable period.

Acid Phosphatase

PRINCIPLE

Acid phosphatase (orthophosphoric monoester phosphohydrolase, 3.1.3.2) is a phosphomonoesterase and hydrolyzes orthophosphoric monoesters [29]. The enzyme has wide specificity and also catalyzes transphosphorylation. α-Naphthyl acid phosphate sodium salt in an acid buffer solution is hydrolyzed by acid phosphatase to naphthol and phosphorus. As the α-naphthol is formed, it couples with the diazonium salt Fast Blue RR to form an insoluble black compound [5]. Electrophoretically the enzyme has been shown to occur in some tissues in multiple molecular forms [4].

EQUIPMENT AND REAGENTS

1. Standard gel electrophoretic equipment (see Standard Gel Electrophoretic Method).
2. Barbital buffer, pH 8.6, ionic strength 0.05.
3. Solution A—buffered gel substrate solution. To 50 ml. of barbital buffer add 250 mg. agarose and bring to a boil. Add 50 mg. of α-naphthyl acid phosphate sodium salt. Keep warm to use for strips.
4. Solution B.

Acetate buffer, pH 4.0		100.0 ml.
$NaC_2H_3O_2$, 0.2M	9 ml.	
$HC_2H_3O_2$, 0.2M	41 ml.	
Distilled water	50 ml.	
Fast Blue RR salt		100.0 mg.
MgCl, 1 percent solution		0.1 ml.

PROCEDURE

1. Prepare strips by trench method and mark midpoint and 3.2-cm. anodic point. Coat strips with 5 ml. of solution A (gel substrate).
2. Inoculate with 15μl. of specimen. Migrate sample at 150 volts until bromphenol blue–albumin marker reaches anodic mark.
3. Remove strips from the cell and place them in solution B for one hour at room temperature.
4. Wash in several changes of distilled water for 30 minutes. Dry.

RESULTS

Areas of enzymatic activity are demonstrated by black bands.

NORMAL

One band is seen in the α_2-globulin area.

Alkaline Phosphatase Isoenzymes

PRINCIPLE

α-Naphthyl acid phosphate sodium salt in an alkaline buffer solution is hydrolyzed by alkaline phosphatase (orthophosphoric monoester phosphohydrolase, 3.1.3.1) to α-naphthol and phosphorus [29]. As the α-naphthol is formed, it couples to the diazonium salt Fast Blue RR to form an insoluble black compound [5] which marks the site of enzyme activity. Multiple molecular forms (isoenzymes) can be identified by starch-gel [13], agarose, or agar electrophoresis [15, 45, 95].

EQUIPMENT AND REAGENTS

1. Standard gel electrophoretic equipment (see Standard Gel Electrophoretic Method).
2. Barbital buffer, pH 8.6, ionic strength 0.05.
3. Solution A—buffered gel substrate solution. To 50 ml. of barbital buffer add 250 mg. of agarose and bring to a boil. Add 50 mg. of α-naphthyl acid phosphate sodium salt.
4. Solution B.

Clark-Lubs buffer, pH 10.0		100.0 ml.
H_3BO_3, 0.2M, and KCl, 0.2M	25.0 ml.	
NaOH, 0.2M	21.95 ml.	
Distilled water	53.05 ml.	
Fast Blue RR salt		50.0 mg.
MgCl, 1 percent solution		0.05 ml.
α-Naphthyl acid phosphate Na		50.0 mg.

PROCEDURE

1. Prepare strips by marking midpoint and 3.2-cm. anodic migration point. Coat strips with 5 ml. of solution A (gel substrate) and use trench method for preparation of inoculation point.
2. Inoculate with 5–15μl. of specimen; migrate at 150 volts until bromphenol blue–albumin marker reaches anodic migration mark.
3. Remove strips from the cell and place them in solution B for one hour at room temperature.
4. Wash in several changes of distilled water for 30 minutes. Dry.

RESULTS

Areas of enzymatic activity are demonstrated by black bands.

NORMAL

Two bands of activity corresponding to α_2- and β-globulin are usually seen. Three bands are not infrequently seen. The additional band is believed to be from intestine. The results of an isoenzyme study must be related to the total enzyme. See Chap. 7, on alkaline phosphatase, for discussion.

NOTE

Diazonium salts are labile compounds that decompose at high temperatures, especially in an alkaline medium. Therefore, it is best to perform the test at room temperature.

The above system may be used for immunoelectrophoretograms, but it is best to place substrate in a gel overlay or as a liquid overlay.

Glutamic-Oxalacetic Transaminase (GOT) Isoenzymes

PRINCIPLE

The enzyme GOT (L-asperate: 2-oxo-glutarate aminotransferase, 2.6.1.1) [29], which catalyzes the transfer of an amino group from aspartic acid to α-ketoglutamate, is detected in an electrophoretogram by diazonium salt (6-benzamide-4-methoxy-m-toluidine diazonium chloride). This couples with the product, oxalacetic acid, to form a purple-red compound [3]. Isoenzymes of the enzyme can be identified in serum [11, 25, 78].

EQUIPMENT AND REAGENTS

1. Standard electrophoretic equipment (see Standard Gel Electrophoretic Method).
2. Barbital buffer, pH 8.6, ionic strength 0.05.
3. Buffered agarose solution. To 50 ml. of barbital buffer add 250 mg. of agarose and bring to a boil.
4. Substrate. Mix equal parts of solution 1 and solution 2.
   Solution 1.
   > *Trans-AC substrate solution* (General Diagnostics Division, Warner-Chilcott, Inc., Morris Plains, N.J.): Composed of alpha-ketoglutamic acid, 5 mM. per liter; L-aspartic acid, 20 mM. per liter; dibasic potassium phosphate (K_2HPO_4), 3.335 gm.; monobasic potassium phosphate (KH_2PO_4), 100 mg.; polyvinylpyrrolidone (PVP), 1.0 gm.; and ethylenediamine tetra-acetic acid (EDTA), tetrasodium salt, 0.1 gm. in 100 ml. of distilled water.

   Solution 2.
   > *Color developer.* (General Diagnostics Division, Warner-Chilcott, Inc., Morris Plains, N.J.): Composed of 6-benzamide-4-methoxy-m-toluidine diazonium chloride, 50 mg., dissolved in 10 ml. of distilled water.

PROCEDURE

1. Prepare strips by marking midpoint and 3.2-cm. anodic point. Coat each strip with 5 ml. of buffered agarose solution and use trench method for preparing inoculation point.
2. Inoculate the strip with 15µl. of material.
3. Place a small dot of bromphenol blue–albumin at the midpoint.
4. Perform electrophoresis at 150 volts until the bromphenol blue marker has reached the anodic mark.
5. Remove strips from the cell and place in substrate for one hour at 37° C.
6. Soak in water 30 minutes to remove buffer salts and dry.

RESULTS

Enzyme activity is demonstrated by the appearance of violet-red bands. One band is between the α_2- and β-globulin area; the other appears in the γ-globulin area.

Normal serum, as a rule, shows only the fast-moving component whereas serum with elevated GOT activity shows both.

IMMUNOELECTROPHORETIC APPLICATION

GOT activity can be demonstrated in immunoprecipitin bands by substrate overlay.

Leucine Aminopeptidase (LAP)

PRINCIPLE

The enzyme leucine aminopeptidase (L-leucyl-peptide hydrolase, 3.4.1.1) splits off N-terminal residue with a free amino group [29]. Detection of the enzyme in serum or other body fluids is based on its reaction with the substrate L-leucyl-β-naphthylamide [58]. β-Naphthylamine is released from the substrate in the presence of the enzyme. The β-naphthylamine couples with the diazonium salt Fast Blue B to form an azo dye [4, 70].

EQUIPMENT AND REAGENTS

1. Standard gel electrophoretic equipment (see Standard Gel Electrophoretic Method).
2. Phosphate buffer, pH 7.5, M/30.

KH_2PO_4	2.8 gm.
Na_2HPO_4	11.8 gm.
Distilled water	3000.0 ml.

3. Buffered agarose solution. To 50 ml. of phosphate buffer add 250 mg. of agarose and bring to a boil.
4. Substrate.

Phosphate buffer, pH 6.5, M/15		10 ml.
KH_2PO_4, M/15	32 ml.	
Na_2HPO_4 M/15	18 ml.	
Sodium chloride, 0.85 percent		2 ml.
L-Leucyl-β-naphthylamide hydrochloride*		8 mg.
Fast Blue B*		10 mg.

PROCEDURE

1. Prepare agarose strips by marking midpoint and 3.2-cm. anodic point. Coat strips with 5 ml. of buffered agarose solution and use trench method for inoculation point.

*See Appendix (Sigma Chemical Co.)

2. Inoculate with 15μl. of the specimen and migrate at 150 volts until bromphenol blue–albumin marker reaches the anodic mark.
3. Remove the strip from the cell and place in substrate for one hour at 37° C.
4. Soak in distilled water to remove buffer salts for 45 minutes. Dry.

RESULTS

Leucine aminopeptidase activity is demonstrated by the appearance of purple-red bands in the α_2-globulin area. Immunoprecipitin bands in the α_2-globulin area show enzyme activity.

Lysozyme Determination

PRINCIPLE

Lysozyme (mucopeptide N-acetylmuramylhydrolase, 3.2.1.17) [29] hydrolyzes β-1,4-mannuronide links which are common components in the wall of some bacteria. The enzyme is a bacteriolytic enzyme found in all human tissues and secretions. It is a basic polypeptide and gives a number of protein reactions. Lysozyme is the principle product of proliferating monocytes in a monocytic or myelomonocytic leukemia. Lysozyme is nondialyzable, soluble in water and in weak saline solutions, insoluble in alcohol and ether, heat resistant, stable at room temperature, and readily absorbed by charcoal and fibrin. The isoelectric point is at pH 10.8, and the molecular weight is 17,000. In an electrophoretic field it migrates cathodic to the γ and has been labeled cationic protein (CP) [68].

Lysozyme readily lyses the carbohydrate structure of the wall of a gram-positive coccus, *Micrococcus lysodeikticus*. Lysozyme, therefore, can be quantitatively measured by simple gel diffusion with the *M. lysodeikticus* incorporated in the gel. As the lysozyme diffuses into the gel, it lyses the organisms. This lysis can be quantitated by measuring the diameter of the clear or lytic zone around the sample well in an otherwise opaque plate. The opacity is caused by the cocci in the agar. Standards of crystalline egg-white lysozyme (10μg. per milliliter to 160μg. per milliliter) are used in each gel-diffusion plate.

Lysozyme activity is increased in monocytic leukemia and myelomonocytic leukemia and may be increased in chronic infections and renal disease.

EQUIPMENT AND REAGENTS

1. Electrophoresis for CP lysozyme. Follow procedure for urine or serum electrophoresis as the case may be.

2. Lysozyme assay.
   a. Phosphate buffer, pH 6.3, M/15.
      KH_2PO_4, M/15, 9.07 gm. per liter—37 ml.
      Na_2HPO_4, M/15, 9.46 gm. per liter—13 ml.
   b. Buffered gel—250 mg. agarose (Bausch & Lomb)* dissolved in 50 ml. of phosphate buffer. Cool slightly.
   c. *Micrococcus lysodeikticus* suspension. Dissolve each of three reagent vials (Worthington Biochemical Corp.)* of *M. lysodeikticus* with 2 ml. of distilled water. Bring to room temperature.
   d. Gel—*M. lysodeikticus* plate. Mix 45 ml. of warm agarose with 5 ml. of *M. lysodeikticus* suspension. Coat four 16-cm.-long strips of photographic film with 10 ml. of the solution. Let harden. Suction out nine wells 2 mm. in diameter and 1.5 cm. apart along the mid-longitudinal axis.
   e. Standard lysozyme. Dilute one vial of standard (Worthington Biochemical Corp.)* with 2.5 ml. of distilled water. This is equal to 160μg. per milliliter of lysozyme. Serially dilute to give standards of 80μg./ml., 40μg./ml., 20μg./ml., and 10μg./ml.
   f. Fixative solution—50 percent (V/V) ethanol.
   g. Photographic equipment—Polaroid MP-3 with 4-inch lens or Solar Immuno-Glo camera (HCS Corp.)* and Immuno-Glo with quantitative grid (HCS Corp. or Hyland Laboratories).*

PROCEDURE

1. Collect a 24-hour urine specimen from the patient. Measure the volume. Concentrate the specimen 100-fold by ultrafiltration.
2. Perform electrophoresis on both the concentrated and the unconcentrated urine specimens and observe for CP lysozyme.
3. Inoculate a lysozyme assay plate with 2μl. of the five standards and the unconcentrated urine specimen.
4. Incubate the lysozyme assay plate for 16 hours in a moist, room-temperature incubator.
5. Fix plate in fixative solution for 30 minutes.
6. Photograph.
7. Measure the diameter of the lytic zones around each standard in millimeters and the unknown well on the Polaroid print from grid pattern (Fig. 15-20 in Quantitation of Immunoglobulins section).
8. Plot the concentration of CP lysozyme standards on the X axis versus the diameter in millimeters of the lysed zones of the Y axis of semilog paper. Interpolate the concentration of the unknowns.

*See Appendix.

RESULTS

1. Example of results shown in positive findings by gel diffusion case of acute leukemia is given in Chap. 5.
2. CP lysozyme is not demonstrable in a normal urine or serum.
3. Lysozyme assay—0.4μg. per milliliter in urine; 0.2μg. per milliliter in serum.

INTERPRETATION

Lysozyme excretion is increased in monocytic and myelomonocytic leukemia. It may also be elevated in certain chronic renal diseases, such as pyelonephritis and nephrosis. The involvement in renal disease is not related to the lysozyme excretion.

Alpha$_1$-Antitrypsin

PRINCIPLE

Normal human serum inhibits the proteolytic activity of trypsin. Electrophoretically, the principle inhibitor travels as an α_1-globulin and has been shown to be a protein. About 90 percent of the neutralizing action of the serum against trypsin is found in the α_1-globulin fraction. The remaining 10 percent of inhibition in serum has been shown to involve α_2-macroglobulins (serum plasmin inhibitors), inter-α inhibitor, and an antitrypsin separate from and slower than the α_1-antitrypsin.

Laurell and Eriksson first found a deficiency of α_1-antitrypsin in several patients and suggested possible relationship to pulmonary disease [47]. In a series of rapid reports, Laurell et al. [48], Eriksson [30], and Axelsson et al. [2] published findings which established α_1-antitrypsin as an important protein with genetic variation. Persons who are homozygous for the α_1-antitrypsin deficiency gene are susceptible to pulmonary emphysema. The mechanism is believed to involve failure of neutralization of trypsin by serum α_1-antitrypsin, with resulting destruction of lung tissue by the trypsin [46]. The source of trypsin may be from dying leukocytes during pulmonary inflammation [46].

The method is based on the ability of α_1-antitrypsin to inhibit trypsin digestion of fibrinogen of a fibrinogen-agarose plate (Fig. 15-17) [15, 35].

EQUIPMENT

1. Oven with temperature to 70° C.
2. Suction water aspirator.
3. Room-temperature incubator.
4. Photographic film strips—DuPont Cronar film, P40B, cut into 16-cm. lengths.

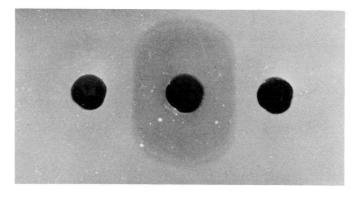

FIGURE 15-17

Assay for the presence or absence of α_1-antitrypsin. The diffusion plate is made with an agarose-fibrinogen mixture. The central well contains trypsin. Normal control and patient serum are in the two outer wells. Digestion in this example is interfered with by the presence of α_1-antitrypsin in both normal control and patient.

REAGENTS

1. Barbital buffer, pH 8.6, ionic strength 0.05.
2. Agarose.
3. Fibrinogen solution, 2 percent (W/V)—1 gm. bovine fibrinogen dissolved in 50 ml. barbital buffer.
4. Calcium chloride, 0.025M.
5. Trypsin solution—trypsin, 5 gm. per liter, 1:300 (Colorado Serum Company, Denver, Colo.).
6. Aqueous acetic acid, 10 percent (V/V).

PROCEDURE

1. Bring to room temperature 4 ml. of fibrinogen solution and 2 ml. of CaCl.
2. Dissolve 250 mg. of agarose in 25 ml. of barbital buffer by bringing to a boil. Cool the solution to 60° C.
3. To the fibrinogen-CaCl solution quickly add 9 ml. of the warm agarose solution and immediately coat two 16-cm. strips of photographic film with 6 ml. of the mixture.
4. After the agarose is firm, place the strips on a damp towel and seal in an airtight tray. Place the tray in a 65–70° C. oven for one hour. During this time the fibrinogen will coagulate and the fibrin plate will be opaque. It should have a "frosted glass" pattern.
5. By suction produce three wells 3 mm. in diameter 1 cm. apart along the mid-longitudinal axis. The first well is for the serum to be tested, the middle well for the trypsin solution, and the last well for the control serum.

6. Fill each well with 25μl. of the above solutions.
7. Incubate the fibrin plate for 16 hours at room temperature.
8. Remove the plate from the incubator and place in 10 percent acetic acid for one hour to stop the trypsin digestion.
9. Photograph the plate as for immunoelectrophoresis (Fig. 15-17).

RESULTS

NORMAL. The trypsin digestion of the fibrinogen will be inhibited when the patient has a normal level of α_1-antitrypsin. This is illustrated in Figure 15-17 as a sharp line of demarcation between the digested fibrinogen (cleared zone) and the undigested fibrinogen (opaque zone).

ABNORMAL. Clearing of fibrinogen about the trypsin well in the form of a complete circle will occur if there is no inhibition by α_1-antitrypsin.

NOTE

Absence of α_1-antitrypsin can be suspected from a serum protein electrophoretogram by the absence or marked decrease of α_1-globulin. Genetic variations of α_1-antitrypsin with different electrophoretic mobility have been noted by Laurell [48]. The genetic variations can be described by three different technics utilizing immunoelectrophoresis. One is a regular serum immunoelectrophoresis using patient and control serum in the antigen wells against the anti-α_1-antitrypsin. Displacement of the immunoprecipitin band from a residual position will denote genetic variation.

The second method involves immunoelectrophoresis using a fibrin plate as described in the α_1-antitrypsin procedure. After preparation of the fibrin plate, an immunoelectrophoretic pattern is cut with two antigen wells and one antiserum trench (open pattern). The wells are inoculated with patient and control serum and migrated according to the standard method for immunoelectrophoresis. The trench is filled with a solution of trypsin as used for a diffusion test and diffused 16 hours (Fig. 15-18). In Figure 15-18 are six studies. The lower three show inhibition. The upper three do not. One of these is a negative control (top), and the other two are from patients with α_1-antitrypsin deficiency. The electrophoretic variations will be noted by the lysed pattern of the immunoelectrophoretogram.

Third, a two-dimensional system with the second dimension performed in agarose saturated with anti-α_1-antitrypsin has been advanced as a useful method. The variations show up as differences in

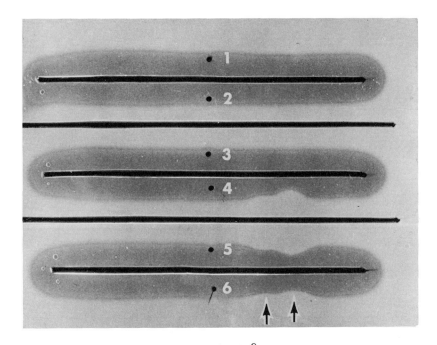

FIGURE 15-18
Immunoelectrophoretic method of identifying position of α_1-antitrypsin. The plate contains agarose-fibrinogen. After the electrophoretic phase trypsin is introduced into the antiserum trenches and leaves a clearing zone where it digests the fibrinogen in its diffusion path. In antigen wells 1, 2, and 3, no antitrypsin is noted. In 4, 5, and 6, zones of inhibition marking the position of α_1-antitrypsin are clearly indicated. Two zones (*arrows*), a slow and a fast, are recognized. See text. Anode is on the right.

immunoprecipitin peaks, and very small differences of electrophoretic mobility can be depicted.

Esterase–Alpha-Naphthyl Acetate Method

PRINCIPLE

Carboxylesterases (carboxylic-ester hydrolase, 3.1.1.1) are classified on the basis of the substrate specificity [29]. Since these enzymes hydrolyze carboxylic acid esters of alcohols, naphthols, and phenols, it is readily apparent that the number of esterases is extensive. This group is usually referred to as ali-esterases or B-esterases. Further substrate specificity is relative, and it appears best to think of these enzymes as a group of heterogeneous molecules with ill-defined specificity. Zymographic studies have shown just how heterogeneous these enzymes are [5, 6, 85]. At the same pH Markert and Hunter [57] demonstrated 10 esterase zones in starch-gel electrophoretograms of adult mouse liver. Keep in mind that variations in pH, temperature, ionic strength, ion species, substrate analogues, etc., would reveal a whole spectrum of zymogram patterns. It seems useless at this point to dwell on classification other than to outline areas where some clarity has already been found useful. Ali-esterases originally were designed to denote those enzymes with aliphatic substrate specificity such as lipases and non-specific esterases. Arylesterases, arom-esterases or A-esterases (aromatic esterases) (aryl-ester hydrolase, 3.1.1.2), are not not inhibited by L-serine. Cholinesterases or pseudocholinesterases (acylcholine acyl-hydrolase, 3.1.1.8) hydrolyze choline esters [5]. The method to be discussed is based on the substrate α-naphthyl acetate and demonstrates primarily the carboxylesterases [5]. The enzymatic activity is demonstrated by the coupling of α-naphthyl acetate with the azo salt Fast Blue RR to form a black precipitate.

EQUIPMENT AND REAGENTS

1. Standard electrophoretic equipment (see Standard Gel Electrophoretic Method).
2. Barbital buffer, pH 8.6, ionic strength 0.05.
3. Buffered agarose solution. To 50 ml. of barbital buffer add 250 mg. of agarose and bring to a boil.
4. Substrate.
   Solution 1.
   Phosphate buffer, pH 7.0, M/15   10 ml.
   Solution 2.
   Fast Blue RR* (1 mg. per milliliter distilled water)   100 ml.
   Solution 3.
   α-Naphthyl acetate* (1 percent in acetone)   2 ml.

PROCEDURE

1. Prepare strips by marking midpoint and 3.2-cm. anodic point. Coat strips with 5 ml. of buffered agarose solution and use trench method for preparation of inoculation point.
2. Inoculate with 15μl. of specimen, migrate at 150 volts until bromphenol blue–albumin marker reaches the anodic mark.
3. Remove strips from cell and place in substrate for one hour at room temperature.
4. Soak in several changes of distilled water to remove buffer salts for 30 minutes. Dry.

RESULTS

Esterase activity is shown by black staining bands.

*See Appendix (Sigma Chemical Co.)

Glucose-6-Phosphate Dehydrogenase (G-6-PD) Isoenzymes

PRINCIPLE

The enzyme G-6-PD (D-glucose-6-phosphate: NADP oxidoreductase 1.1.1.49) is the first enzyme in the hexose monophosphate shunt in the direct catabolism of glucose [29]. In erythrocytes this enzyme has been shown to be responsible for maintaining an appropriate level of reduced glutathione which preserves the erythrocyte from hemolysis. Deficiency of the enzyme occurs in nonspherocytic hemolytic anemia and favism. Electrophoretically, the enzyme is heterogeneous, and isoenzymes A and B have been described. Three common phenotypes are known among Negro males: type A and enzyme deficient (A—), type A and normal enzyme activity (A+), and type B and normal activity (B+) [14, 43, 72]. American Caucasians are type B regardless of the state of enzyme deficiency [14, 43, 72]. In Negro females type AB exists in addition to type A and type B. The gene controlling the synthesis of G-6-PD is located on the X chromosome. The enzyme is best demonstrated in erythrocytes and leukocytes.

EQUIPMENT AND REAGENTS

1. Standard gel electrophoretic equipment (see Standard Gel Electrophoretic Method).
2. Buffer for tank.

Water		1000.0	ml.
Sodium chloride (0.05M)		2.0	gm.
Tris, pH 8.8, 0.05M			
Tris HCl	1.23 gm.		
Tris base	5.13 gm.		
EDTA, 2.7 mM.		1.005	gm.
Nicotine adenine dinucleotide phosphate			
(NADP)* in cathode tank		7.6	mg.

3. Agarose strips.

Tris, pH 8.0, 0.05M		50.0	ml.
Tris HCl	0.444 gm.		
Tris base	0.265 gm.		
Water	100.0 ml.		
EDTA		100.0	mg.
NADP,* 1 mg. per milliliter		0.4	ml.
Agarose		250.0	mg.

4. Substrate.

Water	30.0	ml.
Tris buffer, pH 8.0, 0.5M	5.0	ml.
NADP	3.0	mg.

*See Appendix (Sigma Chemical Co.)

Na cyanide	15.0	mg.
Glucose-6-phosphate (G-6-P)*	6.0	mg.
Phenazine methosulfate (PMS)*		
(1 mg. per milliliter)	0.2	ml.
Nitro Blue Tetrazolium (NBT)*	5.0	mg.

PROCEDURE

1. Collect blood in heparin or EDTA and do not permit tube to become warm. Use beaker with water and ice to control temperature.
2. Wash erythrocytes four times in cold saline solution.
3. Pack red cells and add equal volume of iced water to hemolyze the cells.
4. Centrifuge to remove stroma.
5. Add 1 mg. of G-6-phosphate to each 2 ml. of hemolysate.
6. Prepare agarose strip as for general method and place in electrophoretic cell.
7. Inoculate with 3μl. of the hemolysate using razor blade method.
8. Apply dot of bromphenol blue–albumin as starting marker.
9. Migrate at 150 volts at 4° C. until the albumin has migrated to the 3.2-cm. mark.
10. Remove from the cell and cover with substrate.
11. Incubate in the dark at 30° C. for three hours.
12. Transfer to water to remove buffer salts.
13. Dry.

RESULTS

Purple bands appear in the area of enzyme activity.

B, A, and AB bands can be demonstrated according to Boyer et al. [14].

IMMUNOELECTROPHORETIC APPLICATION

Immunoprecipitin bands can be studied to determine enzyme activity by this method. The substrate overlay is perhaps the most promising approach.

Catalase

PRINCIPLE

The enzyme catalase (hydrogen peroxide: hydrogen-peroxide oxidoreductase, 1.11.1.6) catalyzes the breakdown of hydrogen peroxide to oxygen and water [29]. The reaction rate is very rapid. The enzyme is found in plants [7] and animals. Absence of the enzyme occurs in

*See Appendix (Sigma Chemical Co.)

acatalasia, which is found only in Japanese [19]. The condition is associated with ulcers of the mouth [19, 84]. The trait behaves like an autosomal recessive [19]. The enzyme is found in most tissues, particularly erythrocytes, and can be localized in electrophoretograms [86].

EQUIPMENT AND REAGENTS

METHOD A.

1. Standard electrophoretic equipment (see Standard Gel Electrophoretic Method).
2. Standard electrophoretic reagents (see Standard Gel Electrophoretic Method).
3. Standard agarose-buffer mixture (see Standard Gel Electrophoretic Method).
4. Hydrogen peroxide, 3 percent.

METHOD B.

1. Standard electrophoretic equipment (see Standard Gel Electrophoretic Method).
2. Standard electrophoretic reagents (see Standard Gel Electrophoretic Method).
3. Starch-agarose buffer mixture. Dissolve 250 mg. agarose and 150 mg. of cornstarch in 50 ml. of barbital buffer, pH 8.6, ionic strength 0.05, by bringing solution to a gentle boil.
4. Solution A. (Mix just before use.)
H_2O_2, 3 percent (0.88M)	10 ml.
$Na_2S_2O_3 \cdot H_2O$, 1.5 percent (0.06M)	5 ml.

   Solution B.
KI, 1.5 percent (0.09M)	10 ml.

PROCEDURE

METHOD A.

1. Prepare strips by marking midpoint and 3.2-cm. anodic migration point. Coat strips with 5 ml. of warm buffered agarose solution.
2. Inoculate with 1.5µl. of sample. Carry out electrophoresis at 150 volts until bromphenol blue–albumin marker reaches anodic migration point. The enzyme is usually detectable in small sample size. Erythrocytes are particularly rich in the enzyme.
3. Place strips in 10 ml. of 3 percent hydrogen peroxide for a few minutes.
4. Observe for bubble formation. Photograph with indirect light using Immuno-Glo and Polaroid or Solar camera.
5. Place in fixative of methanol.
6. Dry and stain strips for protein if desired. Usually this is not necessary.

METHOD B.

1. Method B is the same as method A through step 2.
2. Place strips in solution A for 10–20 seconds. Remove, pass once through distilled water, and place in solution B until suitable differentiation develops.
3. Fix in absolute methanol and dry strips.
4. Strips may be photographed directly while wet or after drying can be used as negatives for contact prints or used in an enlarger.

RESULTS

Catalytic activity is demonstrated in areas where bubble formation takes place. In hemolysates this area is just behind the A_1 hemoglobin band. In stained strips the starch is blue-black with an area of clear starch-agarose marking the zone of enzyme activity.

IMMUNOELECTROPHORESIS

Immunoelectrophoretograms soaked in the substrate show bubbles in the immunoprecipitin bands. The staining method may also be used if desired although the clear transparency is reduced because of the starch.

Proteolytic Activity

PRINCIPLE

Exposed x-ray film with a gelatin base is used as a substrate for localization of proteolytic enzymes in electrophoretograms or immunoelectrophoretograms. The electrophoretograms are placed agarose-side down against the x-ray film [18]. Suitable pH and temperatures are used to demonstrate enzyme activity. The enzyme zone shows up as a cleared zone on the x-ray film. Uriel described a procedure using protein stain of electrophoretograms after they had been put into substrate solution [88]. The protein from substrate solution diffused into the gel but was digested where the enzyme was located.

EQUIPMENT AND REAGENTS

1. Standard gel electrophoretic equipment (see Standard Gel Electrophoretic Method).
2. Standard gel electrophoretic reagents (see Standard Gel Electrophoretic Method).
3. Standard agarose-buffer mixture (see Standard Gel Electrophoretic Method).
4. Exposed x-ray film.
5. Phosphate buffer. The pH is determined by the source of the proteolytic enzyme.

PROCEDURE

1. Follow procedure for serum protein electrophoresis or immuno-electrophoresis. In the latter case the unreacted reagents should be removed by soaking strip for 24 hours in saline solution or a filter paper overlay may be used for 30 minutes. Urine and other fluids, organ extracts, and cells may be used.
2. Soak a piece of x-ray film in phosphate buffer of the required pH for about five minutes.
3. Remove strips from cell and place the agarose-coated side of the electrophoretogram face down on the x-ray film.
4. Incubate at 37° C. for 60 minutes in a moist chamber.
5. Separate x-ray film and wash in tap water. The electrophoretogram can be soaked, dried, stained, and used as a marker for enzyme activity although this may not be practical since there is uptake by the agarose of gelatin from the x-ray film which is difficult to remove.

RESULTS

Enzymatic activity shows clear areas on the x-ray film where the gelatin emulsion has been destroyed.

Cawley, Eberhardt, and Goodwin found that bromelin proteolytic activity was best demonstrated at pH 5.5 in phosphate buffer [18].

Malic Dehydrogenase (MDH) Isoenzymes

PRINCIPLE

The enzyme malic dehydrogenase (L-malate: NAD oxidoreductase, 1.1.1.37) catalyzes the oxidation of L-malate to oxaloacetate [29]. It is found in many tissues throughout the body and is present in multiple molecular forms [20].

EQUIPMENT AND REAGENTS

1. Standard gel electrophoretic equipment (see Standard Gel Electrophoretic Method).
2. Phosphate buffer, pH 7.4, M/30.*
3. Buffered gel for strips. Dissolve 250 mg. of agarose in 50 ml. of phosphate buffer, pH 7.4, M/30, by bringing to a boil as quickly as possible.
4. Substrate.
   a. Solution I.

Sodium cyanide	24 mg.
Magnesium chloride	8 mg.
L-Malic acid	535 mg.
Phosphate buffer	40 ml.

*See Appendix.

b. Solution II.
Phenazine methosulfate   1 mg. per milliliter of distilled
water (make fresh just before using).
c. Mix the substrate each day immediately before using.
Solution I   20 ml.
β-Nicotinamide adenine dinucleotide (NAD)*   40 mg.
Nitro Blue Tetrazolium (NBT)*   24 mg.
Solution II   1 ml.

PROCEDURE

1. Prepare strips by marking midpoint and 3.2-cm. anodic point.
Coat strips with 5 ml. of buffered gel and use trench method for
preparation of inoculation point.
2. Inoculate with 15μl. of specimen, migrate at 150 volts until
bromphenol blue–albumin marker reaches anodic mark.
3. Remove strips from cell and place in substrate for one hour at
47° C.
4. Soak in several changes of distilled water for 30 minutes. Dry.

RESULTS

Malic dehydrogenase activity is demonstrated in hemolysates as
one blue-black band which travels toward the anode just behind the
A_1 hemoglobin band.

Lactate Dehydrogenase (LDH) Isoenzymes

PRINCIPLE

Visual localization of electrophoretically separated LDH (L-lactate:
NAD oxidoreductase, 1.1.1.27) isoenzymes of serum or other body
fluid is produced by the reduction of nitrotetrazolium blue to a purple
insoluble formizan as the terminal electron acceptor [15, 29, 93]. A
reagent overlay containing substrate, phenazine methosulfate, and nico-
tinamide adenine dinucleotide (NAD) is placed over the gel electrophoretogram. The linked reactions are as follows:

$$\text{L-Lactate} + \text{NAD} \xrightarrow{\text{LDH}} \text{pyruvate} + \text{NADH}_2$$
$$\text{NADH}_2 + \text{PMS} \longrightarrow \text{NAD} + \text{reduced PMS}$$
$$\text{Reduced PMS} + \text{NBT} \longrightarrow \text{formizan} \downarrow + \text{PMS}$$

At completion the strips are soaked in 10 percent aqueous acetic
acid, dried, and scanned in the Analytrol at 550 mμ.

EQUIPMENT AND REAGENTS

1. Standard gel electrophoretic equipment (see Standard Gel
Electrophoretic Method).

*See Appendix (Sigma Chemical Co.).

2. Phosphate buffer, pH 7.4, M/30.
3. Buffered agarose solution. Dissolve by gently boiling 250 mg. agarose in 50 ml. phosphate buffer.
4. Substrate.
   Solution I.

Sodium cyanide	96.0	mg.
Magnesium chloride	32.0	mg.
Lithium lactate*	1.088	gm.
Phosphate buffer	160.0	ml.

   Solution II.
   Phenazine methosulfate (PMS)†   1 mg. per milliliter of distilled water

   Histochemical overlay.   Mix   substrate   fresh each day just before using:

Solution I	10.0	ml.
Solution II	0.5	ml.
β-Nicotinamide adenine dinucleotide (NAD)†	20.0	mg.
Nitro Blue Tetrazolium (NBT)†	12.0	mg.

5. Aqueous acetic acid, 10 percent (V/V).
6. Marker dye—albumin saturated with bromphenol blue.

PROCEDURE

1. Coat strips marked at midpoint and 2.5-cm. anodic point with 5 ml. of agarose-phosphate buffer solution utilizing the trench method, using 1-cm. lengths of smooth-surfaced wire the size of #1 paper clips in place of 1.5-cm. lengths of capillary tubing. After the gel has solidified, the wire lengths can be removed by a magnet held directly above, but not touching, the gel surface.‡ Figure 15-19 is a diagram of steps using agar-substrate overlay. This is what we originally used; it was initially developed by Wieme [93].
2. Inoculate the strip with 5μl. of serum, concentrated urine, concentrated cerebrospinal fluid, or other body fluid, using a micropipette or Hamilton syringe.
3. Place a small dot of bromphenol blue–albumin at the edge of the strip, on a line with the trench, for a marker.
4. Migrate at 12 ma. per strip (approximately 150 volts), until the albumin has reached the 2.5-cm. point.
5. Take strips from cell, remove 4 cm. of agarose from each end, and cover the remaining agarose with 1 ml. of the substrate mixture.

*See Appendix (G. Fredrick Smith Chemical Co.).
†See Appendix (Sigma Chemical Co.).
‡Suggestion of James B. McCormick, M.D., Hinsdale, Ill.

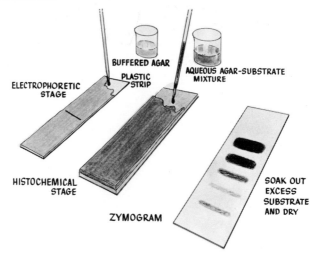

FIGURE 15-19

Outline of steps for demonstration of LDH isoenzymes using agar-substrate overlay mixture [93]. Although current procedure leaves out agar [15], there are times when use of agar or gelatin in the substrate to form a gel may be desirable. (From E. J. Wright, L. P. Cawley, and L. Eberhardt, *Amer. J. Clin. Path.* 45:737, 1966. © 1966, The Williams & Wilkins Co., Baltimore, Md. 21202, U.S.A.)

6. Incubate one hour at 47° C. in total darkness and soak in 10 percent aqueous acetic acid for one hour.
7. Dry.
8. Scan on Analytrol with B-2 cam with 550 mμ filter and calculate percentage.

RESULTS

Lactate dehydrogenase activity is demonstrated by five blue-violet bands, which are numbered, starting with number 1 at the anode (see Chap. 3).

TABLE 15-2

ZYMOGRAMS FROM NORMAL INDIVIDUALS

LDH Fraction	Mean Percent	Percent Range
LDH_1	22.0	15.0–30.0
LDH_2	39.0	33.0–50.5
LDH_3	27.0	20.0–35.5
LDH_4	7.5	2.5–13.5
LDH_5	4.5	0–8.5
Total[a]	36.0 I.U.	19.0–54.0 I.U.

[a]Total LDH in international units (I.U.) normal 60 or less.

IMMUNOELECTROPHORETIC APPLICATION

The gel substrate overlay may be superior to the liquid substrate method for the same reasons advanced elsewhere, namely, the loss of products of the reaction into liquid environment is prevented.

The results of 25 LDH serum zymograms from normal individuals are shown in Table 15-2.

Ferritin Localization

PRINCIPLE

In immunochemical studies involving the electron microscope, antibodies labeled with ferritin are frequently needed. The method to be described is designed to determine effectiveness of the tagging. Prussian blue, a stable blue pigment, is produced by the reaction of ferric iron with potassium ferrocyanide forming the complex ferric ferrocyanide [54, 55]. This stain applied to an electrophoretogram of ferritin antibody distinguishes the coupled ferritin from the ferritin not in union with the antibody. An estimate of the effectiveness of the tagging method can be ascertained from these findings.

EQUIPMENT AND REAGENTS

1. Standard gel electrophoretic equipment (see Standard Gel Electrophoretic Method).
2. Barbital buffer, pH 8.6, ionic strength 0.05.
3. Standard agarose-buffer mixture (see Standard Gel Electrophoretic Method).
4. Prussian blue stain.
   Solution A.
   Potassium ferrocyanide, 2 percent (W/V) solution (keeps several weeks)
   Solution B.
   Hydrochloric acid, 1 percent (W/V) solution
   Solution C.
   To one part of solution A add three parts of solution B (make fresh just before using).

PROCEDURE

1. Prepare strips by marking midpoint and 3.2-cm. anodic migration point. Coat strips with 5 ml. of buffered agarose solution using the trench method for preparation of inoculation point.
2. Inoculate with 15μl. of the specimen, dot midpoint with bromphenol blue–albumin and migrate at 150 volts until the albumin reaches the 3.2-cm. anodic migration point.
3. Remove agarose strips from cell and place in solution C for 15 minutes.

4. Soak in distilled water for 30 minutes to remove buffer salts and dry.

RESULTS

A bright blue color is formed in the area where ferritin is present.

NOTE

The same procedure can be carried out on immunoelectrophoretograms.

Autoradiography

PRINCIPLE

Autoradiography of electrophoretograms and immunoelectrophoretograms is applicable for detection and localization of isotope-labeled compounds. Most frequently the isotope is attached to a protein. Localization in electrophoretograms of thyroxin-binding protein, ceruloplasmin, transferrin, and other important specific carrier proteins has been successfully performed using specific isotope-labeled compound placed in the test sample before electrophoresis and finalized by autoradiography. The localization of radioactive iron as a function of iron-binding capacity has been advocated as a useful clinical laboratory test. Localization of antigen, of antigen-antibody reaction, or of antibody in immunoelectrophoretograms or electrophoretograms has been based on use of I^{131} or I^{125} tag and has found numerous applications in immunochemistry [34]. The procedure to be described is designed to give a basic method of preparing strips or sheets for autoradiography.

A method of immunoelectrophoresis for localization of a specific immunoglobulin against an antigen, as, for example, ragweed, is described below. An I^{131}-tagged specific antigen is mixed with the anti-γG, anti-γA, and anti-γM antisera, which are then used as antisera against a serum suspected of having antibodies to the isotope-tagged antigen. After diffusion, soaking, and drying, the immunoelectro-plate is placed agarose side down in contact with the x-ray film, usually from 8 to 16 hours. Specific reaction sites of I^{131}-labeled antigen with the antibody produced in the serum are demonstrated on the x-ray film as a black area [34].

EQUIPMENT AND REAGENTS

1. Equipment and reagents for serum protein immunoelectrophoresis.
2. X-ray film and x-ray developing solutions.
3. I^{131}-tagged specific antigen.

PROCEDURE

1. Mix I^{131}-tagged antigen with the γG, γA, and γM antisera. The ratio of I^{131} antigen to antisera depends upon the titer of the specific antigen.
2. Prepare, inoculate, and migrate immunoelectro-plate as for serum protein immunoelectrophoresis.
3. Fill the trenches of the immunoelectro-plate with 100μl. of the respective antisera.
4. Allow the immunoelectro-plate to diffuse at moist room temperature for 16 hours.
5. Soak the plates in saline solution for 48 hours to remove excess proteins and buffer salts.
6. Dry the immunoelectro-plates.
7. With the dried gel surface down, leave the immunoelectro-plates in contact with x-ray film for 24 hours. Remove the immunoelectro-plates and develop the x-ray film.
8. The immunoelectrophoretogram can be stained with Thiazine Red R for comparison with autoradiograph.

RESULTS

Immunoprecipitin bands are those normally expected, i.e., γG, γA, and γM. If the serum had anti-ragweed antibodies that were γG, the γG immunoprecipitin band would contain γG anti-ragweed antibodies precipitated by goat anti-γG. The isotope-tagged ragweed theoretically should become attached to the precipitated γG anti-ragweed. Many other approaches can be developed.

Fluorescent Labeling and Scanning of Electrophoretograms

PRINCIPLE

The coupling of fluorescein isothiocyanate to antibodies is a well-established, chemically sound, and sensitive method of labeling protein [61] and is well suited to serum proteins fixed in a dry gel matrix of an electrophoretogram [16, 17]. The principle involves the concept of solid phase labeling of proteins to improve end point detection in contrast to staining. The determination of serum albumin by fluorescent dye has been reported by Reed et al. [76]. (See Table 15-3.)

EQUIPMENT AND REAGENTS

1. Standard gel electrophoretic equipment (see Standard Gel Electrophoretic Method) plus a Turner III fluorometer coupled to a recorder.
2. Standard gel electrophoretic reagents (see Standard Gel Electrophoretic Method).

3. Standard agarose-buffer mixture (see Standard Gel Electrophoretic Method).
4. Fixative solution—absolute methanol or 75 percent ethanol.
5. Fluorescein isothiocyanate (Baltimore Biological Laboratories, Baltimore, Md.)—10 mg. in 100 ml. of carbonate buffer, pH 9.0. This solution must be prepared fresh.

PROCEDURE

1. Prepare strips according to standard protein electrophoresis method.
2. Place in cell filled with barbital buffer, pH 8.6, ionic strength 0.05.
3. Inoculate strips with 2μl. of serum.
4. Migrate until leading edge of marker dye reaches 4.2-cm. migration mark (30–35 minutes).
5. Fix strips for 15 minutes, then dry.
6. Place dried strips in fluorescein isothiocyanate solution for 10 minutes for coupling to protein.
7. Wash excess fluorescein isothiocyanate from the strip in 2 or 3 washes of distilled water.
8. Scan strips (wet or dry) in a Turner III fluorometer coupled to a recorder using the #110-853 blue lamp and filter BG-12 as a primary filter with Wratten filter #15 as a secondary filter.

TABLE 15-3

PERCENT DISTRIBUTION OF PROTEIN FRACTIONS BY THIAZINE STAIN,
FLUORESCEIN, AND GRAVIMETRIC METHODS

Fraction	Thiazine	Fluorescein	Gravimetric
Alb.	57.5	33.5	64.6
α_1	4.5	10.4	6.1
α_2	11.0	18.1	11.3
β	11.5	16.2	5.0
γ	15.5	21.8	13.2

SOURCES OF ERROR

1. Sources of error discussed in the procedure for serum protein electrophoresis are applicable.
2. The two most critical aspects of this procedure are the pH and age of the fluorescein isothiocyanate solution. The solution must be prepared fresh. It is good for only one to two hours after preparation.

Quantitation of Immunoglobulins

Principle

Quantitation of immunoglobulins can be done by electroimmuno-diffusion, double gel diffusion, single gel diffusion, or radial single gel diffusion [80]. Described below is a commercial preparation of radial single gel diffusion [56]. In this method a specific antiserum is incorporated into the gel.

The standards and unknowns are inoculated, diffused for 16 hours, and intensified, and the precipitin rings are measured. The greater the concentration of the antigen, the larger the diameter of the precipitin ring [56, 77]. γM, because of its large size, will usually not diffuse through the gel in proportion to its concentration.

Equipment and Reagents

1. Quantitative Immunoplate (Hyland Laboratories)* (γG, γA, γM). Each kit comes with three standards—high, mean, low— and six plates for one specific antibody.
2. Hamilton syringe, 1μl. to 10μl.
3. Incubation chamber.
4. Solar Immuno-Glo camera and Immuno-Glo with quantitative grid (HCS Corp. or Hyland Laboratories).*

Procedure

1. The plates are labeled according to their antigen specificity.
2. Each plate contains six antigen wells, which are inoculated with 8μl. of sample as follows:
   γG plate antigen wells.
     (1) γG high standard
     (2) γG mean standard
     (3) γG low standard
     (4), (5), (6) Unknowns
   γA antigen wells.
     (1) γA high standard
     (2) γA mean standard
     (3) γA low standard
     (4), (5), (6) Unknowns
   γM antigen wells.
     (1) γM high standard
     (2) γM mean standard
     (3) γM low standard
     (4), (5), (6) Unknowns

*See Appendix.

3. The Immunoplates are covered and placed in a room-tempera-ture, moist incubator for 16 hours (overnight).
4. The plates are intensified by reaction in a 1 percent tannic acid solution for five minutes.
5. The tannic acid is gently removed by a wash of distilled water.
6. The plates are placed on the quantitative grid (1-mm. squares) and a Polaroid print is taken (Fig. 15-20).
7. The millimeter marks can then be counted on the print and recorded.
8. Each immunoglobulin is plotted on semilog paper (Fig. 15-21). It helps to have a clear plastic overlay or envelope on which to draw the curve with wax pencil so that the system may be repeated (Fig. 15-22). The diameters of the standards are plotted on the X axis against log concentration on the Y axis (Fig. 15-23). The curve is usually linear except at low concen-tration. Also the well size causes some error.
9. From the calibration curve interpolate the concentration of the unknowns.

RESULTS

The immunoglobulins are reported in milligrams per 100 ml. of serum. Qualitative judgment of values in reference to electrophoretic pattern of γ-globulin may on occasion suggest a difference which needs to be evaluated. Small monoclonal peaks sometimes cause con-fusion in terms of concentration, and it should always be within the scope of the laboratory's activities to explore such discrepancies until a "best answer" or diagnostic answer is obtained.

NORMALS

Newborn infant, milligrams per 100 ml.
  γG 600–1200
  γA 3
  γM 10
6-Month infant, milligrams per 100 ml.
  γG 300–650
  γA 10–50
  γM 40–60
12-Month infant, milligrams per 100 ml.
  γG 450–900
  γA 20–60
  γM 50–80
6 Years, milligrams per 100 ml.
  γG 600–1100
  γA 50–140
  γM 50–90

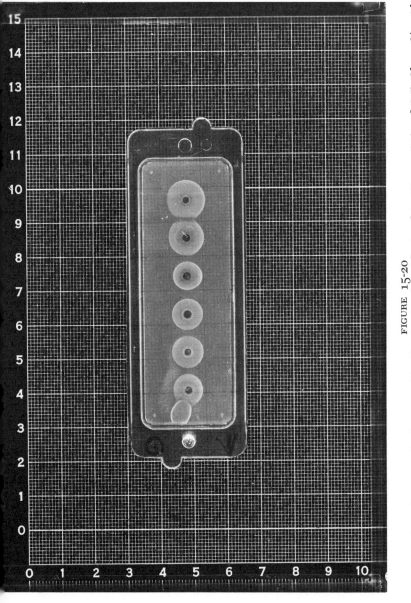

FIGURE 15-20

A view looking down on the Immuno-Glo illumination box (depicted in Figures 2-5 and 2-7, Chap. 2) with the Quantitative Immuno-plate on a transparent grid. Light from the Immuno-Glo box illuminates the Immuno-plates as well as the grid. A Polaroid photograph permits easy and accurate measurement of the diameter of the radial immunodiffusion precipitate by noting the calibrated grid markings in reference to diameter.

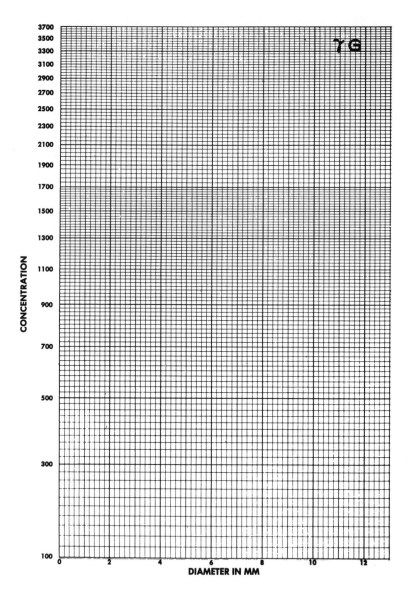

FIGURE 15-21

Semilog paper used to plot diameter in millimeters of the radial immunodiffusion precipitate against concentrations

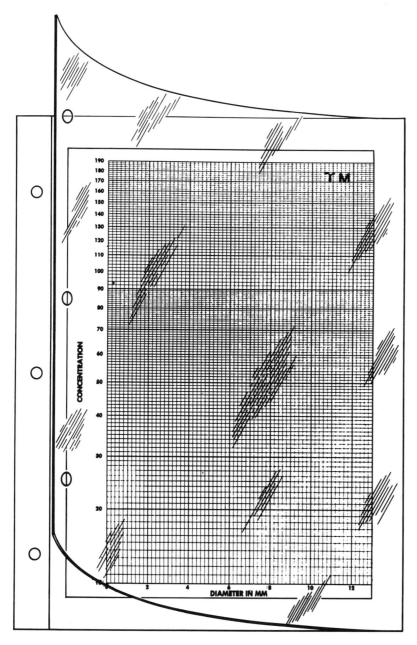

FIGURE 15-22

Method used in our laboratory to permit easier handling of large volumes of immunodiffusion assay. The γM immunoglobulin sheet is inserted in a transparent plastic holder upon which diameters are marked with a wax pencil.

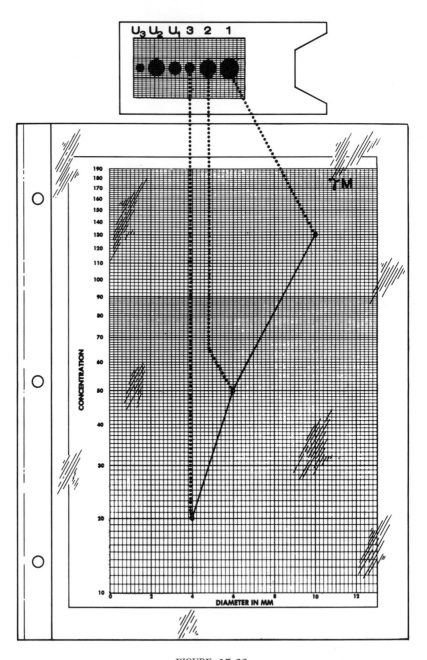

Adults, milligrams per 100 ml.
　γG 600–1200
　γA 50–240
　γM 50–150

Macroglobulin-SIA Test

PRINCIPLE

The SIA test is a *nonspecific* screening test for increased γ-globulin. First recommended as a screening procedure for macroglobulins, it has since been shown to be negative at times for macroglobulins, and it is also positive in other monoclonal and polyclonal gammopathies [77, 91]. Serum protein electrophoresis is always performed *in addition to* the requested SIA test.

PROCEDURE

1. Add 1 or 2 drops of fresh serum to 1 ml. of distilled water.
2. Observe the water for cloudiness or precipitate.

RESULTS

If cloudiness or precipitate forms, the SIA test is positive. The report is released along with the electrophoretic report.

Pyroglobulin

PRINCIPLE

Pyroglobulin is a thermoprotein which precipitates usually irreversibly at 56° C. This abnormal globulin is present in serums of monoclonal and polyclonal gammopathies.

The method described is a screening procedure for the detection of heat-sensitive globulin [77, 91]. In addition, a serum protein electrophoretogram should always accompany the screening procedure.

EQUIPMENT

1. Water bath to 56° C.
2. Test tube.

PROCEDURE

1. Collect fresh centrifuged serum.
2. Into each of two tubes measure 1 ml. of serum.
3. Incubate one tube at 56° C. for 24 hours and the other tube at 37° C. for 24 hours.
4. If there is precipitate or gelling in the tube incubated at 56° C., a pyroglobulin is present in the serum. If there is no precipitation, pyroglobulin is not present.

NORMAL

Negative for pyroglobulin.

Cryofibrinogen

PRINCIPLE

Cryofibrinogen is a thermoprotein of fibrinogen and may represent transition products in the conversion of fibrinogen to fibrin [91]. Cryofibrinogens can be present in some cold sensitivities, prostatic carcinomas, and other specific disorders [77, 91]. The cryofibrinogen can be demonstrated by incubation of plasma (heparinized or oxalated) at 4°C. for 24 hours. Quantitation and specific identification of cryofibrinogen are done using the procedure described for cryoglobulins, below.

EQUIPMENT AND REAGENTS

1. Same as for serum protein immunoelectrophoresis.
2. Barbital buffer, pH 8.6, ionic strength 0.05.
3. Wintrobe sedimentation tubes.

PROCEDURE

1. Centrifuge fresh heparinized plasma and separate from cells.
2. Fill two Wintrobe sedimentation tubes with 1 ml. of plasma each.
3. Incubate one tube erect at 37° C. for 24 hours and the other tube erect at 4° C. for 24 hours.
4. If there is fine white precipitate in the 4° C. tube at the end of 24 hours, a cryofibrinogen is present. If there is no precipitate, the plasma is negative for cryofibrinogen. The 37° C. tube is a control for comparison of the plasma's natural clearness.
5. If it is positive for cryofibrinogen, centrifuge the 4° C. tube at 2500 rpm for five minutes while still cold. A refrigerated centrifuge is preferable.
6. Read the amount of precipitate from the sedimentation scale; 1 mm. of precipitate is equal to 1 percent cryofibrinogen content.
7. To specifically identify the components of the cryofibrinogen, wash the precipitate four times with cold saline solution (4° C.).
8. Dilute precipitate 1:10 in barbital buffer, pH 8.6, ionic strength 0.05. Place suspension of precipitate-buffer in 37° C. water bath until precipitate is dissolved.
9. Inoculate an immunoelectro-plate (as for serum protein immunoelectrophoresis) with cryofibrinogen-buffer suspension

in antigen wells 1, 3, 5, 7 and plasma sample in antigen wells 2, 4, 6, 8.

10. Migrate at 150 volts until albumin reaches the 3.2-cm. anodic migration point.
11. Fill the antibody trenches with 100μl. of the specific antisera: γG, γA, γM, fibrinogen.
12. Diffuse, intensify, and photograph as for serum protein immunoelectrophoresis.

RESULTS

Normal: no cryofibrinogen detected at the end of 24 hours at 4° C.

Cryoglobulin

PRINCIPLE

Cryoglobulin is a thermoprotein; that is, it has the physical characteristics of gelling or precipitating at 4° C. [77, 91]. This physical property of cryoglobulin is usually reversible. Cryoglobulins may be typed according to what immunoglobulins make up the thermoprotein. At present there are at least seven types of cryoglobulins described: (1) γG, (2) γM, (3) γA, (4) BJ, (5) γM-lipoprotein, (6) γG-γM, (7) γG-γA [91]. Cryoglobulins are present in some monoclonal gammopathies and in some polyclonal gammopathies [91].

The procedure described quantitates and types the cryoglobulins by use of immunoelectrophoresis [71].

EQUIPMENT AND REAGENTS

1. Same as for serum protein immunoelectrophoresis.
2. Cold saline solution.
3. Refrigerator at 4° C.
4. Wintrobe sedimentation tubes.

PROCEDURE

1. Fill two Wintrobe sedimentation tubes with 1 ml. of fresh centrifuged serum each.
2. Place one tube erect in a 37° C. water bath and place the other tube erect in the refrigerator at 4° C.
3. Leave tubes undisturbed for 24 hours.
4. At the end of 24 hours observe the 4° C. tube for precipitation or cloudiness. The 37° C. tube is a control for comparison of the serum's natural clearness.
5. If no precipitation or cloudiness is present, the test for cryoglobulins is negative.
6. If there is precipitation or cloudiness, the amount can be quantitated and, if the quantity of precipitate is sufficient, it can be typed.

7. To quantitate, centrifuge the 4° C. tube at 2500 rpm, preferably in refrigerated centrifuge, immediately.
8. Read the amount of precipitate from the Wintrobe scale; 1 mm. of precipitate would represent 1 percent cryoglobulin present.
9. To type the cryoglobulin, wash the precipitate four times with cold (4° C.) saline solution.
10. Dissolve the precipitate in barbital buffer, pH 8.6, ionic strength 0.05, by placing a suspension of precipitate and buffer in a 37° C. water bath until dissolved. A dilution of 1:10 of precipitate-buffer is used.
11. Inoculate the open immunoelectro-plate with the cryoglobulin-buffer solution in the 1, 3, 5, 7, 9, 11 wells with the whole serum sample in the 2, 4, 6, 8, 10, 12 wells (see Serum Protein Immunoelectrophoresis).
12. Migrate the albumin to the 3.2-cm. anodic migration mark.
13. Fill the trenches with 100μl. of antisera: γG, γA, γM, IEP, χ, λ.
14. Diffuse, intensify, and photograph as for serum protein immunoelectrophoresis.

RESULTS

1. Normal: No cryoglobulin would be detected at the end of 24 hours at 4° C.
2. Abnormal: Cryoglobulin present.
3. The quantities in percent and types of cryoglobulin present are reported.
4. It is our practice to do a serum protein electrophoretogram when a test for cryoglobulin is ordered.
5. When typing of cryoglobulins is done, the patient's serum serves as a check or control for comparison. IEP antiserum shows how well the other proteins were washed away. Generally there is still a trace of albumin present.

Creatine Phosphokinase (CPK) Isoenzymes

PRINCIPLE

Isoenzymes of creatine phosphokinase (ATP: L-arginine phosphotransferase, EC 2.7.3.2) [29] can be demonstrated by agarose electrophoresis [9]. The visualization of this isoenzyme is shown by the following reactions [65]:

$$\text{ADP} + \text{creatine phosphate} \xrightleftharpoons{\text{CPK}} \text{ATP} + \text{creatine}$$

$$\text{ATP} + \text{glucose} \xrightleftharpoons{\text{Hexokinase}} \text{glucose-6-phosphate} + \text{ADP}$$

Glucose-6-phosphate $+$ NADP $\xrightleftharpoons{\text{G-6-PD}}$

$\qquad$ 6-phosphogluconic acid $+$ NADPH$_2$

NADPH$_2$ $+$ phenazine methosulfate (PMS) $\longrightarrow$

$\qquad$ NADP $+$ reduced PMS

Reduced PMS $+$ Nitro Blue Tetrazolium (NBT)* $\longrightarrow$

$\qquad$ formizan $\downarrow$ $+$ PMS

EQUIPMENT AND REAGENTS

1. Standard gel electrophoretic equipment (see Standard Gel Electrophoretic Method).
2. Barbital buffer, pH 8.6, ionic strength 0.05, for electrophoretic tanks.
3. Barbital buffer, pH 8.6, ionic strength 0.025, for agarose strips.
4. Buffered gel. To 50 ml. of barbital buffer, pH 8.6, ionic strength 0.025, add 250 mg. of agarose and bring to a boil.
5. Substrate.
   Solution A.

Phosphate buffer, pH 7.5, M/15	200.0	ml.
$\quad$ KH$_2$PO$_4$, M/15 $\qquad$ 7 ml.		
$\quad$ Na$_2$HPO$_4$, M/15 $\qquad$ 43 ml.		
Magnesium acetate	2.0	mM.
Nitro Blue Tetrazolium (NBT)*	54.0	mg.

   (The solution is stable for three weeks.)
   Solution B.

Solution A	15.0	ml.
Glucose	0.11	mM.
Adenosine diphosphate (ADP)*	15.0	μM.
Adenosine monophosphate (AMP)*	0.15	mM.
Nicotine adenine dinucleotide phosphate		
$\quad$ (NADP)*	5.2	μM.
Creatine phosphate	33.0	μM.
Hexokinase solution	10.0	I.U.
Glucose-6-phosphate dehydrogenase (G-6-PD)*	5.0	I.U.
Phenazine methosulfate (PMS) solution*		
$\quad$ (1 mg. per milliliter)	0.4	ml.

PROCEDURE

1. With 5 ml. of warm buffered agarose gel coat strips as for trench method. Mark at the midpoint (origin) and 3.2-cm. anodic migration point.
2. Inoculate with 15μl. of specimen. Migrate at 150 volts until bromphenol blue–albumin marker reaches the anodic migration mark.

*See Appendix (Sigma Chemical Co.)

3. Remove from cell and cover with solution B for one hour at 37° C.
4. Soak in distilled water to remove buffer salts and dry.

RESULTS

Purple-blue bands are formed at the sites of enzyme activity.

Acrylamide Electrophoresis

PRINCIPLE

Most of the preliminary work utilizing acrylamide as a support medium for electrophoresis was performed by two groups. One, at the William Pepper Laboratory of Clinical Medicine, University of Pennsylvania, was made up of Raymond, Weintraub, Wang, and Nakamichi [73, 74, 75]. The second group consisted of Ornstein and Davis at Mount Sinai Hospital, New York City [23, 24, 66]. Others [37, 92] are now reporting their findings utilizing acrylamide.

As mentioned, acrylamide gels fall into the classification of molecular sieves, and, as with starch, separation of proteins is based not only upon electrophoretic mobility of the protein but also upon the size and shape of the protein molecule. Since the pore size in the gel can be controlled, considerable separation is possible. The results are quite compatible with those achieved with starch and in many instances superior. Since acrylamide is perfectly transparent after destaining, the distinctiveness of the patterns is striking. Acrylamide is soluble in buffer solution and polymerizes to form a suitable gel for electrophoresis. When cross-linked with a small proportion of methylene bisacrylamide, it creates a very suitable molecular sieve.

Two basic systems are in use, that described by Raymond et al. [73, 74, 75] which utilizes acrylamide in sheets and blocks, and that of Ornstein and Davis [23, 24, 66], which employs discontinuity in pH and gel size, giving rise to the name *disc electrophoresis*. The experimental conditions outlined by Kohlrausch in 1897 were utilized by Ornstein [66] to formulate the experimental model of disc electrophoresis. The system is formed inside a glass column in three layers; the lower gel (small-pore gel) is separated from the sample by a spacer gel (large-pore gel), and the upper gel is applied over the surface of the spacer gel. In this system the proteins are first separated according to their mobility into highly concentrated discs of protein in the spacer gel. When they enter the small-pore gel, further separation takes place based upon the size of molecule. The system produces extremely sharp separation of proteins. The amount of material necessary for the system is quite small.

We will describe disc electrophoresis and discuss some of the uses and modifications that we have found for it in our laboratories. The reagents and equipment were obtained from Canalco.*

*Canal Industrial Corporation, Bethesda, Md.

EQUIPMENT AND REAGENTS*

1. N,N′-Methylenebisacrylamide (BIS). BIS is a white crystalline solid that should be stored in the refrigerator. After prolonged storage, self-polymerization may take place.
2. 2-Amino-2-(hydroxymethyl)-1,3-propanediol (Tris).
3. N,N,N′,N′-Tetramethylethylenediamine (TEMED).
4. Riboflavin.
5. HCl, 1N.
6. Acrylamide monomer.
7. Ammonium persulfate, reagent grade. Store dry reagent in refrigerator.
8. Glycine (ammonia-free).
9. Standard gel, stock solutions.†

Solution A.

HCl, 1N	48.0	ml.
Tris	36.3	gm.
TEMED	0.23	ml.
Water to make	100.0	ml.

Solution B.

HCl, 1N	48.0	ml.
Tris	5.98	gm.
TEMED	0.46	ml.
Water to make	100.0	ml.

Solution C.

Acryamide	28.0	gm.
BIS	0.735	gm.
Water to make	100.0	ml.

Solution D.

Acrylamide	10.0	gm.
BIS	2.5	gm.
Water to make	100.0	ml.

Solution E.

Riboflavin	4.0 mg. per 100 ml. water

Solution F.

Ammonium persulfate	0.14	gm.
Water to make	100.0	ml.

10. Buffer, Tris glycine. This buffer is made by dissolving 6.0 gm. of Tris and 28.8 gm. of glycine in 1 liter of water. pH of solution is 8.3 (concentrated buffer).
11. Acetic acid solution, 7.5 percent (V/V).
12. Amido black stain. This is made by dissolving 1 gm. of Amido Black 10-B in 100 ml. of 7.5 percent acetic acid.
13. Tracking dye—a few crystals of bromphenol blue dissolved in water.

*The individual reagents described here to make standard gel can be ordered from Canalco or Distillation Products Industries, Rochester, N.Y.
†Premixed reagents are available from Canalco.

PROCEDURE

PREPARATION OF GEL COLUMNS. Three layers of gel are built up in cylindrical glass columns: starting at the bottom, small-pore gel or lower gel, then two layers of large-pore gel or upper gel. Each has a particular function. The lower or small-pore gel serves as the molecular separation medium. The middle layer, or large-pore gel, serves as a spacer gel to concentrate the sample into thin discs of protein. The top layer, or large-pore gel, acts as a stopper to prevent back-diffusion of sample into the buffer reservoir.

The lower gel is formed by mixing one part of solution A, two parts of solution C, one part water, and four parts of solution F. After mixing, the resulting solution is de-aerated by vacuum.

The de-aerated solution is pipetted into glass tubes 5 mm. in inside diameter and 60 mm. in length. The tubes are fitted on the ends with rubber base caps and placed in a vertical position. The lower gel is introduced, as well as the other solutions, into the tubes by means of a syringe. We have found a 10-ml. plastic syringe to be preferable. The syringe is fitted with a small-diameter piece of polyethylene tubing which may then be inserted in the glass column. Filling proceeds from the bottom upward.

The lower gel solution is put into the glass tube to a depth of 40 mm. Immediately after introduction of this gel into the tube, water is layered about 3 mm. in depth over the gel. In order to prevent turbulence or intermixing at this point, the water is introduced by a 1-ml. plastic tuberculin syringe fitted with a 25-gauge needle which is slightly bent near the hub. Fill the syringe with water and remove the plunger. Place the tip of the needle on the wall of the tube just above the small-pore gel and allow the water to flow down by gravity. If there is any mixing between the water and the gel, the tube must be discarded. Once the water layer has been applied, the tube is allowed to remain undisturbed for 45 minutes.

The next layer of gel, spacer gel, is formed by mixing one part of solution B, two parts of solution D, one part of solution E, and four parts water. The mixture is de-aerated as described for lower gel. The water layer in the tubes is removed by inverting the tubes and gently shaking them. The tubes are rinsed once with upper gel solution and then filled with upper gel to about 5 mm. above the lower gel. Again place a layer of water on top of the spacer gel, as before. The upper gel, unlike the lower gel, which is catalyzed chemically, is catalyzed by light. A 15-watt fluorescent light directly behind the tubes causes polymerization in about 20 minutes.

The third layer of gel is applied after polymerization of the spacer gel. This is made exactly as the spacer gel is made. The tube is filled to the top. Before the gel polymerizes, the sample is applied—a modification developed in our laboratories. With a Hamilton microsyringe (10μl.), 5μl. of unmodified sample are then layered between the spacer gel and the upper gel by inserting the tip of the syringe down to the upper level of the spacer gel. The 5μl. are then slowly injected to take up a position at the interphase between the spacer gel and the upper gel. The syringe is gently removed and the upper gel permitted to polymerize with light.

ELECTROPHORESIS.   The electrophoretic stage is now ready to begin. The base caps of the tubes are removed, and the tubes are inserted in the soft plastic mountings of the upper buffer compartment. The lower end is then immersed in the lower compartment. The two buffer compartments are filled with buffer, 1 liter each of concentrated buffer diluted 1:10. One drop of bromphenol blue tracking dye is added to the upper buffer compartment. The electrodes are connected to their respective receptacles—the negative electrode to the top reservoir and the positive electrode to the bottom.

The electrophoretic phase is started by adjusting the power supply to deliver 5 ma. per tube. The distance of migration is not the same in all tubes. Therefore, if reproducibility is desired, a distance of 30 mm. is marked on the outside of the columns from the upper end. When the tracking dye bound to albumin meets this point, that particular tube is withdrawn. (A hard plastic rod is used to plug the upper compartment opening before the tube is withdrawn with an insulated hemostat.) The power is lessened by 5 ma. for each tube removed. At the completion of this phase all tubes, then, should show the same distance of migration.

REMOVING GEL FROM COLUMN, STAINING, AND DESTAINING.   The gels are removed from the glass column by immersing the entire tube in ice water. With the tube under water, a teasing needle is inserted into the glass column and the gel is rimmed. A rubber bulb is then placed over one end of the glass column, preferably the upper end, and pressure exerted. If just the right pressure is exerted, the gel column slides swiftly from the glass tube. Too much pressure will cause the gel to tear. If resistance is met, the gel is rerimmed and pressure reapplied. The columns of gel are then immediately immersed in a tube of stain containing fixative for one hour. It is important to stain promptly since there is diffusion of the proteins after electrophoresis.

Destaining is done electrically. The gel is placed in a destaining tube, which is then connected to the upper and lower tanks as previously described for electrophoresis. A 7.5 percent solution of acetic acid is placed in both reservoirs and 12.5 ma. per tube of current is applied. The entire destaining process takes about an hour. The gels are removed after destaining and placed in glass tubes with 7.5 percent acetic acid as a preservative. The tubes may be stoppered and stored. The gels may also be dehydrated by permitting them to lie exposed to the air in a stainless-steel pan. They can be stored in this condition for a very long time and when rehydrated with 7.5 percent acetic acid will reassume their original size without loss of significant detail.

If special histochemical or other studies are necessary, the gel, after removal from the glass columns, may be placed in a glass tube and frozen in the deepfreeze. When it is later removed for study, staining with special histochemical stains or protein stains is readily accomplished.

In our laboratory, we have found it unnecessary to remove the columns of gel from the glass tube before freezing. We simply take the tube straight from the electrophoretic phase, place it inside a test tube, and freeze in the deepfreeze. Then, at the time of the special histochemical work, we remove the gel from the column of glass as previously described after the electrophoretic stage.

SCANNING. For scanning, a microdensitometer is available from Canalco,.or other systems utilizing photographic reduplication with scanning of the photograph or modification of a Spinco Analytrol are possible. In our laboratory a method using photography proved to be satisfactory. To speed analysis of disc electrophoretograms a 35-mm. color and black-and-white photographic technic was developed for densitometric scanning. A 35-mm. Exakta* camera with a f35 mm. 1:28 lens with a closeup attachment featuring the auto-quinaron proved to be the most satisfactory for this purpose.

PHOTOGRAPHIC TECHNICS. The Exakta 35-mm. camera is placed on a Yashica† copy stand containing a light box approximately 14″ by 18″, with six 25-watt bulbs for illumination. A black mask outlining the shape of the disc electrophoretogram is placed over the surface of the illuminating box. A small opening is made in the mask near the tube for the insertion of numbers, such that each photograph will have a built-in system of identification. High-speed Ektachrome B, ASA 125 is used. The approximate exposure time is 1/200th of a second at f22 for color slides with a green filter #X1. Panatomic X film developed in Hyfinol for $5\frac{1}{2}$ minutes at 68° C. with proper fixation and washing produced excellent black-and-white slides.

DENSITOMETRIC SCANNING. The Beckman Micro Scanner Attachment is employed with the Analytrol as specified by the manufacturer. The slides are scanned at 550 mμ and a B-2 cam is used. The 35-mm. color photograph of the disc electrophoretogram is inserted in the carriage and scanned at a speed of 70. Figure 12-4 in the chapter on analytical technics shows a photograph of a disc electrophoretogram

*Nikon, Inc., New York, N.Y.
†Tokyo, Japan.

of nonhemoglobin erythrocyte protein (NHEP) and scan. The scan and disc electrophoretogram were matched from separate photographs. The resolution of this method of scanning compares favorably with that of a microdensitometer manufactured by Canalco designed specifically for scanning disc electrophoretograms.

Discussion

We have found the system of disc electrophoresis applicable to separation of body fluids containing minute quantities of protein. In our hands the system works quite well for separation of spinal fluid proteins. The procedure utilizes essentially the same mechanism as the standard procedure. The gel solutions remain unchanged in composition. The spacer gel is increased to 15 mm. The lower gel remains the same. Increasing the thickness of the spacer gel makes marked concentration of the protein possible. It is, therefore, unnecessary to perform a preliminary concentration of the spinal fluid by dialysis or ultrafiltration before performing a disc electrophoretic analysis. This is accomplished by increasing the length of the total tubes as well as proportionately increasing the thickness of the spacer gel. The inoculation of the sample is also increased. For spinal fluid, we have found 0.6 ml. of spinal fluid to be satisfactory. To accommodate the increased sample size, the glass tubes are increased to 6 inches in length. The electrophoretic time is increased to 1½ hours. The marker dye is still incorporated to note distance of migration. With such long columns there is some difficulty in removing the gels from the columns. We have, therefore, introduced a modification which overcomes this objection. Two glass columns of the usual length are coupled together by a tight-fitting plastic ring. The gels are then built up inside this elongated double tubing of glass, which then serves as the electrophoretic separating chamber. At the completion of electrophoresis, the two columns are disassociated, and the lower portion, which contains the small-pore gel, is treated exactly as for serum. Using the technic just described, we have been able to identify approximately 16 bands in spinal fluid.

Urinary proteins can also be separated by this technic. However, preliminary concentration by ultrafiltration is necessary. For best results, 10 ml. of urine are concentrated to about 1 ml. by means of an ultrafiltration system. We have found the metal pressure filtration apparatus, MD 50-10, manufactured by Membranfiltergesellschaft MBH,* very satisfactory. The system utilizes a celloidin membrane and high pressure. In our system 150 pounds per square inch is applied to the 10 ml. of urine. In about two hours the 10 ml. has been concentrated to 1 ml. or less. However, the ultrafiltration method described under concentration of cerebrospinal fluid and urine for elec-

*Carl Schleichier & Schuell Co., Keene, N.H.

trophoresis is used routinely. We utilize 0.6 ml. of the concentrated urine and apply to the columns as described for spinal fluid. Urinary protein separation has been reasonably good by this technic. Some blurring of bands is present.

References

1. Afonso, E. On human serum amylase. *Clin. Chim. Acta* 14:195, 1966.
2. Axelsson, U., and Laurell, C. B. Hereditary variants of serum gamma$_1$-antitrypsin. *Amer. J. Hum. Genet.* 17:466, 1965.
3. Babson, A. L., Shapiro, P. O., Williams, P. A. R., and Phillips, G. E. The use of a diazonium salt for the determination of glutamic-oxaloacetic transaminase in serum. *Clin. Chim. Acta* 1:199, 1962.
4. Barka, T. Electrophoretic separation of acid phosphatase in rat liver on polyacrylamide gels. *J. Histochem. Cytochem.* 9:542, 1961.
5. Barka, T., and Anderson, P. J. *Histochemistry Theory Practice and Bibliography.* New York: Hoeber Med. Div., Harper & Row, 1963. Pp. 227–238.
6. Barron, K. D., Bernsohn, J., and Hess, A. R. Separation and properties of human brain esterases. *J. Histochem. Cytochem.* 11:139, 1963.
7. Beckman, L., Scandalios, J. G., and Brewbaker, J. L. Catalase hybrid enzymes in maize. *Science* 146:1174, 1964.
8. *Beckman Methods Manual,* Technical Bulletin 6095A. Palto Alto, Calif.: Spinco Div., Beckman Instruments, Inc., November 1961. Pp. 10–11.
9. Berggard, I. Identification and Isolation of Urinary Proteins. In Peeters, H. (Ed.), *Protides of the Biological Fluids.* Proceedings of the Eleventh Colloquium, Bruges, 1963. Amsterdam: Elsevier, 1964. Pp. 285–291.
10. Berk, J. E., Kawaguchi, M., Zeineh, R., Ujihira, I., and Searcy, R. Electrophoretic behavior of rabbit serum amylase. *Nature* (London) 200:572, 1963.
11. Block, W. D., Carmichael, R., and Jackson, C. E. Quantitative determination of isoenzymes of glutamic oxaloacetic transaminase in human serum. *Proc. Soc. Exp. Biol. Med.* 115:941, 1964.
12. Blumberg, B. S. Clinical Significance of Serum Haptoglobins. In Sunderman, F. W., and Sunderman, F. W., Jr. (Eds.), *Hemoglobin: Its Precursors and Metabolites.* Philadelphia: Lippincott, 1964. Pp. 318–324.
13. Boyer, S. H. Alkaline phosphatase in human sera and placenta. *Science* 134:1002, 1961.
14. Boyer, S. H., Porter, I. H., and Weilbacker, R. G. Electrophoretic heterogeneity of glucose-6-phosphate dehydrogenase and its relationship to enzyme deficiency in man. *Proc. Nat. Acad. Sci. U.S.A.* 48:1868, 1962.
15. Cawley, L. P. *Workshop Manual on Electrophoresis and Immunoelectrophoresis* (rev. ed.). Commission on Continuing Education, Council on Clinical Chemistry. Chicago: American Society of Clinical Pathologists, 1966. Pp. 115–118.

16. Cawley, L. P., Dibbern, P., and Eberhardt, L. Fluorescent labeling of electrophoretograms by fluorescein isothiocyanate. *Clin. Chem.* 13: 701, 1967.

17. Cawley, L. P., and Eberhardt, L. Simplified gel electrophoresis: I. Rapid technic applicable to the clinical laboratory. *Amer. J. Clin. Path.* 38:539, 1962.

18. Cawley, L. P., Eberhardt, L., and Goodwin, W. L. Agar-gel electrophoretic and analytic agar-gel electrophoretic study of bromelin. *Transfusion* 4:441, 1964.

19. Clarke, C. A. *Genetics for the Clinician. Acatalasia.* Oxford, Eng.: Blackwell, 1962. Pp. 175–176.

20. Conklin, J. L., and Edward, J. N. Malic dehydrogenase isoenzymes of the chick embryo. *J. Histochem. Cytochem.* 13:510, 1965.

21. Crowle, A. J. *Immunodiffusion.* New York: Academic, 1961.

22. Davidson, C. K., and Aide, P. A. A colorimetric screening method for cholinesterase using agar gel. *Anal. Biochem.* 12:70, 1965.

23. Davis, B. J. *Disc Electrophoresis,* Part II. Rochester, N.Y.: Distillation Products Industries, 1961.

24. Davis, B. J., and Ornstein, L. A New High Resolution Electrophoresis Method. Delivered at meeting of the Society for the Study of Blood, at the New York Academy of Medicine, March 24, 1959.

25. Decker, L. E., and Rau, E. M. Multiple forms of glutamic-oxaloacetic transaminase in tissues. *Proc. Soc. Exp. Biol. Med.* 112:114, 1963.

26. Dencker, S. J., and Swahn, B. Clinical Value of Protein Analysis in Cerebrospinal Fluid, a Micro-immuno-electrophoretic Study. From the Departments of Neurology and Medicine, Lund University. Lund, Sweden: Gleerup, 1961.

27. Diamond, I., and Aquino, T. I. Myoglobulinuria following unilateral status epilepticus and ipsilateral rhabdomyolysis. *New Eng. J. Med.* 272:834, 1965.

28. Dutcher, T. F. Description of Technique of Molecular Sieving and Application of Technique for Serum Haptoglobin Determination and Qualitative Urinary Myoglobin Determination. In *Manual for Workshop on Chromatography.* Commission on Continuing Education, Council on Clinical Chemistry. Chicago: American Society of Clinical Pathologists, 1965. Pp. 58–66.

29. *Enzyme Nomenclature.* Recommendations of the International Union of Biochemistry on the Nomenclature and Classification of Enzymes, Together with Their Units and the Symbols of Enzyme Kinetics. Amsterdam: Elsevier, 1965.

30. Eriksson, S. Pulmonary emphysema and alpha-antitrypsin deficiency. *Acta Med. Scand.* 175:197, 1964.

31. Ferris, T. G., Easterling, R. E., Nelson, K. J., and Budd, R. E. Determination of serum-hemoglobin binding capacity and haptoglobin-type by acrylamide gel electrophoresis. *Techn. Bull. Regist. Med. Techn.* 36:191, 1966.

32. Fredrickson, D. S., Levy, R. I., and Lees, R. S. Fat transport in lipoproteins—an integrated approach. *New Eng. J. Med.* 276:34, 94, 148, 215, 273, 1967.

33. Grabar, P., and Butin, P. *Immuno-electrophoretic Analysis.* Amsterdam: Elsevier, 1964.
34. Hale, R., Cawley L. P., Holman, J. G., and Minard, B. Radioimmunoelectrophoretic studies of antigen binding capacity of atopic sera. *Ann. Allerg.* 25:88, 1967.
35. Heimburger, N., and Schwick, G. Die Fibrin-agar-elektrophorese, eine Methode zur Immunologischen Charakterisierung des Fibrinolytischen Systems. In Peeters, H. (Ed.), *Protides of the Biological Fluids.* Proceedings of the Ninth Colloquium, Bruges, 1961. Amsterdam: Elsevier, 1962. P. 303.
36. Henry, R. J. *Clinical Chemistry.* New York: Hoeber Med. Div., Harper & Row, 1964. Pp. 231–233.
37. Hermans, P. E., McGuckin, W. F., McKenzie, B. F., and Bayrd, E. D. Electrophoretic studies of serum proteins in cyanogum gel. *Proc. Staff Meet. Mayo Clin.* 35:792, 1960.
38. Holmes, E. W., Malone, J. I., Winegrad, A. I., and Oski, F. A. Hexokinase isoenzymes in human erythrocytes: Association of type II with fetal hemoglobin. *Science* 156:646, 1967.
39. Joseph, R. R., Olivero, E., and Ressler, N. Electrophoretic study of human isoamylases. *Gastroenterology* 51:377, 1966.
40. Kabat, E. A., and Mayer, M. M. *Experimental Immunochemistry* (2d ed.). Springfield, Ill.: Thomas, 1961. Chap. 43, Dialysis and Ultrafiltration.
41. Kaplan, J. -C., and Beutler, E. Hexokinase isoenzymes in human erythrocytes. *Science* 159:215, 1968.
42. Kelsey, R. L., de Graffenried, T. P., and Donaldson, R. C. Electrophoretic fractionation of serum glycoproteins on cellulose acetate. *Clin. Chem.* 11:1058, 1965.
43. Kirkman, H. N., and Hendrickson, E. M. Sex-linked electrophoretic difference in glucose-6-phosphate dehydrogenase. *Amer. J. Hum. Genet.* 15:241, 1963.
44. Klemperer, M. R., Gotoff, S. P., Olper, C. A., Levin, A. S., and Rosen, F. S. Estimation of the serum beta$_1$C globulin concentration: Its relation to the serum hemolytic complement titer. *Pediatrics* 35:765, 1965.
45. Kreisher, J. H., Close, V. A., and Fishman, W. H. Identification by means of L-phenylalanine inhibition of intestinal components separated by starch gel electrophoresis of serum. *Clin. Chim. Acta* 11:122, 1965.
46. Kueppers, F., and Bearn, A. G. A possible experimental approach to the association of hereditary α_1-antitrypsin deficiency and pulmonary emphysema. *Proc. Soc. Exp. Biol. Med.* 121:1207, 1966.
47. Laurell, C. B., and Eriksson, S. The electrophoretic gamma globulin pattern of serum in gamma antitrypsin deficiency. *Scand. J. Clin. Lab. Invest.* 15:132, 1963.
48. Laurell, C. B., and Eriksson, S. The serum alpha-antitrypsin in families with hypo-gamma-antitrypsinemia. *Clin. Chim. Acta* 11:395, 1965.
49. Lees, R. S., and Fredrickson, D. S. The differentiation of exogenous and endogenous hyperlipemia by paper electrophoresis. *J. Clin. Invest.* 44:1968, 1965.
50. Lees, R. S., and Hatch, F. T. Sharper separation of lipoprotein

species by paper electrophoresis in albumin-containing buffer. *J. Lab. Clin. Med.* 61:518, 1963.

51. Lowenthal, A. *Agar-gel Electrophoresis in Neurology.* New York: American Elsevier, 1964.

52. McGeachin, R. L., and Reynolds, J. M. Inhibition of amylase by rooster antisera to hog pancreatic amylase. *Biochim. Biophys. Acta* 39:531, 1960.

53. McGeachin, R. L., and Reynolds, J. M. Serological differentiation of amylase isozymes. *Ann. N.Y. Acad. Sci.* 94:996, 1961.

54. McManus, J. F. A., and Mowry, R. W. *Staining Methods: Histologic and Histochemical.* New York: Hoeber Med. Div., Harper & Row, 1965. Pp. 195–196.

55. Mallory, F. B. *Pathological Technique.* Philadelphia: Saunders, 1942. P. 137.

56. Mancini, G., Carbonara, A. O., and Heremans, J. F. Immunochemical quantitation of antigens by single radial immunodiffusion. *Immunochemistry* 2:235, 1965.

57. Markert, C. L., and Hunter, R. L. The distribution of esterases in mouse tissues. *J. Histochem. Cytochem.* 7:42, 1959.

58. Martinek, R. G., Berger, D., and Broida, D. Simplified estimation of leucine aminopeptidase (LAP) activity. *Clin. Chem.* 10:1087, 1964.

59. Miale, J. B. *Laboratory Medicine—Hematology* (2d ed.). St. Louis: Mosby, 1962.

60. Muller-Eberhard, H. J., Nilsson, U., and Aronsson, T. Isolation and characterization of two beta$_1$-glycoproteins of human serum. *J. Exp. Med.* 11:217, 1960.

61. Nairn, R. C. *Fluorescent Protein Tracing.* Baltimore: Williams & Wilkins, 1964.

62. Naumann, H. N. Differentiation of Bence Jones protein from uroglobulins. *Amer. J. Clin. Path.* 44:413, 1965.

63. Nerenburg, S. T. *Electrophoresis. A Practical Laboratory Manual.* Philadelphia: Davis, 1966.

64. Oger, A., and Bischops, L. Les iso-enzymes de l'amylase. *Clin. Chim. Acta* 13:670, 1966.

65. Oliver, I. T. A spectrophotometric method for the determination of creatine phosphokinase and myokinase. *Biochem. J.* 61:116, 1955.

66. Ornstein, L. *Disc Electrophoresis,* Part I. Rochester, N.Y.: Distillation Products Industries, 1961.

67. Osserman, E. F., and Lawlor, D. Immunoelectrophoretic characterization of the serum and urinary proteins in plasma cell myeloma and Waldenström's macroglobulinemia. *Ann. N.Y. Acad. Sci.* (Art. 1) 94:93, 1961.

68. Osserman, E. F., and Lawlor, D. P. Serum and urinary lysozyme (muramidase) in monocytic and monomyelocytic leukemia. *J. Exp. Med.* 124:921, 1966.

69. Owen, J. A., and Smith, H. Detection of ceruloplasmin after zone electrophoresis. *Clin. Chim. Acta* 6:441, 1961.

70. Pearse, A. G. E. *Histochemistry* (2d ed.). Boston: Little, Brown, 1960. P. 913.

71. Peetoom, F., and van Loghem-Langereis, E. IgM-IgG (beta$_2$M-

$7S_{gamma}$) cryoglobulinaemia. An auto-immune phenomenon. *Vox Sang.* 10:281, 1965.

72. Porter, I. H., Boyer, S. H., Watson-Williams, E. J., Adam, A., Szeinberg, A., and Siniscalco, M. Variation of glucose-6-phosphate dehydrogenase in different populations. *Lancet* 1:895, 1964.

73. Raymond, S., and Nakamichi, M. Electrophoresis in synthetic gels: I. Relation of gel structure to resolution. *Anal. Biochem.* 1:23, 1962.

74. Raymond, S., and Wang, Y. J. Preparation and properties of acrylamide gel for use in electrophoresis. *Anal. Biochem.* 1:391, 1960.

75. Raymond, S., and Weintraub, L. Acrylamide gel as a supporting medium for zone electrophoresis. *Science* 130:711, 1959.

76. Rees, V. H., Filds, J. E., and Laurence, D. J. R. Dye-binding capability of human plasma determined fluorimetrically and its relation to determination of plasma albumin. *J. Clin. Path.* 7:336, 1954.

77. Ritzmann, S. E., and Levin, W. C. Polyclonal and monoclonal gammopathies. *Lab Synopsis* (Sect. II) 2:9, 1967.

78. Romel, W. C., and LaManciesa, S. J. Electrophoresis of glutamic oxaloacetic transaminase in serum, beef heart, and liver homogenates on cellulose acetate. *Clin. Chem.* 11:132, 1965.

79. Schultze, H. E., and Heremans, J. F. *Molecular Biology of Human Proteins*, Vol. 1. New York: American Elsevier, 1966.

80. Schwick, G., and Storiko, K. Qualitative and quantitative determination of plasma proteins by immunoprecipitation. *Lab Synopsis* 1:1, 1965.

81. Smith, I. *Chromatographic and Electrophoretic Techniques*, Vol. II. New York: Interscience, 1960.

82. Snapper, I., and Kahn, A. I. Multiple myeloma. *Seminars Hemat.* 1:87, 1964.

83. Strickland, R. D., Mack, P. A., Guriele, F. T., Podleski, T. R., Salome, O., and Childs, W. A. Determining serum proteins gravimetrically after agar electrophoresis. *Anal. Chem.* 31:1410, 1959.

84. Takahara, S., Hamilton, H. B., Neel, J. V., Kobara, T. Y., Ogura, Y., and Nishimura, E. T. Hypocatalasemia: A new genetic carrier state. *J. Clin. Invest.* 39:610, 1960.

85. Tashian, R. A., and Shaw, M. W. Inheritance of an erythrocyte acetylesterase variant in man. *Amer. J. Hum. Genet.* 14:295, 1962.

86. Thorup, O. A., Strole, W. B., and Leavell, B. S. A method for the localization of catalase on starch gel. *J. Lab. Clin. Med.* 58:122, 1961.

87. Uriel, J. Étude de l'activité enzymatique de la ciruloplasmine de sérum humain après électrophorèse et immunoélectrophorèse en gelose. *Bull. Soc. Chim. Biol.* (Paris) 39 (Suppl. 1):105, 1957.

88. Uriel, J. A method for the direct detection of proteolytic enzymes after electrophoresis in agar gel. *Nature* (London) 188:853, 1960.

89. Uriel, J. The Characterization Reactions of the Protein Constituents Following Electrophoresis or Immunoelectrophoresis in Agar. In Grabar, P., and Butin, P. (Eds.), *Immunoelectrophoretic Analysis*. Amsterdam: Elsevier, 1964. Pp. 46–47.

90. Van der Veen, K. J., and Willebrands, A. F. Isoenzymes of creatine phosphokinase in tissue extracts and in normal and pathological sera. *Clin. Chim. Acta* 13:312, 1966.

91. Wert, E. B. Identification and Measurement of Thermoproteins. In Sunderman, F. W., and Sunderman, F. W., Jr. (Eds.), *Serum Proteins and the Dysproteinemias.* Philadelphia: Lippincott, 1964.

92. Wieme, R. J. An Integrated Procedure for Acrylamide Gel Electrophoresis. In Peeters, H. (Ed.), *Protides of the Biological Fluids.* Proceedings of the Tenth Colloquium, Bruges, 1962. Amsterdam: Elsevier, 1963. Pp. 309–311.

93. Wieme, R. J. *Agar Gel Electrophoresis.* Amsterdam: Elsevier, 1965.

94. Wilding, P. The electrophoretic nature of human amylase and the effect of protein on the starch-iodine reaction. *Clin. Chim. Acta* 12:97, 1965.

95. Yong, J. M. Origins of serum alkaline phosphatase. *J. Clin. Path.* 20: 647, 1967.

APPENDIX

STAINS

323

SOLUTIONS, BUFFER

325

SOURCES OF REAGENTS AND
EQUIPMENT

327

BUFFERS

330

MODIFICATION OF SAMPLE OR
MEDIUM FOR IMPROVEMENT
OF ELECTROPHORESIS

334

ART FORM

339

CALCULATIONS

342

APPENDIX REFERENCES

344

STAINS

1. Thiazine Red R, 1 percent (W/V) for protein.
   a. Fix strips for 20 minutes in 10 percent acetic acid–absolute methanol and dry.
   b. Place dried strips in 1 percent Thiazine Red R for 30 minutes.

Thiazine Red R	1 gm.
Glacial acetic acid	10 ml.
Distilled water	90 ml.

   c. Destain in several changes of 2 percent aqueous acetic acid.
   d. The solutions can be reused several times.
2. Amido Black B (Amido Schwartz, Aniline Black), 1 percent (W/V) for protein.
   a. Fix strips for 20 minutes in 10 percent acetic acid–absolute methanol and dry.
   b. Place dried strips in 1 percent Amido Black B for 30 minutes.

Amido Black B	1 gm.
Glacial acetic acid	10 ml.
Distilled water	90 ml.

   c. Destain in several changes of 2 percent aqueous acetic acid.
   d. Solutions can be used several times.
3. Peroxidase for heme pigments.
   a. After migration is complete, place strip in stain solution for 10 minutes. Peroxidase stain:

Aqueous acetic acid 10 percent (V/V) saturated with benzidine	50.0 ml.
Hydrogen peroxidase 30 percent	0.1 ml.

      Check reactivity by placing a few drops of stain on a drop of blood. Should turn blue-black immediately.
   b. Wash strips in distilled water until buffer salts are removed. Dry strips.
4. Sudan Black B for lipoprotein.
   a. Fix strips in 55 percent reagent alcohol 15–30 minutes and dry.
   b. Place dried strips in Sudan Black B stain for 1 hour.

Sudan Black B	10 gm.
Warm 55 percent reagent alcohol	1 L

      Let sit overnight before using.

Immediately before using, filter-stain. Fast Red 7B may also be used [20].

c. Destain in several solutions of 55 percent reagent alcohol.

5. Periodic acid–Schiff (PAS) for glycoprotein or carbohydrates.

    a. PAS Stain Pak P-403—Fischer. Contains periodic acid, Schiff's reagent, 1N HCl, and potassium metabisulfate.

    b. The following reagents can be made in place of commercial Stain Pak.

      (1) Periodic acid: Dissolve and mix fresh each time used.

Periodic acid	2.0	gm.
Distilled water	75.0	ml.
Reagent alcohol	175.0	ml.
Sodium acetate	0.68	gm.

      (2) Schiff's reagent: Dissolve:

Basic fuchsin	6.0	gm.
Distilled water 90° C.	1.2	L

Cool to 50° C. and filter. Add:

Hydrochloric acid, 2N	30 ml.
Potassium metabisulfite	4 gm.

Stopper and allow to stand in cool, dark place overnight. Add:

Animal charcoal	3 gm.

Mix thoroughly, allow to stand 1 minute, and filter. Add:

Hydrochloric acid, 2N	40 ml.

Keep in refrigerator. Discard when solution turns pink.

      (3) Potassium metabisulfite rinse:

Potassium metabisulfite	4 gm.
Distilled water	1000 ml.

Dissolve and add:

Hydrochloric acid, conc.	10 ml.

Discard rinse after use.

    c. Stain procedure using either set of reagents.

      (1) After migration is complete, fix strip in reagent alcohol 30 minutes.

      (2) Dry.

      (3) Periodic acid, 5 minutes.

      (4) Running water, 10 minutes.

      (5) Schiff's reagent, 30 minutes.

      (6) Potassium metabisulfite, 10 minutes.

      (7) Running water, 10 minutes.

      (8) Potassium metabisulfite, 10 minutes.

      (9) Dry in moderate heat in the dark.

Agarose, being carbohydrate, will retain a background color. Spraying dried, stained pattern with a plastic spray will prevent further discoloration.

SOLUTIONS, BUFFER

1. Barbital buffer, pH 8.6, ionic strength 0.05.
   a. Dissolve one package of commercial packaged barbital buffer labeled pH 8.6, ionic strength 0.075, in 1500 ml. distilled water.
   b. Diethyl barbituric acid   1.84   gm.
   Sodium diethylbarbiturate   10.30 gm.
   • Distilled water   1.0   L
2. Barbital buffer, pH 8.6, ionic strength 0.025.
   Mix equal parts of barbital buffer, pH 8.6, ionic strength 0.05, with distilled water.
3. Phosphate buffer, pH 7.5, M/30.
   a. With 4.5 percent sucrose

Potassium monobasic phosphate	0.94	gm.
Sodium dibasic phosphate	3.94	gm.
Sucrose	45.0	gm.
Distilled water	1.0	L

   b. Without sucrose

Potassium monobasic phosphate	0.94	gm.
Sodium dibasic phosphate	3.94	gm.
Distilled water	1.0	L

4. Phosphate buffer, pH 7.5, M/30.
   a. With 4.5 percent sucrose

Potassium monobasic phosphate	2.8	gm.
Sodium dibasic phosphate	11.8	gm.
Sucrose	45.0	gm.
Distilled water	1.0	L

5. Phosphate buffer, pH 7.4.
   a. Without sucrose
   Dissolve one package of commercial packaged phosphate buffer pH 7.4, M/10, in 1000 ml. of distilled water.
   b. With sucrose
   Dissolve one package of commercial phosphate buffer, pH 7.4, M/10, and 45 gm. sucrose in 1 liter of distilled water.
6. Phosphate buffer, pH 7.5, M/15.

Na_2HPO_4 (M/15, 9.46 gm./L)	43.0	ml.
KH_2PO_4 (M/15, 9.07 gm./L)	7.0	ml.

7. Phosphate buffer, pH 6.5, M/15.

Na_2HPO_4 (M/15, 9.46 gm./L)	18.0	ml.
KH_2PO_4 (M/15, 9.07 gm./L)	32.0	ml.

8. Phosphate buffer, pH 6.3, M/15.

Na_2HPO_4 (M/15, 9.46 gm./L)	13.0	ml.
KH_2PO_4 (M/15, 9.07 gm./L)	37.0	ml.

9. Phosphate buffer, pH 7.0, M/15.

Na_2HPO_4 (M/15, 9.46 gm./L)	31.0	ml.
KH_2PO_4 (M/15, 9.07 gm./L)	19.0	ml.

10. Clark-Lubs buffer, pH 10.0, 0.2M.

Solution A	50.0	ml.
$\quad B(OH)_3$ (0.2M, 12.4 gm./L)		
$\quad KCl$ (0.2M, 14.9 gm./L)		
Solution B	43.9	ml.
$\quad NaOH$ (0.2M, 8.0 gm./L)		
Distilled water to	200.0	ml.

11. Acetate buffer, pH 4.0, 0.1M.

$NaC_2H_3O_2$ (0.2M, 16.4 gm./L)	9.0	ml.
$HC_2H_3O_2$ (0.2M, 11.55 ml./L)	41.0	ml.
Distilled water	50.0	ml.

12. Acetate buffer, pH 5.6, 0.1M.

$NaC_2H_3O_2$ (0.2M, 16.4 gm./L)	45.2	ml.
$HC_2H_3O_2$ (0.2M, 11.55 ml./L)	4.8	ml.
Distilled water	50.0	ml.

13. Tris buffer, pH 8.8, 0.05M.

Tris HCl	1.23	gm.
Tris base	5.13	gm.
Distilled water	1.0	L

14. Tris buffer, pH 8.0, 0.05M.

Tris HCl	0.444	gm.
Tris base	0.265	gm.
Distilled water	100.0	ml.

15. Sucrose–barbital buffer for CPE.

  a. Stock buffer solution:
     Dissolve contents of one package of Beckman buffer B-2 in 1 liter of distilled water.

  b. Working buffer solution:
     Dissolve 110 gm. of sucrose in 1 liter of distilled water containing 20 ml. of the stock buffer solution. After sugar is dissolved, add distilled water to a final volume of 3 liters.

SOURCES OF REAGENTS
AND EQUIPMENT

Agarose:
    1. Bausch & Lomb, Inc.
        Rochester, New York
    2. Fisher Scientific Co.
        Pittsburgh, Pennsylvania
    3. Certified Blood Donor Service, Inc.
        Woodbury, New York

Barbital buffer, pH 8.6, ionic strength 0.075:
    1. Fisher Scientific Co.
        Pittsburgh, Pennsylvania
    2. Beckman Instruments, Inc.
        Fullerton, California

Phosphate buffer, pH 7.4:
        Fisher Scientific Co.
        Pittsburgh, Pennsylvania

Specific anti–human precipitating antisera made in goat and/or in rabbit:
    1. Hyland Laboratories
        Los Angeles, California
    2. Certified Blood Donor Service, Inc.
        Woodbury, New York

Thiazine Red R:
        Allied Chemical Corp.
        Morristown, New Jersey

Bromphenol blue:
        Allied Chemical Corp.
        Morristown, New Jersey

Sudan Black B:
        Allied Chemical Corp.
        Morristown, New Jersey

Tannic acid:
        Mallinckrodt Chemical Works
        St. Louis, Missouri

Lithium lactate:
>   G. Frederick Smith Chemical Co.
>   Columbus, Ohio

Periodic acid:
>   G. Frederick Smith Chemical Co.
>   Columbus, Ohio

Schiff (basic fuchsin):
>   Beckman Instruments, Inc.
>   Fullerton, California

PAS Stain Pak P-403:
>   Fisher Scientific Co.
>   Pittsburgh, Pennsylvania

Lysozyme—standard and reagent:
>   Worthington Biochemical Corp.
>   Freehold, New Jersey

Enzyme reagents such as NAD, PMS, TPN, NADP:
> 1. Sigma Chemical Co.
>   St. Louis, Missouri
> 2. Calbiochem
>   Los Angeles, California
> 3. Mann Research Laboratories
>   New York, New York

Immuno-Plates—γG, γA, γM, and complement:
>   Hyland Laboratories
>   Los Angeles, California

Cronar Film, P40B, 35 mm., unperforated, 0.004" thickness:
> 1. E. I. duPont de Nemours & Co. (Inc.)
>   Dallas, Texas
> 2. HCS Corporation
>   Wichita, Kansas

Mylar D 5" x 7" sheets, uncoated, 0.005" thickness:
> 1. Tommy Tucker Plastics
>   Dallas, Texas
> 2. HCS Corporation
>   Wichita, Kansas

Solar camera with 4" lens and Polaroid Land Pak:
>   HCS Corporation
>   Wichita, Kansas

Polaroid MP-3 camera with 4" lens and Polaroid Land Pak:
>   Polaroid Corporation
>   Cambridge, Massachusetts

Immuno-Glo:
> 1. HCS Corporation
>   Wichita, Kansas
> 2. Hyland Laboratories
>   Los Angeles, California

Immunoboard and agar scoop:
1. HCS Corporation
   Wichita, Kansas
2. Hyland Laboratories
   Los Angeles, California

Incubation chamber:
1. HCS Corporation
   Wichita, Kansas
2. Hyland Laboratories
   Los Angeles, California

Electro-Gel:
1. HCS Corporation
   Wichita, Kansas
2. Hyland Laboratories
   Los Angeles, California

Electrophoretic equipment:
1. Spinco Durrum Cell
   Spinco Division
   Beckman Instruments, Inc.
   Fullerton, California
2. LKB Instruments, Inc.
   Rockville, Maryland
3. Gelman Instrument Co.
   Ann Arbor, Michigan
4. Colab Labs
   Chicago Heights, Illinois
5. Buchler Instruments, Inc.
   Fort Lee, New Jersey
6. Hyland Laboratories
   Los Angeles, California
7. HCS Corporation
   Wichita, Kansas

BUFFERS

The buffers used in electrophoresis and immunoelectrophoresis make up a predominant share of the environment of proteins during separation and serve to maintain a constant pH, which ensures adequate and consistant net negative charge density of the protein and keeps the protein in a freely soluble form. As the concentration of buffer is decreased, proteins become less soluble and bind with the medium (paper, starch, agar, cellulose acetate, etc.). Conductivity flow is reduced. Low current flow is desirable since the wattage (amps × volts) or heat production is also low. A balance between these two is desirable; i.e., the buffer must be concentrated enough to maintain an environment of adequate and constant pH and at the same time not favor production of excessive heat (high current), which destroys many proteins.

Buffers for electrophoresis are many and are usually made up on the basis of molarity, pH range, and ionic strength. The latter term is an expression of the number of ions resulting from ionization. In idealized situations where complete ionization occurs, ionic strength and conductivity are comparable. Buffers ionize to a different degree, and calculated ionic strengths cannot fully take into consideration the degree of ionization. Activity calculations could increase accuracy of the ionic strength values; however, as discussed by Crowle, measurement of conductivity is directly related to current flow [3]. Table 1 lists buffers used in many laboratories for a variety of electrophoretic studies, along with stated pH, molarity (if available), measured pH (pK_a for organic buffers described by Good et al. [5]), ionic strength, osmolarity, and conductivity. Measurement of osmolarity is done on an osmometer made by Advanced Instruments, Inc., Newton Highlands, Massachusetts, which reads directly in milliosmoles. This measurement is used for buffers which are employed in cell electrophoresis. Serum osmolarity is about 290.8 mOsm, and normal saline osmolarity is 290 mOsm (Table 1). A buffer for cell electrophoresis should have an osmolarity value close to that of serum. Tris base-8 percent sucrose has osmolarity of 275 mOsm, a low conductivity, and a high pH—all features which result in good separation of cells. Why some buffers are better than others cannot be explained by physical measurements alone. There appear to be many biologic properties of

TABLE I

BUFFER CHARACTERISTICS

Buffer	List pH	Molarity[a] (M)	pH	pKa	Ionic Strength[a]	Osmolarity[b] (mOsm)	Conductivity[c] (mMHO)
Veronal[d]	8.6		8.4		0.075 V	164.5	3.0
Veronal[d]	8.6		8.1		0.050 V	114.0	2.2
Veronal[d]	8.6		8.2		0.025 V	56.0	1.42
Phosphate, M/30	7.5	0.033	7.25			90.0	3.55
Phosphate, M/15	7.5	0.066	7.4			155.0	6.0
Phosphate, M/10	7.5	0.1	7.2			235.0	9.0
Phosphate, M/30 with sucrose		0.033	7.1			215.0	3.1
Phosphate, M/10 with sucrose		0.1	7.05			349.5	8.25
Phosphate[a]	7.4	0.033	7.5			89.0	3.9
Phosphate[a] with sucrose		0.033	7.0			223.0	3.3
Phosphate, M/15	6.5	0.066	6.6			136.0	5.25
Phosphate, M/15	6.3	0.066	6.45			150.0	5.0
Phosphate, M/15	7.0	0.066	6.9			154.0	5.75
Clark-Lubs	10.0	0.2	9.4			164.0	6.25
Acetate	4.0	0.1	3.9			237.0	2.01
Acetate	5.6	0.1	5.3			184.0	4.5
Tris base + Tris HCl[e]	8.8	0.05	8.55			62.0	0.56
Tris base + Tris HCl[e]	8.0	0.05	7.6			82.0	1.83
Tris base[e]		0.1	9.45			106.5	0.2
Tris base[e]		0.025	8.55			24.5	0.031
Aaronson (Tris-EDTA-boric)		0.5	8.8			512.0	3.15
Tris-EDTA-boric		0.1	8.8			111.0	0.575
Veronal-1.4% glycine			7.8		0.025 V	246.0	1.35
Veronal-0.3% EDTANa4			8.85		0.025 V	91.0	2.7
Glycylglycine		0.1	5.35	8.4		101.5	0.11
Glycylglycine with NaOH added		0.1	7.55			127.0	1.25
Glycylglycine-Veronal (1:1)		0.1 G	7.35		0.05 V	103.0	1.12

TABLE 1 *(Continued)*

Buffer	List pH	Molarity[a] (M)		pH	pK$_a$	Ionic Strength[a]		Osmolarity[b] (mOsm)	Conductivity[c] (mMHO)
Tris-glycylglycine (1:1)		0.1		7.72				102.0	0.78
Tris-glycylglycine		0.05							**0.45**
Tris-glycylglycine		0.025							**0.23**
Tris-Veronal (1:1)		0.1	T	8.72	8.15	0.05	V	110.0	1.24
Tricine		0.1		4.12				102.0	0.08
Tricine-Veronal (1:1)		0.1	Tr	7.3		0.05	V	108.0	1.2
Bicine		0.1		4.15	8.35			105.0	0.0143
Bicine-Veronal (1:1)		0.1	B	7.4		0.05	V	180.0	1.15
Glycine		0.1		5.05				105.0	0.575
Glycine-Veronal (1:1)		0.1	G	8.05		0.05	V	108.0	1.15
Saline		0.025		5.5				60.2	2.65
Saline		0.05		5.45				111.5	4.3
Saline, normal, 0.85%		0.145		5.3				290.0	10.6
Saline		0.15		5.6				301.0	11.0
Saline		0.3		5.3				582.0	21.0
Serum (pooled stored)[f]				8.6				290.8	8.25
Veronal, 0.3M sucrose				6.8		0.0005	V	275.0	0.033
Tris base, 0.3M sucrose		0.025		8.9				383.0	0.0275
Tris base, 1% sucrose		0.025		8.9				55.4	0.0262
Tris base, 6% sucrose		0.025		8.9				212.0	0.0232
Tris base, 8% sucrose		0.025		8.9				275.0	0.0265

[a] V = Veronal; T = Tris; Tr = Tricine; B = Bicine; G = Glycine.
[b] Freezing point depression measured with osmometer.
[c] Conductivity = 1/OHMS per cm. = MHO.
[d] Fisher Scientific Co., Pittsburgh, Pa.
[e] Sigma Chemical Co., St. Louis, Mo.
[f] Total protein = 6.9 gm./100 ml.

protein which are adversely influenced by certain buffer systems, and at the moment only a test of separation will establish the biologic compatibility of a buffer. An example is the separation of serum proteins in two buffers: glycylglycine-Veronal (1:1), osmolarity 103 mOsm and conductivity 1.12 mMHO, and Veronal (ionic strength 0.050), osmolarity 114 mOsm and conductivity 2.2 mMHO. Both patterns are sharp, but the glycylglycine-Veronal (1:1) buffer selectively increases the electrophoretic mobility of α_1-globulin so that it is masked by albumin. Good et al. [5] stress the importance of selecting buffers on the basis of several factors, a number of which are related to how the buffer affects biologic systems. Many who are using electrophoresis and immunoelectrophoresis are now beginning to review the physical chemical properties of buffers and to add to their knowledge concerning the biologic properties of buffers, which should extend our knowledge considerably.

MODIFICATION OF SAMPLE OR MEDIUM FOR IMPROVEMENT OF ELECTROPHORESIS

It is often desirable to study the effect on the electrophoretic mobility of single or multiple components by changing either the sample itself or the environment. The effect on the electrophoretic mobility of chemicals added to serum as well as chemicals added to the medium is not well known. Electrophoresis may be looked at as consisting of approximately five parts: the electrophoretic field, sample (particle), support (medium), buffer, and detecting system (Table 2).

Electrophoretic Power

The driving force of electrophoresis is usually derived from direct current although pulsating current may be used and one may alter the program in such a way that the voltage changes during electrophoresis. These changes have not necessarily been very helpful in improving electrophoretic separation.

Modification of Sample

As in gas chromatography, separation may be improved on occasion by creating derivatives of the sample. This is just beginning to become a popular area of investigation in protein chemistry. Effort to increase electrophoretic resolution by creating derivatives of proteins was at one time discounted, primarily because most derivatives of protein resulted in loss of charge density or caused inactivation of the protein. In either event, separation by electrophoresis was not possible. The introduction of new protein derivatives which did not disturb the integrity of the proteins has stimulated renewed interest in creation of derivatives. Fluorescein, ferritin, and enzyme-tagged antibodies are types of derivatives of protein which do not inactivate the proteins [11, 12, 18]. Such derivatives have been studied briefly in our laboratory by electrophoresis, and in general there is a change in the electrophoretic mobility of some proteins and an increase in the mo-

TABLE 2

COMPONENT PARTS OF ELECTROPHORESIS

Electrical field	
Direct current	Pulsating current
Low voltage	
High voltage	
Sample (particles)	
Native charge	Chemical derivatives
Support	
Nonselective	Selective
Liquid	Molecular sieve (starch-gel acrylamide)
Agarose	Immunochemical
Paper	Chemical derivatives
Cellulose acetate	Ion exchange
Starch block	Molecular exclusion
Pivakon	Adsorption
Buffer	
Nonadditive	Additives
Continuous	Immunologic
Discontinuous	Chemical groups
Detection	
No stain	Stains
Ultraviolet adsorption	Dyes, fluorescent stains, etc.
	Enzyme

bility of tagged antibodies (fluorescein, ferritin, and enzymes). These types of derivatives are prepared as chemical bonds between the protein and the marker (tag).

DERIVATIVES OF PROTEIN BY ADSORPTION OR NONCHEMICAL BINDING

Such derivatives are of some interest and have been applied indirectly in many instances. For example, the addition of tannic acid to serum proteins results in a complex causing some inactivation and precipitation. However, with minimal doses, changes too subtle to see visually can nevertheless be recorded electrophoretically. Inactivation of γ-globulin occurs at low doses. With progressive increase of dose, inactivation of β, α_2, α_1, and albumin occurs, albumin being the most resistant to the action of tannic acid. Another approach is to prestain serum proteins for lipoproteins and proteins with stains such as Sudan Black B, Thiazine Red R, Evans blue, bromphenol blue, Amido Black B, or toluidine blue. In the case of protein, there is a difference in the electrophoretic mobility resulting from binding of dye to protein.

Some stains are more effective than others, and the proteins may be visualized during separation. Electrophoretic mobility at pH 8.6 of some protein stains in decreasing mobility is as follows: Evans blue, Thiazine Red R, Amido Black, and bromphenol blue. All migrate faster than albumin. Evans blue results in an increased number of serum protein components with gross distortion of the electrophoretic pattern. It has been noted by some investigators that the electrophoretic mobility of a certain α_2 protein increases with Evans blue [13], which is probably identical to α_2D, a breakdown product of β_1C complement as described by West [22]. Keflin, alpha methyldopa, formaldehyde, and paraformaldehyde do not result in satisfactory preparations for electrophoresis when added directly to the serum, but when incorporated into the support (agarose) the results are more clear cut. Derivatives between an antibiotic or chemical tag and a protein can be formed in the test tube before electrophoresis. These reactions often result in aggregate particles which remain at the point of application during electrophoresis.

DERIVATIVE FORMATION DURING ELECTROPHORESIS

In this technic the chemical to be studied is incorporated within the substance of the agar or agarose during preparation of the gel. The protein sample migrates in an environment containing the agent under study; for example, tannic acid when added to agarose creates a much better pattern for study than when added to the serum sample before electrophoresis. There is a gradual reaction between the tannic acid and many of the protein fractions during the course of electrophoresis. As mentioned, γ-globulin is the first fraction adversely affected by the presence of the tannic acid, and there is a decrease in electrophoretic mobility. With this technic we have studied the effect on serum protein electrophoretic mobility of formaldehyde, paraformaldehyde, dyes including Thiazine Red R, Amido Black B, bromphenol blue, toluidine blue, trypan blue, eosin, and Evans blue. In addition, serum proteins have been subjected to electrophoresis in agarose containing alpha methyldopa, Keflin, gelatin, starch, glass beads, proteolytic enzymes, EDTA, glutamine, glycine, Polybrene, polyglycine, Methocel, and a number of other detergents. In general, the most rewarding results came with the study of alpha methyldopa, Keflin, formaldehyde, paraformaldehyde, and tannic acid. In these instances a predictable and consistent change in electrophoretic mobility of one or more of the important serum proteins was observed. γ-Globulin in all instances seemed to be most susceptible to change, and except with tannic acid showed an increase in electrophoretic mobility suggesting that the positive charge of the amino groups was neutralized by the chemical agents. On the other hand, tannic acid causes denaturization of the γ-globulin and no migration occurs. A continuing study of the reaction between proteins, antibiotics, and drugs appears to be worthwhile; the methods outlined above are well suited to this type of investigation.

Modification of the Support

Since agarose is used so extensively as a support medium for electrophoresis, it occurred to us that the system could be much more flexible if the nature of the agarose could be changed. Originally, change was effected by adding different substances to the agarose directly, such as starch (see Chap. 8), but a sounder method involving chemical coupling of any desired substance to the agar or agarose seemed to offer more widespread use. In one instance the agarose (made up of repeating units of galactose) was oxidized with periodic acid and then coupled to the chosen substance by N-ethyl-N′-3-dimethyl-amino propyl carbodiimide [1, 6, 17]. The principles involved were essentially those used for making cellulose derivatives [21]. The separation obtained from electrophoresis could be altered by the addition of either a negative or a positive charge to the agarose. Negatively charged groups are largely responsible for the net cathodic liquid flow in agarose. Positively charged groups would be responsible for an opposite effect. In order to alter our system in reference to charges on the agarose medium we coupled the agarose to trypan blue, eosin Y, glycine, lycine, glutamine, polyglycine, and Girard's T reagent. While Ragetli et al. [14] had earlier succeeded in changing endosmotic flow by coupling differently charged groups to agar, our results were disappointing; even though a difference in endosmosis was demonstrated, the electrophoretic serum protein patterns were unsatisfactory.

In another approach used to change the nature of the agarose medium the matrix of the fluid gel was altered to create more of a "sieving" effect. The agarose was not oxidized but coupled directly to other substances by use of N-ethyl-N′-3-dimethyl-amino propyl carbodiimide. The substances used in this study were starch and gelatin. Coupling starch to agarose or gelatin to agarose should make possible a sieve comparable to starch or acrylamide [4, 19]. The results, although not satisfactory, ought nevertheless to be explored further. Coupling antiserum to the gel [2] or incorporating it into the gel offers a method of creating an immunologic atmosphere against which the antigen is driven by the electromotive force. The latter is a form of cross electrophoresis introduced by Nakamura et al. [10] and used and perfected by Laurell [7] and Ressler [15]. Our results from chemically labeling the electrophoretic support with antisera at this point have not been satisfactory; however, we believe they justify more exploration. The principles of solid-phase synthesis of peptides introduced by Merrifield [9] and labeling of solid phase with antigen or antibody [16, 21] are applicable to creation of a selective support for electrophoresis (Table 2).

Detection

Most commonly used to detect proteins in an electrophoretogram are stains such as Thiazine Red R, Amido Black B, and bromphenol blue. A slightly different approach is offered in this section.

DETECTION OF PROTEINS IN ELECTROPHORESIS

An approach was examined which did not alter the agarose but attempted to change the end point detecting system for proteins from the usual stain methods by tagging the proteins after separation with fluorescent stain, ferritin, or enzymes. We tried to create serum protein in derivatives after electrophoresis with substances which would magnify the end point. One coupling reagent was fluorescein (see Chap. 12) and another coupling agent was ferritin, commonly used with antibodies in electron microscopy [18]. After ferritin has been coupled to the protein, a stain for ferritin is performed. The end point can be magnified to a greater degree if an enzyme is coupled to the protein after electrophoresis. When the coupling has taken place, a stain for the enzyme is performed. We chose peroxidase and LDH to couple to the protein. Coupling was performed with slight success, but enough to show the reliability of the principle.

ART FORM

Although the text has been primarily concerned with the clinical applications of electrophoresis and immunoelectrophoresis, I believe it is worth accentuating the beauty frequently inherent in electrophoretograms and immunoelectrophoretograms. To the experienced laboratory investigator or worker, such patterns have often been a source of pleasure and delight. In some instances their intricacy is striking, and one never ceases to be thrilled at looking at them.

Figure 1 is a multiple electrophoretogram done with different sera at one time on a single 5″ by 7″ plastic sheet covered with agarose. The series was performed at 4° C. by a special cooling device. The individual protein zones are quite sharp. The lower pattern is plasma and shows fibrinogen.

Figure 2 is an immunoelectrophoretogram made in 1964 for an investigative study dealing with the effect of bromelin on serum proteins. Serum was prepared in four repetitive duplicate sets, each set containing an upper pattern from the result of the action of bromelin on the serum and a lower pattern from normal untreated serum. The reaction was carried out with rabbit polyvalent anti–human serum. Bromelin obviously caused some change in the serum protein since a number of bands either are missing or have changed positions in reference to untreated serum. The pattern, however, because of its repetitive nature and the unusualness of the lines is a thing of beauty which has for a number of years hung in the laboratory and been a source of attraction to visitors as well as to the laboratory staff. No one has tired of viewing this unusual pattern or attempting to decipher some of the changes. Done in color, many such patterns are extraordinarily striking.

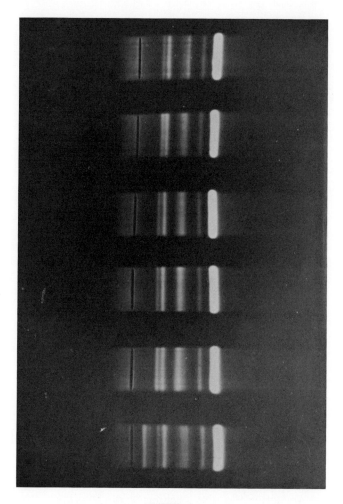

FIGURE 1

Six electrophoretic patterns obtained on single 5″ × 7″ sheet of agarose-coated mylar plastic. Run took 18 minutes at 400 volts (about 20 volts/cm.) and 70 ma. Temperature was controlled to 4° C. by a special cooling device. Note the sharp, narrow protein zones for each fraction, caused by lack of diffusion resulting from rapid separation and low temperature. Lower pattern is plasma and shows fibrinogen zones. Anode is on the right.

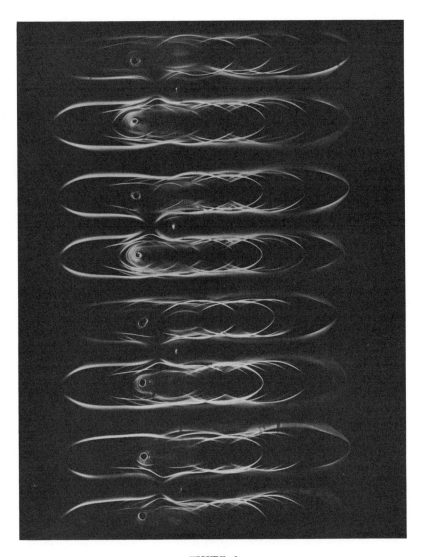

FIGURE 2

Four sera arranged into two sets each of bromelin treated (*upper*) and non–bromelin treated (*lower*), reacted against rabbit anti–normal human serum. Bromelin, a mixture of proteases and other proteins, causes some changes in serum proteins, as noted in the upper electrophoretograms in each of the four sets. Anode is on the right.

CALCULATIONS

Little attention has been paid in the text to the problem of obtaining quantitative values for each of the six protein fractions in an electrophoretogram. It has become the custom to report either percent of each fraction to the total, grams of protein in each fraction, or both. Despite the extreme genetic heterogeneity of proteins from one individual to another, which considerably limits the reliability of quantitative values, and the inaccuracy of protein determination other than by weighing, the clinical laboratories still report the total protein and the quantity of protein per fraction. Qualitative judgment of the electrophoretogram and the densitometric scan is far more informative than the digital data. However, all the information that can be obtained about an electrophoretogram extends the usefulness of the interpretation. Thus, qualitative and quantitative reporting is desirable. Quantitation of an electrophoretogram can theoretically be fully automated with on-line or off-line electronic integrators which determine the area beneath each peak of a densitometric scan by recognizing the beginning and the end of a peak through use of the second derivative. Experience has shown, however, that few such systems can match the eye in detecting fractions. Thus, Lotito et al. [8], using an on-line analog computer, graphed the values of each fraction from paper electrophoretograms marked for specific fractions by visual judgment.

A slightly less automated method is one we currently employ based on a desk computer, the Olivetti Underwood Programma 101 (Fig. 3). This computer operates from a program which is stored on a magnetic card. The computer in our laboratory is programmed to accept six integration values for albumin, α_1, α_2, β_1, β_2, and γ-globulin obtained from a suitable densitometric instrument and also the total protein. From these values the computer prints out the sum, the percent of each fraction of the total, and the grams per 100 ml. of each fraction. The time required for the entire operation is less than 1 minute. The computer has many other programs common and usual to every laboratory, such as computation of standard deviation from grouped and ungrouped values, conversion of raw data from spectrophotometric readings into specific terms for multiple tests, and a number of useful mathematical calculations.

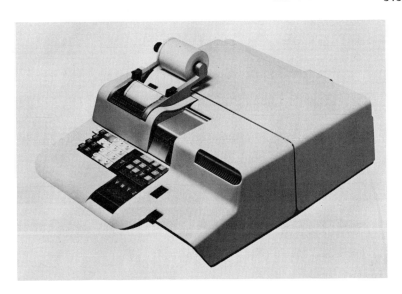

FIGURE 3

Photograph of Olivetti Underwood Programma 101 desk computer. (Courtesy of Olivetti Underwood Corp., New York, N.Y.)

APPENDIX REFERENCES

1. Carpenter, R. R., and Reisberg, M. A. Carbodiimide-induced bentonite-antigen complexes: Readily prepared immunoadsorbents. *J. Immun.* 100:873, 1968.
2. Cawley, L. P. Future of Electrophoresis. Travel Tapes, The Audio Newsletter for Pathologists, Commission on Continuing Education. Chicago: American Society of Clinical Pathologists, 1968.
3. Crowle, A. *Immunodiffusion.* New York: Academic, 1961.
4. Davis, B. J., and Ornstein, L. A New High Resolution Electrophoresis Method. Delivered at meeting of the Society for the Study of Blood, March 24, 1959, at the New York Academy of Medicine.
5. Good, N. E., Winget, G. D., Winter, W., Connolly, T. N., Izawa, S., and Singh R. M. M. Hydrogen ion buffers for biological research. *Biochemistry* 5:467, 1966.
6. Johnson, H. M., Brenner, K., and Hall, H. E. The use of a water-soluble carbodiimide as a coupling reagent in the passive hemagglutination test. *J. Immun.* 97:791, 1966.
7. Laurell, E. B. Antigen-antibody crossed electrophoresis. *Anal. Biochem.* 10:358, 1965.
8. Lotito, L. A., McKay, D. K., and Seligson, D. Analog computation of electrophoresis patterns. *Clin. Chem.* 11:386, 1965.
9. Merrifield, R. B. Automated synthesis of peptides. *Science* 150:178, 1965.
10. Nakamura, S., Tanaka, K., and Murakowa, S. Specific protein of legumes which react with animal proteins. *Nature* (London) 188:144, 1960.
11. Nakane, P. K., and Pierce, G. B. Enzyme-labeled antibodies for the light and electron microscopic localization of tissue antigens. *J. Cell Biol.* 33:307, 1967.
12. Nakane, P. K., and Pierce, G. B. Enzyme-labeled antibodies: Preparation and application for the localization of antigens. *J. Histochem. Cytochem.* 14:929, 1967.
13. Rabinovitz, M., and Schen, R. J. Observations on an α_2-globulin having increased electrophoretic mobility with Evans Blue dye: Its relationship to the $\beta_1 C$ component of complement. *Clin. Chim. Acta* 20:227, 1968.
14. Ragetli, H. W. J., and Weintraub, M. Agar derivatives for electrophoresis requiring reduced endosmotic liquid flow. *Biochim. Biophys. Acta* 112:160, 1966.
15. Ressler, N. Two-dimensional electrophoresis of protein antigens with an antibody containing buffer. *Clin. Chim. Acta* 5:795, 1960.

16. Robbins, J. B., Haimovich, J., and Sela, M.   Purification of antibodies with immunoadsorbents prepared using bromoacetyl cellulose. *Immunochemistry* 4:11, 1967.

17. Sheehan, J. C., and Hess, G. P.   A new method of forming peptide bonds. *J. Amer. Chem. Soc.* 77:1067, 1955.

18. Singer, S. J., and Schick, A. F.   The properties of specific stains for electron microscopy prepared by the conjugation of antibody molecules with ferritin. *J. Biophys. Biochem. Cytol.* 9:519, 1961.

19. Smithies, O.   Zone electrophoresis in starch gels: Group variations in serum proteins of normal adults. *Biochem. J.* 61:629, 1955.

20. Straus, R., and Wurm, M.   New staining procedure and a method for quantitation of serum lipoproteins separated by paper electrophoresis. *Techn. Bull. Reg. Med. Techn.* 28:89, 1958.

21. Weliky, N., and Weetall, H. H.   The chemistry and use of cellulose derivatives for the study of biological systems. *Immunochemistry* 2: 293, 1965.

22. West, C. D., Davis, N. C., Forristol, J., Herbst, J., and Spitzer, R.   Antigenic determinants of human β_1C- and β_1G-globulins. *J. Immun.* 96:650, 1966.

INDEX